cpo science

A member of
School Specialty
Science

CPO Focus on Physical Science Teacher's Guide
First Edition

ISBN-10: 1-58892-261-8
ISBN-13: 978-1-58892-261-8
4 5 6 7 8 9 - QWE- 11 10 09 08 07

CPO Science
80 Northwest Boulevard
Nashua, New Hampshire 03063
(866)588-6951
http://www.cposcience.com
Printed and Bound in the United States of America

Credits

Principal Writers

Sonja L. Taylor
B.S., Chemistry, Stephen F. Austin State University, Nacogdoches, Texas

She has worked as an analytical chemist and taught high school chemistry, biology, and physical science for four years. While teaching in an innovative, academy-based high school, Sonja was the team leader of her academy, overseeing integration across curricula, and serving as the liaison between students, faculty and members of the corporate advisory board.

Catherine C. Reed
B.S., Secondary Education, Akron University; M.Ed., Administration, Curriculum from Baldwin-Wallace College

Consulting writer and Curriculum Consultant. She is a Middle School Science Curriculum Specialist.

Patsy Eldridge
B.S., Biology, Grove City College; M.Ed., Tufts University.

Experienced high school Physical Science teacher and national hands-on science trainer and presenter. As an adjunct professor for Endicott College in Beverly, MA, and the College of Charleston, developed content-intensive Physical Science courses for pre- and in-service educators. Technical background in the medical device field. Patsy has developed curriculum and training materials with CPO for six years.

Senior Editor

Lynda Pennell – Executive Vice President
B.A., English, M.Ed., Administration, Reading Disabilities, Northeastern University; CAGS Media, University of Massachusetts, Boston

Nationally known in high school restructuring and for integrating academic and career education. Served as the director of an urban school with 17 years teaching/ administrative experience. Acted in a variety of administrative positions in the Boston Public Schools.

Investigation Dialog Writers

Dr. Sara Anantharaman
Biology & ELL Science Teacher
Dr. Phillips High School, FL

Louis Reyes
Marine Biology & Chemistry Teacher
Lake Mary High School, FL

Sharon Swalina
Physical Science & Chemistry Teacher
Lake Mary High School, FL

Lisa Hamilton
Physical and Life Science Teacher
Williston High School, FL

Adaptive Support Specialists

Leslie Sheen – ELL Specialist

Has a BA in English from Gordon College; M.A.T., Teaching English as a second language, Georgetown University taught ESOL for 10 years. Taught elementary school and adult education in the US. Taught secondary school and at University of Helsinki in Finland.

Tia L. Shields – ELL Teacher
Position: 7th Grade Life Science/Health and English Language Learner Teacher
Nicolas Junior High School, CA

Assessment Data/Answers

Mary Ann Erickson – Project Coordinator
She has a degree in Naval Architecture and Marine Engineering from the Massachusetts Institute of Technology.

She ran a technical writing consulting business, writing process control manuals for water treatment plants, software design documentation for simulation software, and operator manuals for mining equipment

Erik Benton – Data and Verification
B.F.A., University of Massachusetts his minor was in Physics

Principle Investigation Editor for CPO Science. Erik taught for 8 years in public and private schools, focusing on inquiry and experiential learning

David Bliss – Writer

David has taught for years in all levels of Physics. He has also taught 8th Grade Physical Science Mohawk Central School, NY

Kelly A. Story – Reviewer
B.S, Chemistry and English from Gordon College; Masters in Chemistry from the University of Massachusetts Lowell

Taught chemistry and maintains a position as lab instructor at Gordon College, Wenham, MA.

Reviewers

Nancy Joplin
Position: English Language Arts Department Chair
Ray Wiltsey Middle School, CA

Jodye Selco
Position: Professor
California State Polytechnic University, Pomona

Brian E. Goodrow
Position: Physical Science Teacher
Apple Valley Middle School, CA

Philip L. Hunter
Position: Science Department Chair
Johnson Middle School, CA

Bianca N. McRae
Position: Science Teacher, Department Chair
Menifee Valley Middle School, CA

Tia L. Shields
Position: 7th Grade Life Science/Health Teacher
Nicolas Junior High School, CA

Kelly McAllister
Position: Science Teacher
Gage Middle School, CA

Brad Joplin
Position: 8th Grade Science Teacher
Ray Wiltsey Middle School, CA

Tony Heinzman
Position: Teacher
Apple Valley, CA

Sylvia Gutman
Position: Science Teacher/Department Head
David A Brown Middle School
Lake Elsinore Unified School District, CA

Illustration/Design

Polly Crisman -Designer and Illustrator
B.F.A., University of New Hampshire

Graphic artist with expertise in advertising, marketing design, freelance illustrating, and caricature art.

Bruce Holloway - Cover Designs
Pratt Institute, N.Y.; Boston Museum School of Fine Arts

Expertise in product design, advertising, and three-dimensional exhibit design. Commissioned for the New Hampshire Duck Stamp for 1999/2003.

Jesse Van Valkenburgh - Illustrator
B.F.A., Rochester Institute of Technology

Worked in prepress film production and design. He worked at PC Connection in Merrimack, NH as a graphic designer doing catalog and direct mailing design, logo design, as well as illustration work.

Technical Support

Tracy Morrow – Framemaker Expert, Technical Editor, Technical Trainer
B.A., English, Texas A&M University; M.A., English, Sam Houston State University

Taught middle school in Klein ISD, a suburb of Houston, Texas, for 9 years preparatory and English at Tomball College for 5 years. After this, Tracy worked as a technical writer in the oil and gas, airlines, and real estate industries. Currently she offers consulting services and technical training. Tracy's expertise lies in the editing program Framemaker.

Resource Planner Pages

The Resource Planner is a two-page table that demonstrates how the CPO materials work together to teach chapter content.

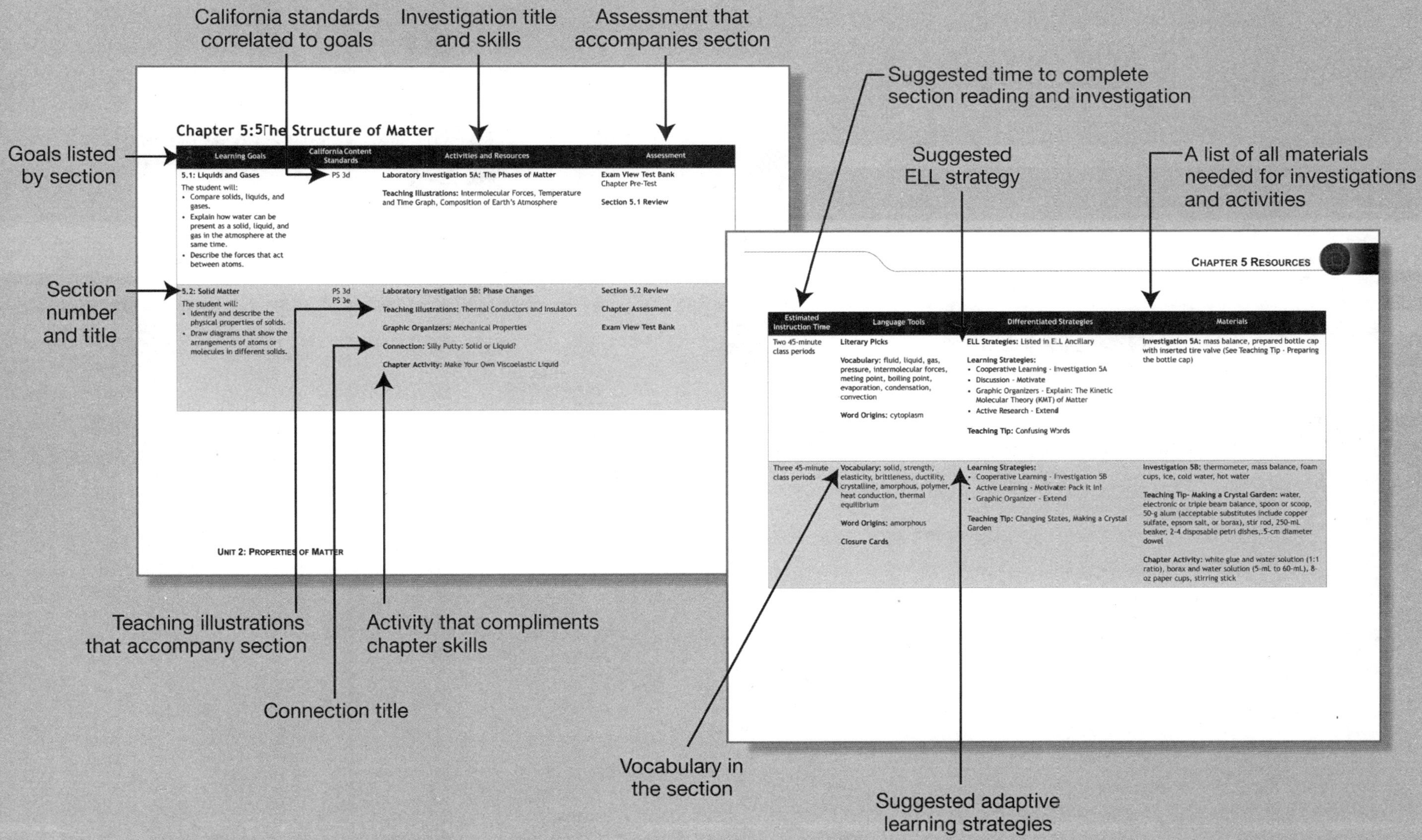

Teaching the Lesson Pages

The layout of this section follows this approach:
Motivate, Explore, Explain, Extend, and **Assess.**

Brief explanation of lesson content

Suggested supplemental student reading

MOTIVATE: A motivational activity for engagement

EXPLORE: Complete the investigation

How to set up for the investigation

EXPLAIN: Suggested explanation for the student

Teaching tips, safety, and other teaching suggestions

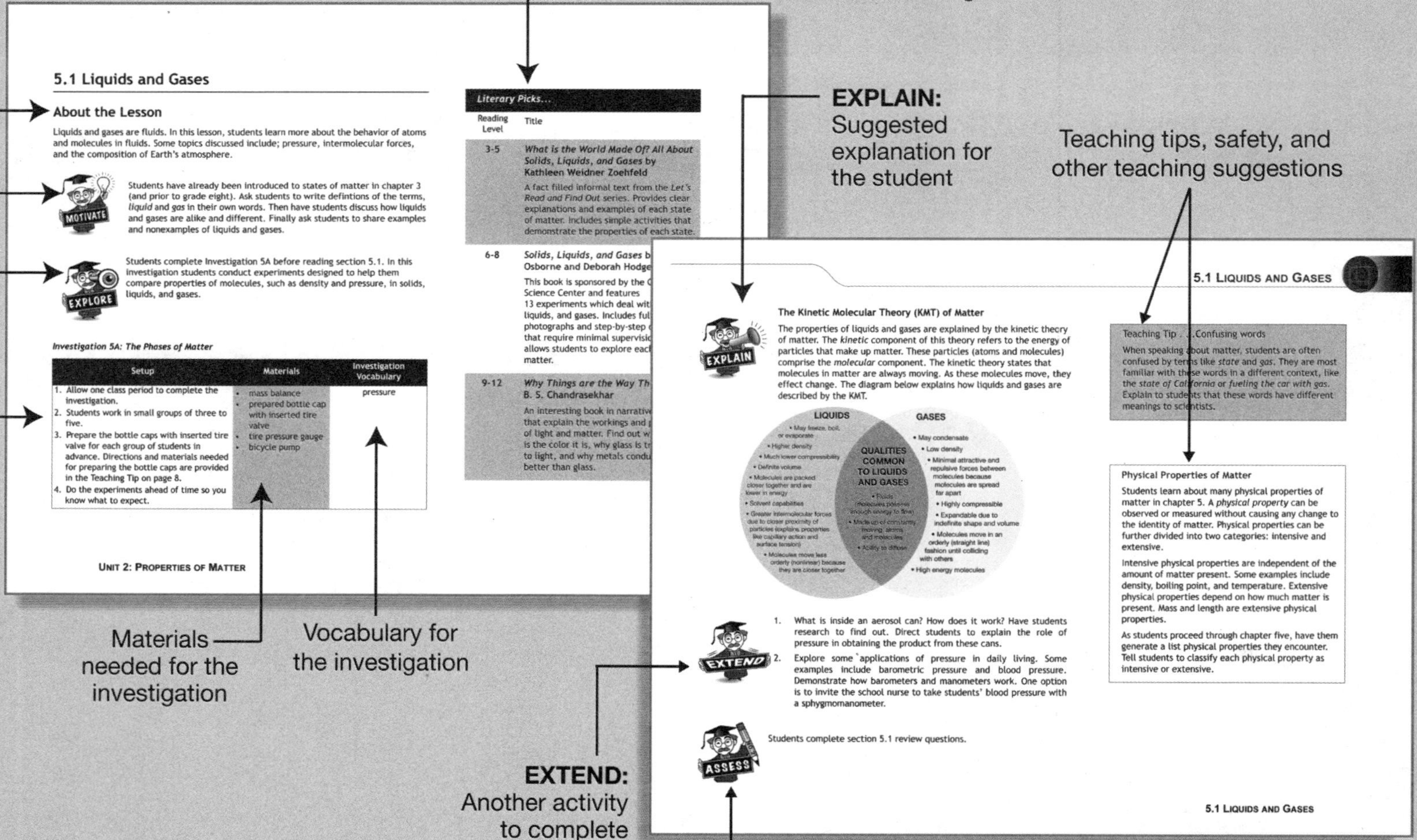

5.1 Liquids and Gases

About the Lesson

Liquids and gases are fluids. In this lesson, students learn more about the behavior of atoms and molecules in fluids. Some topics discussed include; pressure, intermolecular forces, and the composition of Earth's atmosphere.

MOTIVATE: Students have already been introduced to states of matter in chapter 3 (and prior to grade eight). Ask students to write defintions of the terms, *liquid* and *gas* in their own words. Then have students discuss how liquids and gases are alike and different. Finally ask students to share examples and nonexamples of liquids and gases.

EXPLORE: Students complete Investigation 5A before reading section 5.1. In this investigation students conduct experiments designed to help them compare properties of molecules, such as density and pressure, in solids, liquids, and gases.

Investigation 5A: The Phases of Matter

Setup	Materials	Investigation Vocabulary
1. Allow one class period to complete the investigation. 2. Students work in small groups of three to five. 3. Prepare the bottle caps with inserted tire valve for each group of students in advance. Directions and materials needed for preparing the bottle caps are provided in the Teaching Tip on page 8. 4. Do the experiments ahead of time so you know what to expect.	• mass balance • prepared bottle cap with inserted tire valve • tire pressure gauge • bicycle pump	pressure

Unit 2: Properties of Matter

Literary Picks...

Reading Level	Title
3-5	*What is the World Made Of? All About Solids, Liquids, and Gases* by Kathleen Weidner Zoehfeld A fact filled informal text from the *Let's Read and Find Out* series. Provides clear explanations and examples of each state of matter. Includes simple activities that demonstrate the properties of each state.
6-8	*Solids, Liquids, and Gases* b… Osborne and Deborah Hodge… This book is sponsored by the … Science Center and features 13 experiments which deal wit… liquids, and gases. Includes ful… photographs and step-by-step … that require minimal supervisi… allows students to explore eac… matter.
9-12	*Why Things are the Way Th…* B. S. Chandrasekhar An interesting book in narrativ… that explain the workings and … of light and matter. Find out w… is the color it is, why glass is t… to light, and why metals condu… better than glass.

5.1 Liquids and Gases

EXPLAIN: The Kinetic Molecular Theory (KMT) of Matter

The properties of liquids and gases are explained by the kinetic theory of matter. The *kinetic* component of this theory refers to the energy of particles that make up matter. These particles (atoms and molecules) comprise the *molecular* component. The kinetic theory states that molecules in matter are always moving. As these molecules move, they effect change. The diagram below explains how liquids and gases are described by the KMT.

Teaching Tip . . . Confusing words

When speaking about matter, students are often confused by terms like *state* and *gas*. They are most familiar with these words in a different context, like *the state of California* or *fueling the car with gas*. Explain to students that these words have different meanings to scientists.

Physical Properties of Matter

Students learn about many physical properties of matter in chapter 5. A *physical property* can be observed or measured without causing any change to the identity of matter. Physical properties can be further divided into two categories: intensive and extensive.

Intensive physical properties are independent of the amount of matter present. Some examples include density, boiling point, and temperature. Extensive physical properties depend on how much matter is present. Mass and length are extensive physical properties.

As students proceed through chapter five, have them generate a list physical properties they encounter. Tell students to classify each physical property as intensive or extensive.

EXTEND:

1. What is inside an aerosol can? How does it work? Have students research to find out. Direct students to explain the role of pressure in obtaining the product from these cans.
2. Explore some applications of pressure in daily living. Some examples include barometric pressure and blood pressure. Demonstrate how barometers and manometers work. One option is to invite the school nurse to take students' blood pressure with a sphygmomanometer.

ASSESS: Students complete section 5.1 review questions.

5.1 Liquids and Gases

Materials needed for the investigation

Vocabulary for the investigation

EXTEND: Another activity to complete

ASSESS: Evaluation and assessment references

A variety of teaching alternatives are provided as suggestions on these pages.

Investigation Dialog Pages

Title of the investigation

Reduced image of first investigation page

Steps of the investigation are matched to the dialog

Teaching tips that will help complete the investigation

Investigation 5A: The Phases of Matter

Have you ever heard sayings like, "He is an airhead!" or "She is full of hot air!" What comes to mind when you hear these phrases?

Both imply emptiness, lack of value, or that nothing of worth is there.

Air is a gas. Although it may feel like "nothing", it actually has considerable mass. In this part of the investigation, you will observe how changes in pressure affect the mass of air in a bottle. What do you think of when I say the word pressure?

Students answers may vary. Some common responses include being under stress or pushing against something.

Yes, those are examples of how humans often describe pressure. For instance, you may apply pressure to a wound or a friend may ask you to stop pressuring him. When scientists think of pressure, they have a different way of defining it. Pressure is the result of applying force to fluids. Fluids have the ability to flow. Which states of matter would qualify as fluids?

Liquids and gases are fluids.

Good answer. Therefore air is a fluid. What happens if you apply pressure to a block of steel by pushing down on it?

Push down on a block of steel or some other solid matter.

By pushing down on the block I am applying force to it. Can you see any change to the steel block as I push down on it?

No change is visible.

How about if I push down on this balloon? What do you see?

The walls of the balloon expand outward in all directions.

Pressure is the force applied to matter per unit of area. It acts in all directions, not just the direction in which it is applied (like the steel block). The pressure readings you collect will be in units of psi, which is an abbreviation for pounds per square inch. It is very important that the pressure not exceed 70 psi.

Write this safety note on the board and continue to remind students of this throughout this part of the investigation.

Each investigation team has a prepared cap, with the tire valve inside, and a one liter carbonated soda bottle. Fill the bottle with water to the very top. Transfer the water into a graduated cylinder to measure its volume. Record the volume of the bottle. Replace the cap on the empty bottle and measure its mass. The liquid water has been removed from the bottle, therefore it is full of air.

Students prepare the bottles.

Use your bicycle pump to raise the pressure in your bottle to 10 psi. Use the gauge to check the pressure in the bottle. Use the balance to measure and record the mass of the bottle and air at 10 psi. Repeat the pumping and mass measurement for pressures between 10 psi and 70 psi. Record your observations in Table 1. Remember that the pressure must not exceed 70 psi.

Demonstrate the proper technique. Students can better observe the relationship between pressure and mass with ample measurements, so increase the pressure by 10 psi each time.

UNIT 2: PROPERTIES OF MATTER

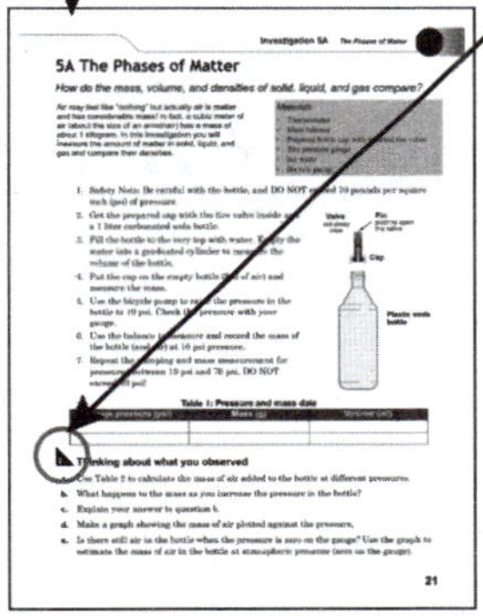

5.1 INVESTIGATION 5A: THE PHASES OF MATTER

1 Thinking about what you observed

Use the data you recorded to help you answer the questions in part two. Your results showed that as pressure in the bottle increased the mass also increased. Why do you think this happened?

Pressure is one factor that affects the density of a gas. Density is the mass per unit of volume. In this part of the investigation, more and more molecules were added to the bottle, increasing the mass each time the gauge pressure was increased. The volume of the bottle remained constant, meaning the air density within the confines of the bottle also increased.

Transfer the gauge pressure readings (in psi) and masses recorded in Table 1 to the first two columns of Table 2. Record the mass of the bottle when the gauge pressure is zero. Subtract the mass of the bottle at zero gauge pressure from the mass of the bottle and air to determine the mass of air added to the bottle. Repeat this calculation for each pressure reading and write your answers in the last column of Table 2.

Model the calculation for the first pressure reading.

Make a graph showing the mass of air (from column 4 in Table 2) plotted against the pressure. Be sure to include an appropriate title, axis labels, and range of data so the trends are easily recognizable. Once you have completed the graph, use it to estimate the mass of air in the bottle at zero gauge pressure.

Explain to students that many tire pressure gauges measure the difference between atmospheric pressure outside and the pressure inside the gauge. Therefore, when gauge pressure reads zero, this is equal to the atmospheric pressure.

2 Liquid and solid phases

Carefully read and follow each step of the procedure in part three. How do you think the volumes of equal masses of solid, liquid, and gas will compare?

Encourage student participation in the discussion. Students should conclude that equal masses of gas, liquid, and solid have very different volumes. They should also notice that the space occupied (volume) by the equal masses of substance varies greatly. For example, the amount of space occupied by the mass of pennies equal to the mass of zero gauge pressure is much less than the space occupied by an equal mass of air.

Teaching Tip . . .Preparing the bottle caps

The materials needed are as follows:

- cap from a one liter plastic bottle for each group
- valve stem for each group (available for purchase at an automotive supply store)
- electric drill and large drill bits
- household silicone adhesive/sealant

Drill a hole in the center of the bottle cap large enough to accommodate the black neck and shoulder of the valve stem. Test-fit the valve stem, then remove it. Place a bead of silicone sealant around the hole inside the bottle cap. Replace the valve stem. Allow the silicone to cure 24 hours before using the cap.

5.1 LIQUIDS AND GASES

Investigation 5A: Data and Answers

Gauge pressure (psi)	Mass (g)	Volume (ml)
0	50.0	1075.0
10	51.2	1075.0
20	51.9	1075.0
30	52.8	1075.0
40	53.6	1075.0
50	54.5	1075.0

1 Thinking about what you observed

a. See Table 2.

b. As the pressure increases, the mass increases.

c. The mass increases because adding pressure is actually adding more material in the form of air. While air seems to be very light, using the pump allows us to put a great deal more air in the bottle than would normally be in there. As pressure is added, air is added, and the mass of the additional air is seen when the mass of the bottle and its contents is measured.

d.

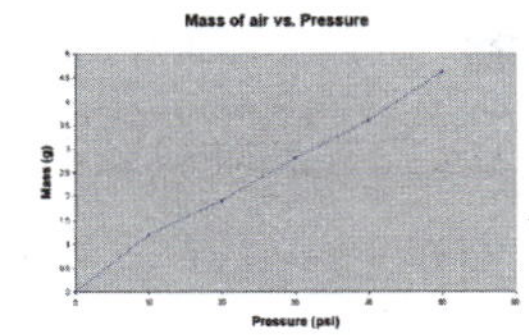

e. There is definitely air in the bottle when the gauge reads zero. The graph shows increases of 0.8 g to 1.0 g of air with each increase of 10 psi, so there is probably around 1.0 g of air in the bottle.

Pressure (psi)	Mass of bottle and air (g)		Mass of bottle at zero gauge pressure (g)		Mass of air added to bottle (g)
0	50.0	-	50.0	=	0
10	51.2	-	50.0	=	1.2
20	51.9	-	50.0	=	1.9
30	52.8	-	50.0	=	2.8
40	53.6	-	50.0	=	3.6
50	54.5	-	50.0	=	4.5

2 Liquid and solid phases

3 Thinking about what you observed

a. These objects all have about the same amount of matter.

b. The density is around 1g/ l. The density is about 1/1000th of water.

c. Answers of the mass will vary. Usually around 250 kg. The air in the room has more mass than the mass of a person.

d. Gas is usually the least dense, followed by liquid and then solid.

UNIT 2: PROPERTIES OF MATTER

Suggested dialog between teacher and students for teaching the investigation

Investigation answers including charts, graphs and diagrams

Assessment Pages

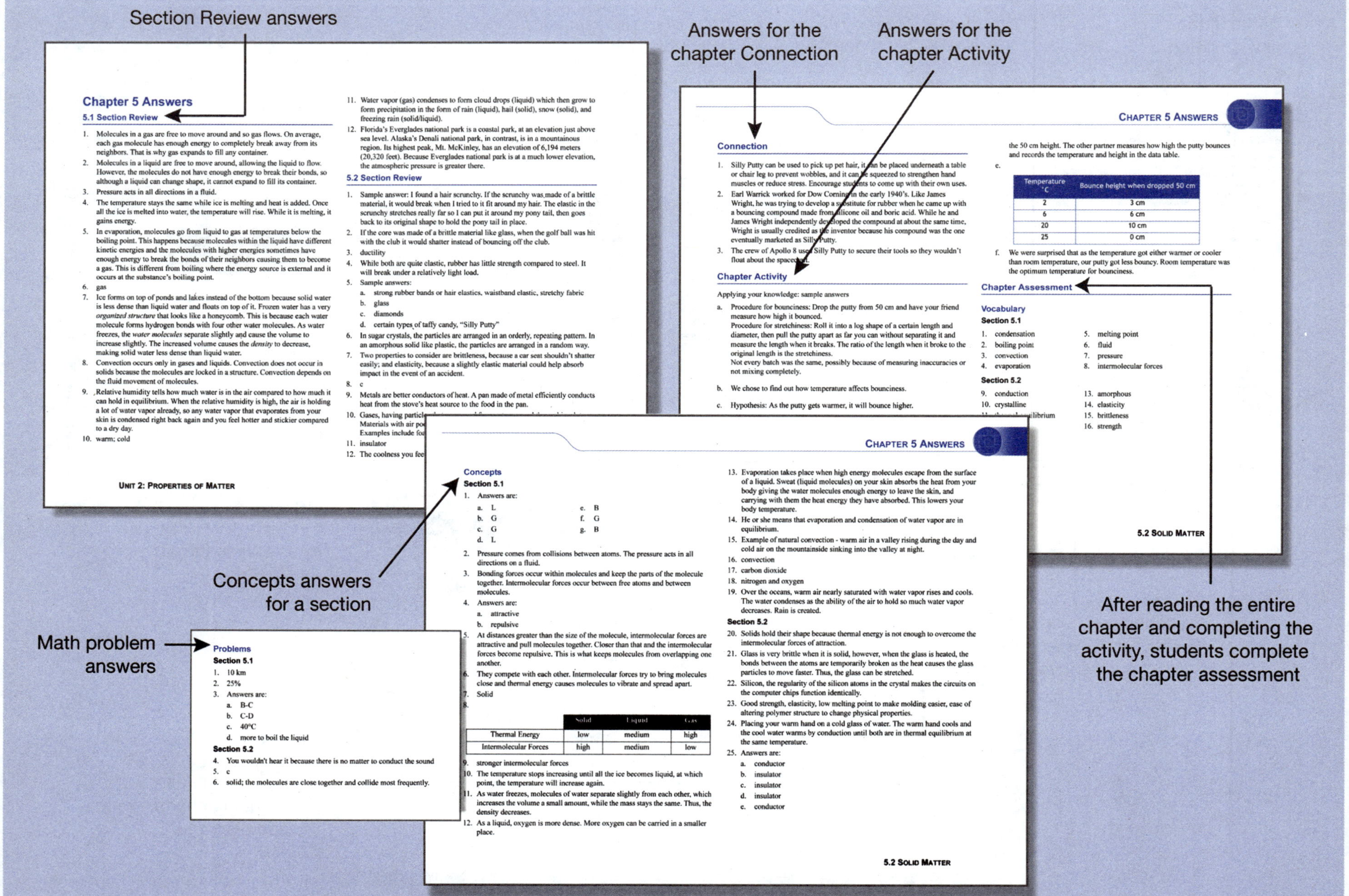

Table of Contents

Appendices

STANDARD SET 1. Motion		Student Text Chapters	Teacher's Guide pages*
1. The velocity of an object is the rate of change of its position. As a basis for understanding this concept:	1.a. Students know position is defined in relation to some choice of a standard reference point and a set of reference directions.	12	192-193
	1.b. Students know that average speed is the total distance traveled divided by the total time elapsed and that the speed of an object along the path traveled can vary.	12	194-199
	1.c. Students know how to solve problems involving distance, time, and average speed.	12	194-199
	1.d. Students know the velocity of an object must be described by specifying both the direction and the speed of the object.	12	194-199
	1.e. Students know changes in velocity may be due to changes in speed, direction, or both.	12, 14	194-199, 234-238
	1.f. Students know how to interpret graphs of position versus time and graphs of speed versus time for motion in a single direction.	12	200-205

STANDARD SET 2. Forces			
2. Unbalanced forces cause changes in velocity. As a basis for understanding this concept:	2.a. Students know a force has both direction and magnitude.	13, 14	218-222, 228-238
	2.b. Students know when an object is subject to two or more forces at once, the result is the cumulative effect of all the forces.	13, 14	212-222, 228-233
	2.c. Students know when the forces on an object are balanced, the motion of the object does not change.	13, 14	218-222, 234-238
	2.d. Students know how to identify separately the two or more forces that are acting on a single static object, including gravity, elastic forces due to tension or compression in matter, and friction.	13, 14	212-222, 234-238
	2.e. Students know that when the forces on an object are unbalanced, the object will change its velocity (that is, it will speed up, slow down, or change direction).	12, 14	194-199, 234-238
	2.f. Students know the greater the mass of an object, the more force is needed to achieve the same rate of change in motion.	14	228-233
	2.g. Students know the role of gravity in forming and maintaining the shapes of planets, stars, and the solar system.	14, 15, 17	228-233, 244-249, 278-283

*** See Chapter Resources for additional points of exposure to the content standards.**

STANDARD SET 3. Structure of Matter		Student Text Chapters	Teacher's Guide pages*
3. Each of the more than 100 elements of matter has distinct properties and a distinct atomic structure. All forms of matter are composed of one or more of the elements. As a basis for understanding this concept:	3.a. Students know the structure of the atom and know it is composed of protons, neutrons, and electrons.	6	94-105
	3.b. Students know that compounds are formed by combining two or more different elements and that compounds have properties that are different from their constituent elements.	3, 8	44-49, 126-136
	3.c. Students know atoms and molecules form solids by building up repeating patterns, such as the crystal structure of NaCl or long-chain polymers.	5, 8	82-87, 132-136
	3.d. Students know the states of matter (solid, liquid, gas) depend on molecular motion.	3, 5	44-49, 76-81
	3.e. Students know that in solids the atoms are closely locked in position and can only vibrate; in liquids the atoms and molecules are more loosely connected and can collide with and move past one another; and in gases the atoms and molecules are free to move independently, colliding frequently.	5	76-81
	3.f. Students know how to use the periodic table to identify elements in simple compounds.	7	110-121

STANDARD SET 4. Earth in the Solar System (Earth Sciences)			
4. The structure and composition of the universe can be learned from studying stars and galaxies and their evolution. As a basis for understanding this concept:	4.a. Students know galaxies are clusters of billions of stars and may have different shapes.	17	278-283
	4.b. Students know that the Sun is one of many stars in the Milky Way galaxy and that stars may differ in size, temperature, and color.	15, 16, 17	244-249, 262-272, 278-283
	4.c. Students know how to use astronomical units and light years as measures of distance between the Sun, stars, and Earth.	15, 16, 17	244-255, 262-267, 278-283
	4.d. Students know that stars are the source of light for all bright objects in outer space and that the Moon and planets shine by reflected sunlight, not by their own light.	16	262-272
	4.e. Students know the appearance, general composition, relative position and size, and motion of objects in the solar system, including planets, planetary satellites, comets, and asteroids.	15, 16	244-255

*** See Chapter Resources for additional points of exposure to the content standards.**

GRADE 8 CALIFORNIA STANDARDS

STANDARD SET 5. Reactions		Student Text Chapters	Teacher's Guide pages*
5. Chemical reactions are processes in which atoms are rearranged into different combinations of molecules. As a basis for understanding this concept:	5.a. Students know reactant atoms and molecules interact to form products with different chemical properties.	9, 10	148-152, 158-163
	5.b. Students know the idea of atoms explains the conservation of matter: In chemical reactions the number of atoms stays the same no matter how they are arranged, so their total mass stays the same.	10	158-163
	5.c. Students know chemical reactions usually liberate heat or absorb heat.	10, 11	164-168, 174-189
	5.d. Students know physical processes include freezing and boiling, in which a material changes form with no chemical reaction.	5	76-81
	5.e. Students know how to determine whether a solution is acidic, basic, or neutral.	9	142-152

STANDARD SET 6. Chemistry of Living Systems (Life Sciences)			
6. Principles of chemistry underlie the functioning of biological systems. As a basis for understanding this concept:	6.a. Students know that carbon, because of its ability to combine in many ways with itself and other elements, has a central role in the chemistry of living organisms.	7, 11	116-121, 174-185
	6.b. Students know that living organisms are made of molecules consisting largely of carbon, hydrogen, nitrogen, oxygen, phosphorus, and sulfur.	11	174-185
	6.c. Students know that living organisms have many different kinds of molecules, including small ones, such as water and salt, and very large ones, such as carbohydrates, fats, proteins, and DNA.	11	174-185

STANDARD SET 7. Periodic Table			
7. The organization of the periodic table is based on the properties of the elements and reflects the structure of atoms. As a basis for understanding this concept:	7.a. Students know how to identify regions corresponding to metals, non-metals, and inert gases.	7	110-121
	7.b. Students know each element has a specific number of protons in the nucleus (the atomic number) and each isotope of the element has a different but specific number of neutrons in the nucleus.	6, 7	94-99, 110-115
	7.c. Students know substances can be classified by their properties, including their melting temperature, density, hardness, and thermal and electrical conductivity.	5, 7	82-87, 116-121

*** See Chapter Resources for additional points of exposure to the content standards.**

STANDARD SET 8. Density and Buoyancy		Student Text Chapters	Teacher's Guide pages*
8. All objects experience a buoyant force when immersed in a fluid. As a basis for understanding this concept:	8.a. Students know density is mass per unit volume.	4	60-65
	8.b. Students know how to calculate the density of substances (regular and irregular solids and liquids) from measurements of mass and volume.	4	60-65
	8.c. Students know the buoyant force on an object in a fluid is an upward force equal to the weight of the fluid the object has displaced.	4	66-70
	8.d. Students know how to predict whether an object will float or sink.	4	66-70

STANDARD SET 9. Investigation and Experimentation			
9. Scientific progress is made by asking meaningful questions and conducting careful investigations. As a basis for understanding this concept and addressing the content in the other three strands, students should develop their own questions and perform investigations. Students will:	9.a. Plan and conduct a scientific investigation to test a hypothesis.	1, 2	10-23, 28-39
	9.b. Evaluate the accuracy and reproducibility of data.	2	28-39
	9.c. Distinguish between variable and controlled parameters in a test.	2	28-39
	9.d. Recognize the slope of the linear graph as the constant in the relationship $y = kx$ and apply this principle in interpreting graphs constructed from data.	2, 12	28-33, 200-205
	9.e. Construct appropriate graphs from data and develop quantitative statements about the relationships between variables.	2, 12	28-39, 200-205
	9.f. Apply simple mathematical relationships to determine a missing quantity in a mathematic expression, given the two remaining terms (including speed = distance/time, density = mass/volume, force = pressure × area, volume = area × height).	2, 3, 4, 12, 13,	28-33, 50-54, 60-70, 194-205, 212-217
	9.g. Distinguish between linear and nonlinear relationships on a graph of data.	2, 12	28-33, 200-205

* See Chapter Resources for additional points of exposure to the content standards.

SAFETY IN THE SCIENCE LABORATORY

The science laboratory is an exciting place for students to discover interesting facts about our world. It is important that everyone work together to create a safe laboratory environment. Safety concerns were considered as the CPO Science Investigations were developed. Investigation activities and experiments have been written with safety in the laboratory as the primary concern.

Safety first

Each CPO Science learning system introduces students to general safety guidelines for all laboratory environments. The *Safety Skills Activity* located in the Investigations Manual emphasizes basic safety rules, discusses special precautions needed when students are working with heat and electricity, explains proper techniques for disposing of materials and supplies, and stresses the importance of diligence in keeping others safe.

The teacher is responsible for ensuring that all students know the location and proper use of all safety equipment, such as fire extinguishers, showers, blankets, and eye wash stations, as well as the emergency exits. The *Safety Skills Activity* also includes a safety quiz, which assesses whether students are familiar with their science lab, understand the rules, and know how to respond in the event of an accident.

The final component of the *Safety Skills Activity* is a contract. The student signs this contract and brings it home for a parent or guardian signature. Keep the original signed copy of the contract for your records and provide a copy for the student.

General safety practices

The teacher plays the most important role in ensuring that students have a safe environment in which to learn. Always check the condition of laboratory equipment and supplies before students begin an investigation. Do not use any damaged glassware or electrical equipment with frayed wiring. Keep the laboratory area clean and free of clutter. Always store and dispose of chemicals according to the supplier's instructions or as directed by your local or state safety authority. Keep the first aid kit in an easily accessible location and replenish it as supplies are used.

Accident preparedness

Even with the best safety implementations accidents may occur. If an accident happens, the teacher is the first responder. Always remain calm as not to alarm students or increase their level of fear or anxiety. Send for help from the school nurse or an administrator if needed. Become familiar with your school district's policy for how to handle and report an accident.

Laboratory classroom management

Students should be monitored in the laboratory at all times. Students are more likely to follow safety guidelines when they know the reason for the guideline and know the teacher is watching. CPO Science investigations employ cooperative learning strategies. It is important for the teacher to maintain a manageable group size.

Note: A complete description of the safety guidelines and procedures for all science courses, including safety regulations for students, is located in the Science Safety Handbook for California Public Schools. Copies of this document may be obtained through the California Department of Education.

SAFETY RULES FOR SCIENCE STUDENTS

he laboratory is a place where you will have the opportunity to "do science." However, you must always act responsibly while working in he science laboratory. The safety tips provided below are to protect you and your classmates from injury.

Start the investigation only after your teacher has given you permission to do so.

Always ask questions if you are unsure about the procedure or need help at any time during the investigation.

Follow your teacher's instructions before, during, and after the investigation.

Eating or drinking in the laboratory or from laboratory equipment is not permitted.

Tell your teacher about any accidents, spills, or injuries, even if it seems to be a small matter.

Your work area must be cleaned and all materials returned to their proper place when the investigation is completed.

irst Aid Checklist

any person in your lab group is injured, immediately notify your teacher. Follow the instructions given below:

Accident or Injury	What to Do
burns	Flush the affected area with large amounts of water.
eye injuries	Escort your classmate to the eye wash station. Flush out the eye with water for at least 15 minutes.
exposure to poisons	Write the name of substance ingested.
bruises or cuts	Apply pressure to stop the bleeding. Apply a cold compress to bruises and a clean bandage or gauze to cuts.
hemical spills on the skin	Use the safety shower or flush the area with large amounts of water.

Chapter 1: Studying Physics and Chemistry

Learning Goals	California Content Standards	Activities and Resources	Assessment
1.1: The Physical Science in Your Life The student will: • Describe everyday applications of physical science. • Distinguish solid, liquid, and gaseous states of matter. • Explain the importance of studying physical science.	PS 3d PS 9a	**Laboratory Investigation 1A:** Inquiry and Scientific Evidence **Investigation Skills:** Lab Safety, Lab Report Format **Reading Skills Sheet:** Using Your Textbook	**Exam View Test Bank** Chapter Pre-Test **Section 1.1 Review**
1.2: Time and Length The student will: • Convert between English and metric measurements. • Explore multiple meanings of time. • Explain the significance of units when reporting measurements. • Estimate distances.	PS 9a	**Laboratory Investigation 1B:** Measuring Time and Distance **Teaching Illustrations:** Metric Prefixes, Converting Between Units of Length **Graphic Organizer:** Time and Length **Investigation Skills:** Recording Observations **Math Skills Sheet:** Stopwatch Math, SI Units, SI English, Dimensional Analysis Conversions **Reading Skills Sheet:** SQ3R - A Reading Study Method	**Section 1.2 Review**
1.3: Inquiry and the Scientific Method The student will: • Apply deductive reasoning skills to solve problems. • Contrast hypotheses, theories, and laws. • Explore the scientific process and apply steps of the scientific method.	PS 9a	**Graphic Organizer:** Inquiry **Teaching Illustrations:** Steps in Learning Through Inquiry **Reading Skills:** Study Notes **Biographies:** James Joule **Connection:** The Art and Sole of Scientific Thinking **Chapter Activity:** Make a Water Clock	**Exam View Test Bank** Chapter Pre-Test **Section 1.3 Review** **Chapter Assessment** **Exam View Test Bank**

Estimated Instruction Time	Language Tools	Differentiated Strategies	Materials
Two 45-minute class periods	**Literary Picks** **Vocabulary:** mass, matter, phases of matter, chemical change **Word Origins:** inquiry	**ELL Strategies:** Listed in the ELL Ancillary **Learning Strategies:** • Cooperative Learning - Investigation 1A, Motivate: Secret Places • Paired Learning - Explain: Introducing Physical Science **Teaching Tip:** Career Day, Research Questions	**Investigation 1A:** small cardboard boxes which may be closed and sealed, selection of objects to be placed in the box
Two 45-minute class periods	**Vocabulary:** distance, astronomical unit, light year	**Learning Strategies:** • Active Learning - Motivate • Cooperative Learning - Investigation 1B **Teaching Tip:** Measuring Time and Length	**Investigation 1B:** CPO timer and two photogates, metric tape, centimeter ruler or meter stick with millimeter gradations, several different books, preferably made with different types of paper **Teaching Tip - Measuring Time and Length:** different tools like meter stick, hourglass, stopwatch, caliper, watch, clock, sundial, etc.
Two 45-minute class periods	**Vocabulary:** natural laws, inquiry, deduce, hypothesis, theory, scientific method, objective, repeatable **Closure Cards**	**Learning Strategies:** • Active Learning - Motivate • Graphic Organizer - Explain • Paired Learning - Extend **Teaching Tip:** Solving Logic Puzzles, Communicating Results **Addressing Misconceptions:** Hypotheses, Theories, and Laws	**Chapter Activity:** Two empty 2-liter soda bottles, 400-mL beaker, water, duct tape, stopwatch

1.1 The Physical Science in Your Life

About the Lesson

How is physical science relevant to humans and their daily activities? In this lesson, students learn how principles of physical science are applicable to their lives. For example, students examine how physics and chemistry principles may be used to describe automobile mechanics, reactions within living organisms, and how energy from the sun reaches and sustains life on Earth.

Secret Places

Divide the class into small groups of students. Assign each group a secret place. Be sure that students are unaware of the places assigned to other groups. Then have each group provide descriptions of their secret place, one clue at a time. Each place should be described in terms of the senses. For example, one place may be a movie theater. Have students describe what they may hear, smell, see, feel, and taste at the theater. Are other groups able to guess the identity of each place?

Students complete Investigation 1A before reading section 1.1. In this investigation students rely on their senses to determine the identity of objects they cannot see. The purpose of this investigation is to introduce students to the process of scientific inquiry.

Investigation 1A: Inquiry and Scientific Evidence

Setup	Materials	Investigation Vocabulary
1. Allow one class period to complete the investigation. 2. Students work in small groups. 3. Setup a variety of boxes before the investigation. Include different items in each box.	• small cardboard boxes which may be closed and sealed • a selection of objects to be placed in the boxes	scientific inquiry evidence hypothesis

Literary Picks...

Reading Level	Title
3-5	***Vision of Beauty: The Story of Sarah Breedlove Walker*** **by Kathryn Lasky** The inspiring story of a first freeborn child of former slaves who rises to the top as queen of beauty products for African-American women. Provides details about her working in a small laboratory using the scientific process and reasoning to develop formulas.
6-8	***Motel of the Mysteries*** **by David Macaulay** This book is a humorous science fiction story about archaeologists who try to explain the past by using present-day knowledge. The book is well illustrated and demonstrates the use of scientific inquiry and deductive reasoning.
9-12	***Stephen Hawking: Solving the Mysteries of the Universe*** **by Ron Cole** Stephen Hawking is known as one of the 20th century's most brilliant physicists. This book discusses his ability to overcome physical limitations in order to promote his own scientific theories.

Introducing Physical Science

In section 1.1 students are introduced to the relevance of physical science in their daily living. Many of the concepts discussed in this lesson may already be familiar to students. Have students read section 1.1. Then have students work in pairs to discuss what they have read.

Distribute three index cards to each pair of students. Have students work together to answer each of the following questions on its own card.

1. What is physical science?
2. How does physical science apply to living organisms?
3. Explain how concepts like force, matter, mass, and energy are important to studying physical science.

It is important to accept the answers students provide. Use this activity to determine what students already know about physical science. Discuss the answers to these questions in a manner that sets the stage for introducing important concepts that integrate the physical sciences: force and energy; the laws of conservation; atomic theory; and kinetic theory.

1. Students choose an object they use each day, like a blow dryer or toothbrush, and explain how physical science applies to its structure, design, or function.
2. Have a career day. Invite a number of scientists to be a part of a panel and explain how physical science is applicable to their field of study. Some scientists to consider are meteorologists, chemists, astronomers, geologists, and physicists. Also, include panelists who apply principles of physical science in their careers, like engineers. Help students to write questions in advance. When having panel discussions, it is important for students to rehearse in order to limit anxiety about speaking before a group of adults. Ask panel members to bring brochures and other literature for your students to keep.

Students complete section 1.1 review questions.

Word Origins...inquiry (from Latin *quaerere* meaning "to seek")

The words, inquiry, query, and question have the same origin. Learning science by inquiry means to seek the answers to questions. Have students use a dictionary to look up the meanings of each term, and then use each word in a sentence.

Teaching Tip . . .Career Day

Ask your school counselor to participate in the career day. Eighth grade is a great time to get students thinking about their future careers as well as the courses they will need to take to prepare for those careers.

Investigation 1A: Inquiry and Scientific Evidence

Complete the *Secret Places* activity before the investigation.

Think about how you were able to guess the secret places your classmates described earlier. Even though you were not physically there, you were able to identify the places. How was this possible?

Each place described by the groups of students had distinctive characteristics that made them identifiable. For example, the smell of popcorn is characteristic of a movie theater.

Do you think that your methods for determining the identity of each place qualify as science?

Students share opinions.

Each of you acted as a scientist when using clues about each place to discover its identity. Scientists solve problems or discover mysteries of nature in a similar way. They make observations and use their investigative skills to draw important conclusions about the world in which we live.

1 Setting up

In today's investigation you will use your mystery solving skills to figure out the identity of an object or objects that you cannot see. Each group should have a box. Before you touch the box, write down at least one thing you know for sure about the object or objects hidden inside.

Give students time to think and write. One immediate observation may be related to the object's size (relative to the size of the box).

Pretend your box is a gift you are not allowed to open. Is there anything that you could possibly do to help you guess what the present might be?

Possible answers would include feeling its weight, shaking it to see what is sounds like, or smelling for any distinct aromas.

Before picking up the box, write down at least three observations you could make that might help you figure out what is inside. These observations may be based on you being able to hold the box and manipulate it in any way you like. Of course, you are not allowed to open it or damage it in any way.

Students make initial observations about the box and its contents.

2 Conducting your inquiry

What comes to mind when you hear the word, inquiry?

Students may or may not be familiar with the meaning of the term. See the Word Origins box for more about the meaning of inquiry.

An inquiry is conducted to find the answers to questions. Police detectives conduct inquiries to solve crimes. A scientific inquiry is an investigation to solve science problems. It includes all the things you do and thought process applied to help you arrive at a solution. Scientific investigations may start with an observation that leads to a research question. What is your research question in this investigation?

The research questions is, "What is in the box?"

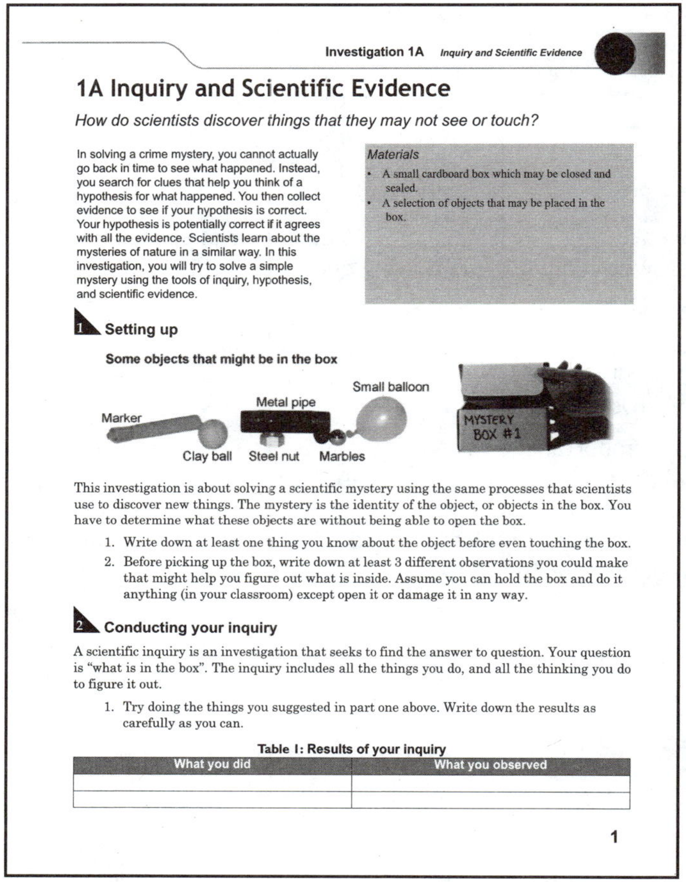

Investigation 1A *Inquiry and Scientific Evidence*

1A Inquiry and Scientific Evidence

How do scientists discover things that they may not see or touch?

In solving a crime mystery, you cannot actually go back in time to see what happened. Instead, you search for clues that help you think of a hypothesis for what happened. You then collect evidence to see if your hypothesis is correct. Your hypothesis is potentially correct if it agrees with all the evidence. Scientists learn about the mysteries of nature in a similar way. In this investigation, you will try to solve a simple mystery using the tools of inquiry, hypothesis, and scientific evidence.

Materials
- A small cardboard box which may be closed and sealed.
- A selection of objects that may be placed in the box.

1 Setting up

Some objects that might be in the box

This investigation is about solving a scientific mystery using the same processes that scientists use to discover new things. The mystery is the identity of the object, or objects in the box. You have to determine what these objects are without being able to open the box.

1. Write down at least one thing you know about the object before even touching the box.
2. Before picking up the box, write down at least 3 different observations you could make that might help you figure out what is inside. Assume you can hold the box and do it anything (in your classroom) except open it or damage it in any way.

2 Conducting your inquiry

A scientific inquiry is an investigation that seeks to find the answer to question. Your question is "what is in the box". The inquiry includes all the things you do, and all the thinking you do to figure it out.

1. Try doing the things you suggested in part one above. Write down the results as carefully as you can.

Table I: Results of your inquiry

What you did	What you observed

1

Teaching Tip . . .Research Questions

It is helpful to write the research question on the board so that students can refer to it at various stages of the investigation. Seeing the question repeatedly keeps students focused on gathering data and making observations that will help them answer the question by the end of the investigation.

n part one you wrote down ideas of how you might figure out what is in the box. Try each of those ideas. Write what you did in the first column and what you observed in the second column.

Walk around to facilitate student progress.

3 Your hypothesis

Did the observations you made about the characteristics of the mystery object lead you to an idea about what might be in the box?

Allow students to share their ideas and opinions about what they think may be in their box and how or why they arrived at their conclusions.

An important part of scientific inquiry is developing a hypothesis. Can anyone tell me what a hypothesis is?

Students may define hypothesis in a number of ways based on their experiences before grade eight. Common responses include an educated guess or a prediction based on observations; or a possible explanation to a problem. Tips for developing a hypothesis are shown at right.

Use your observations to help you write a hypothesis. Remember that your hypothesis is an attempt to answer the research question. It is a possible explanation. Although it may or may not be correct, it must be based on what you actually observed or on your existing knowledge. First, focus on what you think is in the box. How did you arrive at this guess?

Students may respond by describing what they heard, smelled, or felt.

Think about what you observed. Was there anything that you may have heard, felt, or smelled that could have lead you to a different conclusion?

4 Testing your hypothesis

A hypothesis must be testable. When scientists conduct experiments they are actually testing a hypothesis. Now it is your turn to experiment. Think of at least one way you can test your hypothesis without opening or damaging the box. Be sure to record your test and observations on your paper.

5 Stop and think

It is important to remember that experimentation is a tool designed to test your hypothesis. A hypothesis is only as good as its observations. What should you do if one of your observations does not support your hypothesis?

The hypothesis must be changed. Explain to students that scientists are rarely correct the first time. The more they conduct experiments, the more they learn and are able to make a new hypothesis based on these new observations. Students complete questions a-d.

6 Exploring on your own

Think about setting up a mystery box for one of your family members. What would be some of the tests you could have them perform to correctly identify the object(s) in the box?

Students make mystery boxes and four tests that could be used to identify what is in the box.

Hypotheses and Experiments

By the time students reach grade eight, they should already know how to write a hypothesis. Students should also be familiar with the characteristics of a hypothesis. Review these important points about hypotheses with students.

1. A hypothesis may offer an explanation for observations.
2. A hypothesis is often based on prior experiences or knowledge, facts, research, or other observations.
3. A hypothesis leads to or includes a testable prediction.
4. A hypothesis often leads to further inquiry.
5. A hypothesis must be revised or discarded if any observations do not support it.

Another important part of experimentation centers around the *incorrect hypothesis*. It is important for students to realize that scientists' initial hypotheses are often incorrect. Explain to students that scientists may learn as much (or more) from an incorrect hypothesis as they do when they are right. This is because an incorrect hypothesis leads to more research and experimentation.

Investigation 1A: Data and Answers

1 Setting up

2 Conducting your inquiry

What you did	What you observed
shook the box	heard and felt item(s) tumble around
put magnet near box	felt and heard metal object inside move toward magnet
spun box on table	box wobbled somewhat irregularly
smelled box	smelled like a balloon

3 Your hypothesis

My hypothesis is that there is a steel bolt and a small balloon in my box. When I shook the box it felt unevenly weighted, as if there was one heavy item and one light item in the box. When I put the magnet near the box I could feel and hear an object move inside the box. After spinning the box, it seemed there was something taking up a lot of room which did not weigh much, and that item kept the metal item from moving around much. The light item must be a balloon because it is so light and does not weigh very much, and the heavy item must be a metal bolt because it felt long and was attracted to the magnet. I smelled the box and I could smell a rubbery smell that reminded me of a balloon.

4 Testing your hypothesis

- Tip the box over from different sides and see if it falls over.
- The box is clearly much heavier on one side, and falls over rapidly when the heavy side is down.

5 Stop and think

a. My hypothesis is that there is a steel bolt and a small balloon in my box.

b. (see the table for letter c that follows)

c.

What you observed	Support hypothesis?
heard and felt item(s) tumble around	yes
felt and heard metal object inside move toward magnet	yes
box wobbled somewhat irregularly	yes
box tips over irregularly	yes

d. My hypothesis was right.

6 Exploring on your own

- Examine weight of box
- Shake box
- Put magnet near box
- Spin box on table
- Tip box over
- Listen to item(s) bouncing around inside box
- Smell box

Notes:

1.2 Time and Length

About the Lesson

Time and distance are important quantities used to describe scientific measurements. In this lesson students learn how time and distance are measured, identify acceptable units for expressing those measurements, and how to convert from one unit to another.

Place a number of tools before students. Point to each item and ask students to tell what each tool is used to measure. All answers will be either time or distance. Some tools to include are a meter stick, ruler, hourglass, stopwatch, caliper, watch, clock, sundial, and tape measure. Then ask students to generate a list of units that are appropriate for measuring length and time. Additional activities are shown at right.

Students complete Investigation 1B before reading section 1.2. In this investigation students examine how time and distance are measured and described in physical science. Students are introduced to the CPO Timer in this investigation. This device will be used in many other experiments so it is important that students become comfortable using it. Take the time to become familiar with the timer features before the investigation.

Investigation 1B: Measuring Time and Distance

Setup	Materials	Investigation Vocabulary
1. Allow one class period to complete the investigation. 2. Students work in small groups of three to five. 3. The timer will work fine when powered by a 9-volt battery for several hours. Batteries do have a certain life span, and when they begin to run low, the timer may begin to run slow or have trouble giving a reading when used for an extended period of time. Be sure to replace the batteries as needed, or use the AC adapter included with the timer. Using the adapter will ensure the timer will always work properly.	• CPO Timer and two photogates • metric tape • centimeter ruler or meter stick with millimeter gradations • several different books, preferably made with different types of paper	stopwatch time interval reaction time photogate distance precise conversion factor

Teaching Tip . . .Measuring Time and Length

Many students are familiar with units of time and length. However, they are often unable to visualize specific lengths, like one centimeter. Have students look around the classroom and locate items of different lengths. Say to students, "Who can find an item that is one centimeter long?" Repeat this type of questioning for various units (inch, meter, millimeter, foot, etc).

Another option is to present a certain item to students and have them guess its length in a specified unit. The student whose guess is closest to the actual length without going over wins a prize.

Then ask students to think about units of time. Have students remain quiet for a designated period of time. Ask students to guess how much time has elapsed since anyone talked. Be sure to keep a record of the elapsed time and reward the student whose guess is closest to the actual time.

Unit Conversions

Prior to grade eight, students may have learned to convert from one unit to another by using a variety of methods. In this lesson students are introduced to conversions by the *factor-label method*. Students may need to review metric prefixes and conversions using other methods before moving on to factor-label method. Review the steps shown below and then introduce students to factor-label method (Figure 1.1).

TWO WAYS TO TEACH METRIC CONVERSIONS

Line Method

Example: How many kilometers are in 25 decimeters?

Step 1: Arrange the metric prefixes on a line.

Step 2: Identify the given measurement. *(25 decimeters)*

Step 3: Locate the position of the unit to which you are converting. *(kilometers)*

Step 4: Starting with the given measurement, count the number of spaces between the two units. *(4)*

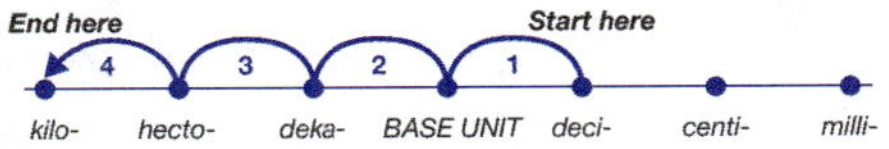

Step 5: Which direction did you move? *(from right to left)*

Step 6: Move the decimal point four spaces to the left.

25. decimeters → .0025 kilometers

Answer: 25 decimeters = .0025 kilometer

Stair Method

Example: Find the number of milligrams in 1.4 grams.

Step 1: Arrange the metric prefixes as steps on stairs.

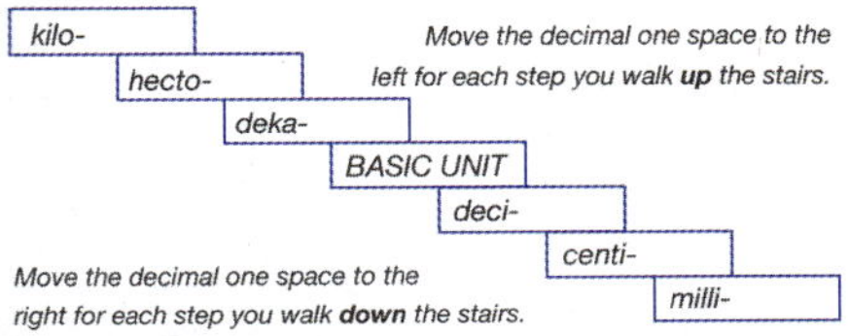

Step 2: Mark your starting position on the stairs. *(grams)*

Step 3: Walk up or down the stairs to your ending position. *(milligrams)*

Step 4: Count the number of steps you took. *(3)*

Step 5: Which direction did you move: up or down? *(down)*

Step 6: Move the decimal point three spaces to the right.

1.4 grams → 1400. milligrams

Answer: 1.4 grams = 1400 milligrams

1. Have students determine the distance in miles from their home to school, then convert the distance from miles to kilometers.
2. Create a "*Time Timeline*." Direct students to track how humans have kept track of time from the periods of ancient civilizations to modern man. Have students offer suggestions for how they think time may be kept in the future.

Students complete section 1.2 review questions.

THE FACTOR LABEL METHOD

The Factor label method is used to convert from one unit of measurement to another.

Example: A student recorded the volume of a liquid as 2.650 liters (L). What is the volume of the liquid in milliliters (mL)?

Step 1: State the relationship between units.

1000mL = 1L

Step 2: Write the relationship as a conversion factor.

$$\frac{1000\text{mL}}{1\text{L}} \quad \text{Or} \quad \frac{1\text{L}}{1000\text{mL}}$$

Step 3: Identify the given quantity in the problem and the unit you want to eliminate.

The given quantity is 2.650 L.
The unit to eliminate is Liters (L).

Step 4: Set up the problem to eliminate the unwanted unit. *Hint: Write the conversion factor with the unit you wish to eliminate in the denominator.*

$$2.65\ \text{L} \times \frac{1000\text{mL}}{1\text{L}} = ?$$

Step 5: Cancel units and complete the calculation.

$$2.65\ \cancel{\text{L}} \times \frac{1000\text{mL}}{1\cancel{\text{L}}} = 2650\ \text{mL}$$

Figure 1.1: The factor-label method.

Investigation 1B: Measuring Time and Distance

I will pay 20 to anyone who can stay after school tomorrow to help clean the classroom. Do I have any volunteers?

Most students should be interested.

Think for a minute. What am I willing to pay for your help after school?

Students may respond $20. Others may think 20 extra credit points. Still others may realize that they are unsure what the 20 is. Accept all responses.

You really do not know what I am offering. It could be 20 of anything. It could even be something you do not want, like 20 spoonfuls of broccoli or 20 homework assignments. Why was there uncertainty about the pay?

Students should have observed that the unit was not specified.

Exactly. The unit is missing. Without a unit, a number is meaningless. This is true with all measurements. Measurements have two parts: the number and the unit. The number is a quantity or answers the questions, "How much?" or "How many?" The unit is the way the number is measured. It is important to know what each unit represents or measures.

Write the following statements on the board. (1) Units give numbers meaning. (2) Measurements come with numbers and meanings. Then write several examples of measurements on the board, like 125 grams or 15 feet, and have students identify the number and unit.

1 Using the timer as a stopwatch

In this investigation, you are going to use a stopwatch. The stopwatch is one of the functions of the photogate timer and it will measure a time interval. Who can tell me what a time interval is?

It is the amount of time between events.

The stopwatch is started with the "A" button and reset with the "O" button. Take a minute and practice starting and resetting the stopwatch function.

Allow time for students to become familiar with the stopwatch function. Be sure you have used the stopwatch function so that you can help troubleshoot areas where students may experience difficulty.

2 Observing reaction time

Now that everyone knows how to correctly use the stopwatch function of the timer, let's play a game. Our game will involve reaction time. Who can tell me what reaction time means?

Reaction time is the amount of time needed to respond to a stimulus.

It takes two people to play this game. One person in each group needs to be the watcher. As the watcher it is your responsibility to pick a number between 5 and 10 seconds. The other person will be the starter/stopper and will need to start and stop the stopwatch without looking at the display. When the stopwatch gets to the amount of time that the watcher has chosen, the watcher will say stop. The starter/stopper will stop the stopwatch as quickly as possible. Do this a few times until you are able to predict your reaction time.

Allow enough time for each student to practice being the watcher and the starter/stopper.

1B Measuring Time and Distance

How are time and distance measured and described in physical science?

We describe the universe with measurements. A measurement includes a unit and the quantity. For example, 3 seconds is a measurement of time that includes a unit (seconds) and a quantity (3). This investigation will explore the measurement and units for time and distance.

Materials
- CPO Timer and 2 photogates.
- Metric tape, centimeter ruler or meter stick with millimeter gradations
- Several different books, preferably made with different types of paper.

1 Using the timer as a stopwatch

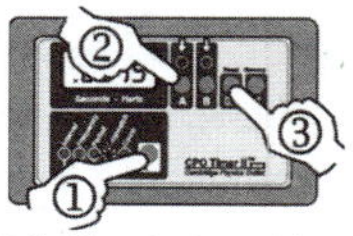

1. Set the timer to **stopwatch**.
2. Start and stop using the "**A**" button.
3. Reset the stopwatch to zero with the "**O**" button.

The photogate timer allows you to make accurate, precise measurements of time. The timer does different functions and the first one to try is **stopwatch**. Use the button (1) to move the light under the word stopwatch.

A stopwatch measures a **time interval**. The stopwatch is started and stopped with the "**A**" button (2). The display shows time in seconds up to 60 seconds, then changes to show **min:sec** for times longer than one minute.

2 Observing reaction time

The time it takes a signal from your brain to move a muscle is called **reaction time**.

1. This experiment takes two people. One person (the watcher) watches the stopwatch and the other person pushes the buttons. The watcher should think of a number between 5 and 10 seconds and keep the number secret.
2. The second person starts (and stops) the stopwatch *without looking at the display*. The watcher looks at the display and says STOP at the secret number. For example, if the secret number is 6 they should say STOP when the display reaches 6.00 seconds.
3. Repeat the experiment several times and estimate reaction time.

3 Mixed units for time

Time is often given in mixed units including hours, minutes, and seconds. Convert (a) and (b) to seconds then arrange the three measurements from smallest to largest:

a. 16,000 seconds

b. 250 minutes

c. 4 hours, 23 minutes and 15 seconds (4:23:15)

3

3 Mixed units for time

What are some units for time? How many seconds are in one minute?

Answers may include hours, minutes, days, weeks, seconds, etc. There are 60 seconds in a minute. Write 60 seconds = 1 minute on the board.

Who can tell me how to find out how many seconds there are in 6 minutes?

Students answers may vary. Stop at this point and model the conversion. Use the steps outlined in Figure 1.2. Be sure to vocalize each step to reinforce it. Students then complete questions a-c. Assist as needed and review the answers to each question before moving on.

4 Using the photogates

Students read each step of the procedure carefully. Then answer the questions that follow.

5 Using the timer with two photogates

Students follow each procedural step in part five and use their observations to fill in Table 1.

6 Another test to try

Think about how the timer and photogates work. Do not tell me your answer but ask yourself this question, "Did the time add up or did the time reset itself?" How do you know this is true? Complete the sentence in part six of the investigation.

7 Measuring small distances

What are some units used to measure distances in the US?

Sample responses may include feet, inches, miles.

Scientists use the SI or *Systeme Internationale*. It is based on the metric system. What are some acceptable units for measuring distance in this system?

Sample responses may include meters, kilometers, etc.

Many measurements are based on the meter. Some are used for very large measurements, like the kilometer and others are used for very small measurements, like the millimeter. Some things are so small that scientists are unable to measure them with a ruler. Can anyone think of something that is too small to measure it with a ruler?

Answers will vary. Some examples are a grain of sand, an atom, and germs.

You will measure the thickness of a piece of paper. How can you accomplish this?

Answers will vary. One possibility is to measure a large stack of papers. Count how many papers are in the stack and divide the measurement by the number of pages.

8 Converting between metric units of distance

Students complete the conversions in part eight.

9 Measuring distances in different units

Students complete the measurements in part nine. Answers may vary from one group to the next if the desks are not identical.

How many seconds are in 6 minutes?

Step 1: State the relationship between seconds and minutes.

60 seconds = 1 minute

Step 2: Write the relationship as a conversion factor.

$$\frac{60\text{ sec}}{1\text{min}} \quad \text{Or} \quad \frac{1\text{min}}{60\text{ sec}}$$

Step 3: Identify the given quantity in the problem and the unit you want to eliminate.

Given the quantity is 6 minutes.
The unit to eliminate is minutes (min).

Step 4: Set up the problem to eliminate the unwanted unit.
Hint: Write the conversion factor with the unit you wish to eliminate in the denominator.

$$6\text{ min} \times \frac{60\text{ sec}}{1\text{min}} = ?$$

Step 5: Cancel units and complete the calculation.

$$6\,\cancel{\text{min}} \times \frac{60\text{ sec}}{1\,\cancel{\text{min}}} = 360\text{ sec}$$

Figure 1.2: Applying the factor-label method.

Investigation 1B: Data and Answers

1 Using the timer as a stopwatch

2 Observing reaction time

3 Mixed units for time

a. 16,000 seconds

b. 15,000 seconds

c. 15,795 seconds

The order from smallest to largest is b, c, a.

4 Using the photogates

a. The clock is started by blocking the light beam.

b. The clock is stopped by letting the light beam reconnect.

c. The clock has measured the time that the beam was blocked.

5 Using the timer with two photogates

A light	B light	How do you start the clock?	How do you stop the clock?	What time interval does the clock measure
On	Off	Block A beam	Stop blocking	Time A is blocked
Off	On	Block B beam	Stop blocking	Time B is blocked
On	On	Block A beam	Block B beam	Time from A to B
Off	Off	can't start it	not measuring	none

6 Another test to try

The timer resets itself every time the beam is blocked because when a time is displayed from a previous measurement and the beam is blocked again, the readout on the timer starts counting from zero.

7 Measuring small distances

a. By measuring the thickness of all the pages of a book with a ruler and dividing the total thickness by the number of pages, the average thickness of each page can be determined in millimeters.

b. 23.0 mm/ 470 pages = 0.05 mm per page.

8 Converting between the metric units of distance

Original measurement	Conversion factor	Converted measurement
0.05 mm	$\frac{1\text{ m}}{1000\text{mm}} = 1$ $\frac{1\text{ km}}{1000\text{m}} = 1$	.00000005 km
0.05 mm	$\frac{1\text{ m}}{1000\text{mm}} = 1$	.00005 m
0.05 mm	$\frac{1\text{ cm}}{10\text{mm}} = 1$	.005 cm

9 Measuring distances in different units

1. 165.1 cm x 61.0 cm
2. 1651 mm x 610 mm
3. 65 in x 24 in

a. The length measurements have different numerical values because each unit used is a different size.

b. 165.1 cm/ 61.0 cm = 2.71
1651 mm/ 610 mm = 2.71
65 in / 24 in = 2.71

Notes:

1.3 Inquiry and the Scientific Method

About the Lesson

Students have already learned that scientists make observations, formulate hypotheses, and conduct experiments to learn more about the universe. In this lesson, students use inquiry to learn more about the process of science. They learn to distinguish hypotheses, theories, and laws. Students also discover how to examine and communicate evidence.

Try this activity to introduce your class to the concept of evidence. Ask students to create a list of five things they know for sure. Have a few student volunteers share their lists. Then have them explain why they are certain of the facts stated on their lists. What is the evidence to support these facts?

Share with students that scientists rely on evidence to help them learn more about the natural world. Ask students to give examples of other careers where evidence is important.

Logic Puzzle: TV Time with the Josephs

Logic puzzles are a fun way to sharpen your students' deductive reasoning skills. Try the tips shown at right first so students can become familiar with how to solve the puzzle. Then have students work in small groups to solve the puzzle described below.

The Joseph family loves to watch television. Each family member has a favorite television show that airs on a different day of the week. The shows the Josephs enjoy watching are *Laugh Really Hard, Reality TV, Crime Solvers, Keep Talking,* and *The Sports Desk*. The family members are Dad, Mom, two daughters (Morgan and Mia), and one son, Nicholas.

Here are the clues to help solve the puzzle. The solution is shown at right (Figure 1.3).

1. The females do not like to watch television on Wednesday.
2. One of the males loves to watch *The Sports Desk*.
3. Nicholas' favorite show airs earlier in the week than Mom's favorite show; but Mom's favorite show is on before *Crime Solvers*.
4. Morgan does not like *Keep Talking*.
5. One adult likes to watch *Reality TV*. The other adult's favorite show airs on Monday.
6. *Laugh Really Hard* is on a day other than Tuesday.

Teaching Tip . . .Solving Logic Puzzles

Try these steps to help students solve the puzzle.

Use a table to mark answers. Do the first clue with students and model how to use the table. Use an x to denote disqualifications and a check mark to denote a correct answer. Have students read each clue and decide if any answers are given in the clue. Remind students that excluding a person from a specific TV show or day is just as important as confirming the person's favorite show or viewing day.

Solution:

1. Dad watches *The Sports Show* on Monday.
2. Mia watches *Keep Talking* on Tuesday.
3. Nicholas enjoys *Laugh Really Hard* on Wednesday.
4. Mom watches *Reality TV* on Thursday.
5. Morgan enjoys watching *Crime Solvers* on Friday.

	Laugh Really Hard	Reality TV	Crime Solvers	Keep Talking	The Sports Desk	Monday	Tuesday	Wednesday	Thursday	Friday
Dad	x	x	x	x	✔	✔	x	x	x	x
Mom	x	✔	x	x	x	x	x	x	✔	x
Morgan	x	x	✔	x	x	x	x	x	x	✔
Nicholas	✔	x	x	x	x	x	x	✔	x	x
Mia	x	x	x	✔	x	x	✔	x	x	x
Monday	x	x	x	x	✔					
Tuesday	x	x	x	✔	x					
Wednesday	✔	x	x	x	x					
Thursday	x	✔	x	x	x					
Friday	x	x	✔	x	x					

Figure 1.3: Solution to the logic puzzle.

"What happens when scientists collect data and make observations that confirm a hypothesis?" This is a common question posed by students. Some options for what may happen when results appear to confirm a hypothesis are shown below. The incorrect relationship among a hypothesis, theory, and law is also explained.

CORRECT

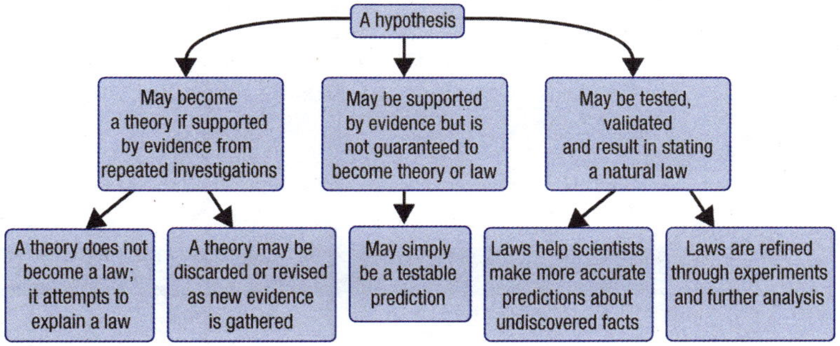

INCORRECT

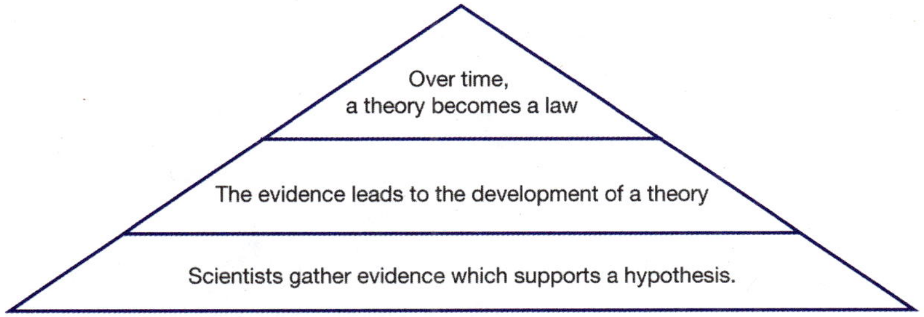

1. Have teams of students work together to write a short story (mystery). Instruct each group to pass their story along to a different group to see whether they can solve the mystery. Tell each group to discuss the evidence that lead to their conclusions.
2. Have students use the Internet or other resources to learn about the *Piltdown Man*. Write a report or make a presentation about the importance of honesty when communicating results.

Students complete section 1.3 review questions.

Addressing Misconceptions: Hypotheses, Theories, and Laws

As students' investigative skills improve they infer that science is a process that usually involves many revisions. Students must realize that many hypotheses are initially incorrect. If so, the hypothesis must be revised and tested further. If the results of an investigation validate a hypothesis, students may think the hypothesis is "advanced to the next level" and becomes a theory. This is a common misconception. Explain to students that a hypothesis must be tested repeatedly by different scientists in order to be validated. Even if the hypothesis withstands the review of other scientists, there is no guarantee that it will become a theory.

A theory explains how or why something occurs and is verified by known facts. However as new evidence is gathered, theories may also be modified or even discarded. Students often ask, "How long does it take for a theory to become a law?" Embedded in this question is the belief that a definite pattern exists between hypotheses, theories, and laws (*see left*). A law states a scientific fact. A law is a precept in nature; and a theory attempts to explain the precept.

Teaching Tip . . .Communicating Results

When conducting investigations, students are sometimes tempted to *force* the results to match their initial hypothesis. Simply stated, students at this age often equate being wrong with inadequacy or failure. Take advantage of this time to explain to students that doing science is a process. Part of the process is truthfully reporting their experimental techniques and results. Emphasize that objectivity is the responsibility of all scientists.

Chapter 1 Answers

1.1 Section Review

1. True
2. False
3. False
4. True
5. False
6. False
7. False; the increase in speed of the car is due to the exchange of energy between the gasoline and the motion of the car.
8. False
9. False
10. True

1.2 Section Review

1. Example answers are:
   a. Both types of time use clocks for measurement.
   b. Historic time identifies a particular moment in time, while a time interval describes a quantity of time.
2. b. 2 minutes; a. 160 seconds; c. 2 minutes 50 seconds
3. 3, 927 seconds
4. d
5. b
6. There are no units on the quantity of rope.
7. 1 mile is longer; 1 mile = 1.61 kilometers
8. 1/4" plywood is 6.3 mm thick, and would be the closest thickness to 7 mm plywood.

1.3 Section Review

1. b
2. A hypothesis is an unproven or preliminary explanation that can be tested.
   A theory is an explanation of why or how something occurs that is supported by evidence
   A natural law is a rule that is obeyed by everything in the universe.
3. b
4. The information in (a) is evidence that could be used to evaluate Julie's hypothesis. Julie has made an observation about the moon. The information in (b) is not evidence, it is a different hypothesis made by Julie's sister.
5. Repeatable means that anyone else who performs the same experiment in the same way observes the same results.
6. b

Connection

1. observation
2. A good hypothesis should propose an explanation to an observation, and make a prediction about how a phenomenon will react when exposed to variables. A good hypothesis can be tested experimentally.
3. Well-cushioned heels and flexible soles would reduce running injuries.
4. They use impact machines to test how long heel cushioning will last, and they use flex testing machines to test sole flexibility.

Activity

Trial	Time (seconds)
#1	5.3
#2	4.8
#3	5.1
#4	5.2
#5	4.9
#6	4.7
Average	5.0

Discussion Questions

a. The moving part used for our clock was water.
b. We named our new unit of time the Sodasecond (Ss).
c. One Sodasecond is equal to approximately 5 seconds.
d. (9.8 s × (1 Ss/5 s) gives a Maurice Green a time of 1.96 Sodaseconds for his 100-meter dash.
e. This clock can only measure one small unit of time, the Sodasecond. It cannot measure multiple Sodaseconds, or smaller or larger units of time, like our modern day clocks. Also, this clock is flawed because it requires quite a lot of movement to be started. The human error in that movement could lead to significant error in trying to take a time with this clock.

Chapter 1 Assessment

Vocabulary

Section 1.1

1. phases of matter
2. energy
3. mass
4. chemical change

Section 1.2

No vocabulary words in this section

Section 1.3

5. objective
6. scientific method
7. hypothesis
8. inquiry
9. natural laws
10. theory

Concepts

Section 1.1

1. Answers are:
   a. When heat is added to the mixed ingredients, a chemical reaction occurs and the batter is turned into a cake.
   b. The strength of iron is due to its physical properties.
   c. Physics explains that the smaller wagon has less mass than the pickup truck, and therefore will move faster when the same force applied.
   d. The bodies of sea otters are less dense than the water, therefore they float.
2. Answers are:
   a. G b. S c. G d. L
3. (c) the sun

Section 1.2

4. Historical time identifies a specific moment in the past, present, or future. Examples will vary and may include any specific date or time: 9/11/01; 2:00 pm; May 1, 2006.
   A time interval identifies a quantity of time. Examples will vary and may include any quantity of time: 2 hours; 49 seconds to run 400 meters; 1 year.
5. A uniform measuring system would make international business more convenient. The SI system is generally considered easier to learn and use.
6. Answers may vary. Reasonable units for each length would be:
   a. meters (m) b. astronomical units (AU) c. microns (μ) d. light years e. kilometers

Section 1.3

7. Answers are:
   a. T b. L c. T d. H
8. A scientific theory (1) must be supported by significant evidence, (2) cannot have any evidence that proves the theory false, and (3) must be unique.

Problems

Section 1.1

1. water
2. The ping pong ball will initially move faster. The bowling ball has more mas and takes more force to get it moving.
3. .

Phase	Description	Examples
Solid	maintains shape	wood, rock, ice
Liquid	flows; has no definite shape	milk, water, gasoline
Gas	expands freely	air, oxygen, helium

Section 1.2

4. Answers are:
   a. 0.0011416 years
   b. 0.41666 days
   c. 600 minutes
5. Answers are:
   a. 4.8 kilometers
   b. 1980 meters
   c. 2.5 meters

Section 1.3

6. Answers are:
   a. b b. a c. d d. e e. c

Chapter 2: Experiments and Variables

Learning Goals	California Content Standards	Activities and Resources	Assessment
2.1: Variables and Relationships The student will: • Define variable. • Discuss commonly used variables in physical science. • Create graphs to infer a relationship between variables. • Contrast direct and inverse relationships. • Apply a formula to solve for volume.	PS 9a PS 9b PS 9c PS 9d PS 9e PS 9f PS 9g	**Laboratory Investigation 2A:** Systems and Variables **Teaching Illustrations:** Relationships Between Variables, Direct and Inverse Relationships, Dependent and Independent Variables, Basic Math Operations Chart, Formula for Volume of a Rectangular Box **Graphic Organizer:** How To Make a Graph **Investigation Skills:** Measuring Mass, Measuring Length, Measuring Temperature **Math Skill Sheets:** Creating Line Graphs, Measuring Angles, Solving Equations with One Variable, What's the Scale **Reading Skill Sheets:** Internet Research	**Exam View Test Bank** Chapter Pre-Test **Section 2.1 Review**
2.2: Experiments and Data The student will: • Design and conduct an investigation. • Distinguish control and experimental variables. • Analyze data from an experiment.	PS 9a PS 9b PS 9c PS 9e	**Laboratory Investigation 2B:** Making Predictions From a Graph **Math Skill Sheets:** Averaging **Reading Skill Sheets:** Bibliographies **Connection:** Testing the Power of Suggestion **Chapter Activity:** Guess the Gram	**Section 2.2 Review** **Assess:** Designing an Experiment **Chapter Assessment** **Exam View Test Bank**

Estimated Instruction Time	Language Tools	Differentiated Strategies	Materials
Four 45-minute class periods	**Literary Picks:** **Vocabulary:** variable, value, graph, independent variable, dependent variable	**ELL Strategies:** Listed in the ELL Ancillary **Learning Strategies:** • Cooperative Learning - Investigation 2A • Demonstration - Motivate: A Tasty System • Guided Practice - Explain: Linear and Nonlinear Relationships	**Investigation 2A:** energy car and track, timer and photogates, meter stick, mass balance **Motivate - Teacher Demonstration:** pitcher of water, sugar, ice, cups, drink mix
Five 45-minute class periods	**Vocabulary:** experiment, procedure, data, analysis, conclusion, experimental variable, control variable, accuracy, error, averaging, significant **Closure Cards**	**Learning Strategies:** • Cooperative Learning - Investigation 2B • Guided Practice - Accuracy and Precision • Graphic Organizers - Flowcharts **Teaching Tip:** Dart Game **Addressing Misconceptions:** Accuracy vs. Precision	**Investigation 2B:** energy car and track, timer and photogates, meter stick, mass balance, simple calculator, graph paper **Teaching Tip - Dart Game:** dart boards **Chapter Activity:** Balance capable of measuring 300 +/- 0.1 grams, assorted items such as coins, paperclips, rubber bands, washers, marbles, popcorn kernels, etc.

2.1 Variables and Relationships

About the Lesson

In this lesson students learn about variables and how scientists examine the relationships among variables to better understand the natural world.

Teacher Demonstration: A Tasty System

Students have had some exposure to systems and variables in grades six and seven. Reintroduce students to systems by preparing a drink. Place the following items on your desk: pitcher of water, sugar, ice, cups, and a package of flavored drink mix. Students may immediately recognize what you are preparing.

Tell students that a system is a group of objects that are related. Systems contain variables. When variables are changed the system is affected. Have students identify the variables in the system. Prepare the drink mix according to the package instructions. Read the instructions aloud as you prepare the drink.

Students complete Investigation 2A before reading section 2.1. In this investigation students use the energy car and track to create a system and learn how scientists find relationships in nature. Students also discover how variables affect the system. Students are introduced to the energy car and track in this investigation, but will also use it in other experiments. Be sure that students become familiar with its components and comfortable with its setup.

Investigation 2A: Systems and Variables

Setup	Materials	Investigation Vocabulary
1. Allow one class period to complete the investigation. 2. Students work in small groups of three to five. 3. Do a trial run of the experiment before presenting it to students.	• energy car and track • timer and photogates • meter stick	system variable controlled experiment experimental variable control variable

Literary Picks...

Reading Level	Title
3-5	***Eureka! Great Inventions and How They Happened*** **by Richard Platt** This book will appeal to all ages with its explanations of how many things that we use daily came to be. The book includes 29 examples with brief biographies of each scientist along with descriptive accounts of their discoveries.
6-8	***June 29, 1999*** **by David Wiesner** This book won the School Library Journal Best Book Award in 1992. It is a humorous science fiction tale about a special science project. Includes illustrations and is a good example of investigation and experimentation.
9-12	***Lucky Science: Accidental Discoveries from Gravity to Velcro, with Experiments*** **by Jeanie and Royston Roberts** Learn about the accidental discoveries and inventions of Silly Putty, penicillin, rubber, Pluto's moon, and others. A brief history and biography along with suggested activities for each chapter are included.

Guided Practice: Linear and Nonlinear Relationships

Use the *Blank Graph* Teaching Illustration from your Teacher Tool Kit to create graphs which represent linear and nonlinear relationships (Figure 2.1). Explain to students that linear means "in a line." Use this activity to discuss dependent and independent variables and direct and inverse relationships.

Graph 1: Helena visited a local pizza parlor. The menu showed that a single topping pizza costs five dollars. Each additional topping costs 50 cents. Create a graph that shows the cost of pizza as the number of toppings increases from one to ten. Ask students to decide which variable is dependent and independent and to choose an appropriate scale. Is the relationship linear? Is it direct or inverse? (See Figure 2.2.)

Graph 2: A scientist conducted an experiment to determine the relationship between the pressure and volume of a gas at constant temperature. His data is shown in the table below. What type of relationship exists between these two variables? (inverse nonlinear.)

Pressure and Volume of a Gas at Constant Temperature

Pressure (mm of mercury)	Volume (liters)
724	1.50
869	1.33
951	1.22
998	1.16
1230	0.94
1893	0.61
2250	0.51

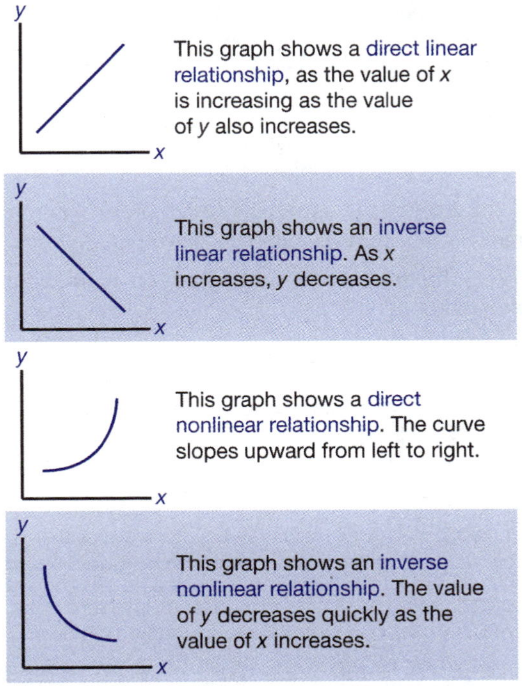

Figure 2.1: Linear and nonlinear relationships.

1. Display different types of graphs. Select graphs that exhibit examples of both linear and nonlinear relationships. Then ask students to explain the trends represented by each graph.
2. Divide students into groups and have them observe trends (like daily high temperatures or gasoline prices) in their area each day for a week. Have each group present a graph of their data to the class.

Students complete section 2.1 review questions.

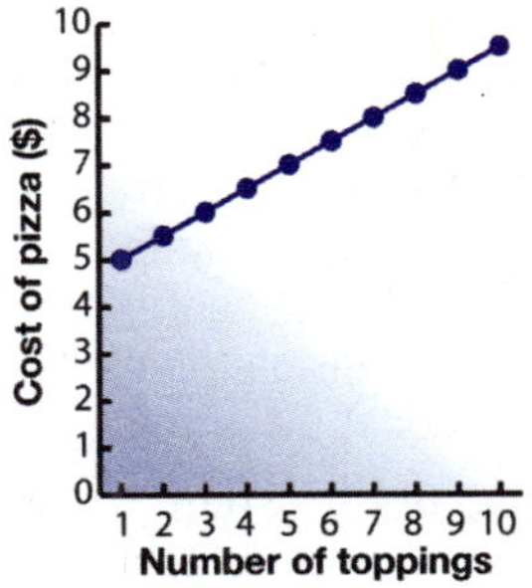

Figure 2.2: Direct linear relationship.

Investigation 2A: Systems and Variables

Complete the teacher demonstration, A Tasty System, before beginning the investigation. Do not serve the drinks to your students without first getting permission from parents or checking the policies of your school district. If your students are allowed to have the drink, then pour each student a cup.

Once again, let's identify the variables in the Tasty System.

Students identify the variables.

What would you expect to happen if I added more sugar, or served it without ice?

The drink would become sweeter or it would be warm.

You are correct, the drink would be sweeter or warm. Suppose I doubled the amount of drink mix. What would happen then?

The drink would taste very bitter.

You are right. I do not think it would taste very good that way. Think about what you have observed thus far. Do you think you have seen evidence of a relationship between systems and variables? If so, what is it?

There is a relationship between systems and variables. Systems are comprised of variables. As variables change the system is affected.

Good answer. What is the easiest way to change a system?

The easiest way to change a system is to change a variable.

Each time you change a variable, the system is also changed. However, you should only change one variable at a time. Can anyone tell me why this is so?

Changing only one variable at a time allows you to know if that change has really affected the system. When multiple variables are changed at once, it is difficult to identify the effect of changing each variable.

1 Making a system

In today's investigation you will use the energy car and track to make another system with different variables you can manipulate.

Have all the supplies for the investigation already laid out at the lab table for easy inventory. Go through the materials list with students and hold up each piece of equipment as you say its name. It would be helpful to identify the physics stand hole you want each group to use on an index card that is easily visible. Walk through the directions with students. If you have an extra setup available, model it for students. Assist as needed.

Set up your energy car and track system as instructed in part one.

Students complete the setup. As students collect data, allow them to put the photogates wherever they choose. Some groups might push the car down the ramp. The data will not make much sense, and this is okay – it will help students see later in the investigation that controlling all the variables is necessary and important.

2A Systems and Variables

How do scientists find relationships in nature?

A system is a group of related objects and influences that you are trying to understand. Variables describe things in the system that can change.

Materials

- Energy car and track
- Timer and photogates
- Meter stick

1 Making a system

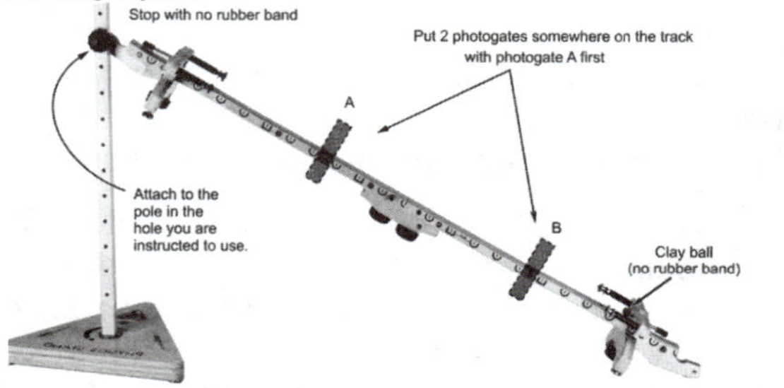

1. Set up the track with a long straight section. Your teacher will tell you which hole in the stand to attach the track. Each group will have a different angle.
2. Put a clay ball on the stop at the bottom.
3. Place two photogates on the track with photogate A higher than photogate B.
4. Roll the car down and record the time it takes the car to pass between the photogates.

2 Stop and think

a. Which track should have the fastest car? Which track should have the shortest time between photogates?

b. Write a one sentence hypothesis that relates the time between photogates to the angle of the track.

c. Use Table 1 to record the results from each group in your class. Record the times in the column labeled "First Trial". Leave the column labeled "Second Trial" blank. How do the results compare with your hypothesis? Can you give a reason why they did or did not behave as you expected?

6

2 Stop and think

Everyone has had a chance to roll the car down the track and time how long it takes for the car to pass through the photogates. Which track do you think should have had the shortest time between photogates?

The track using the highest hole in the stand should have the shortest time.

Why do you think this makes a difference?

This gives the ramp its steepest slope.

Think about the relationship between the slope of the ramp and the time between the photogates. Use your observations to write a hypothesis that relates the time between the photogates to the angle of the track.

Students complete their hypotheses and share results with the class.

3 Variables

Students complete the answers to questions a through d. Discuss the answers to each question with students and use this time to define the terms, *experimental variable* and *control variable*. The experimental variable is adjusted in order to find out how it influences the system. Control variables are kept constant. This is important so each group is able to relate its results to other groups and establish the effect of the experimental variable on the system.

4 A controlled experiment

You are going to redo the experiment. This time you will control all other variables except the experimental variable. Which variable should be adjusted and how should those adjustments be made?

The order of adjustments really does not matter. What is important is that all the controls are in place before students are allowed to continue collecting data. You may need to lead students to the proper adjustments. As students mention each possible adjustment, stop and have the class make a decision about where to place each item. Remember that the greater distance between photogates will give better overall results. Once students are finished and have the data for their individual ramp, have them record it in the "second trial" column.

5 Stop and think

Now that you have repeated the experiment did you get results that agree with your hypothesis? Why did you have better agreement between your hypothesis and the second trial?

All variables except the experimental were controlled. This ensured that the data was influenced by only the angle of the ramp, the only variable mentioned in the hypothesis.

What steps can you take to find the problem if something does not work?

The hypothesis may be modified, or the experimental and control variables may be adjusted.

As scientists it is our job to keep experimenting until we find a solution. Bear in mind that there are many variables that we can change or manipulate in order to reach a solution.

Word Origins...variable (from Latin, *varius*, which means "different")

Ask students to think about what the prefix, *vari-* (or *vario*) means. Then give examples of other words with similar origin, like various or variety. Use these terms in sentences to help students understand that *vari-* means different.

1. There are various science courses offered at the high school.
2. The gardener was happy to see the variety of plants available at the nursery.

Teaching Tip . . .Stop and think (part two)

Draw a chart on the board (similar to the Table 1) that is large enough to accommodate the data for each group, or make a transparency and use it as an overhead to display the class data. Have each group share its data for the time between the photogates. Use this time to discuss students' hypotheses and any discrepancies in data.

Investigation 2A: Data and Answers

1 Making a system

2 Stop and think

a. The track at the steepest angle should have the fastest car. The steepest track should have the shortest time between photogates.

b. The steepest track will have the shortest time between photogates.

c. The results do not agree with our hypothesis. The photogates on the tracks were all at different places so they were not measuring the same thing.

	First Trial	Second Trial
Attached at (holes from bottom)	Time from A to B (seconds)	Time from A to B (seconds)
3	0.4697	0.3355
6	0.5109	0.2273
9	0.3211	0.1833
12	0.5965	0.1588

3 Variables

a.

1 The angle of the hill

2 The mass of the car (how many marbles in car)

3 The angle of the hill

4 The distance between the photogates

5 The position of the photogates on the track

6 Release technique used

7 The friction of each car (different or the same)

b. The angle of the hill is the experimental variable. That is the variable that we carefully set up and compared the effect of on the time from A to B.

c. All the other variables should be kept the same for all the groups in the class. If this is done then we can measure the effect of changing only the angle of the hill on the time from A to B.

d. Answers may vary. Sample answers: time of day, day of the week, color of the car, height of lab table.

4 A controlled experiment

5 Stop and think

a. Yes, the second trial does agree with our hypothesis.

b. Because we controlled all possible variables and experimented with only the angle of the hill.

c. First, try to identify all variables in the system. Second, keep track of each variable. Third, perform an experiment and change only one variable, which will be the experimental variable. All other variables will be control variables and should remain unchanged for the entire experiment. Fourth, analyze the results of all experiments to try and determine which variable, or variables had the greatest effect on the system.

Notes:

2.2 Experiments and Data

About the Lesson

In this lesson, students learn how to design an experiment in order to investigate the relationship among variables. Students explore the components of a well-designed experiment, learn how to analyze data, and communicate the results of an investigation.

Ask students to think about how they experiment and collect data in their daily lives. Is setting up lab equipment always involved? What are some alternative ways to conduct experiments? Have students share an experience about a time when they were curious or puzzled about something. Which problems solving techniques did they apply to arrive at a conclusion?

Students complete Investigation 2B before reading section 2.2. In this investigation students use the energy car and track to collect data. Students use the data they collect to create a graph which serves as a model of the relationship between two variables: distance and time.

Investigation 2B: Making Predictions from a Graph

Setup	Materials	Investigation Vocabulary
1. Allow one class period to complete the investigation. 2. Students work in small groups of three to five. 3. Display one car and track setup as a reference for students.	• energy car and track • timer and photogates • meter stick • simple calculator • graph paper	percent error model

Addressing Misconceptions: Accuracy vs. Precision

Students often use the terms accurate and precise interchangeably. Remind students that these two terms have different meanings to scientists. *Accuracy* refers to how close a measurement is to an accepted or true value. *Precision* means how close a set or group of measurements are to one another. Use the examples at right to help students differentiate the terms.

Guided Practice: Accuracy and Precision

1. Suppose a student is asked to measure the length of her desk. She measures the desk three times and records these measurements: 14.1 cm, 14.8 cm, and 13.9 cm. Are these measurements precise? Are the measurements accurate? (Answers: The measurements are not precise. It is difficult to determine whether any of the measurements is accurate without knowing the true measurement of the desk.)
2. The teacher referred to the manufacturer's guide to determine the actual length of the desk. He then measured the desk to compare the values. According to the manufacturer, the desk is 14.4 cm. The teacher's measurement was 14.43 cm. Is the teacher's answer accurate? (Answer: The measurement is accurate.)

Teaching Tip . . .Dart Game

Playing with suction-cup or foam darts is a fun and easy way to teach students the difference between accuracy and precision. Divide the class into teams of three or four students. Of course the goal of the game is to hit the bullseye, which represents the true or accepted value.

Explain to students that hitting the bullseye is an example of accuracy. However students may demonstrate precision by throwing a group of darts close together, even if they are not near the bullseye. After each team throws its darts, have its members explain whether their darts are accurate, precise, both, or neither. Be sure to remind students to use caution when throwing the darts. Discuss the consequences for violating safety rules before beginning the activity.

1. Instruct students to create a flowchart that explains the steps involved in designing and conducting an experiment. Figure 2.3 shows one example.
2. Have students work with a partner to create note cards to review important concepts in the chapter. Allow class time for students to quiz one another.

Students complete section 2.2 review questions. Then complete the performance-based assessment described below.

Designing an Experiment

Instruct small groups of students to design and carry out an experiment. Students may choose their own hypothesis to test (and have it approved by you) or you may assign specific experiments to students. Some ideas are shown at right. Students may need assistance gathering materials for their investigations. Use the rubric shown below to score each team's experimental design. The total possible points is 40.

The research question is clearly stated.	Yes 5 points	Partial yes 4 points	Partial no 3 points	No 2 points	
The hypothesis is testable.	Yes 5 points	Partial yes 4 points	Partial no 3 points	No 2 points	
One experimental variable is identified. (Other variables are controlled.)	Yes 5 points	Partial yes 4 points	Partial no 3 points	No 2 points	
A materials list is included.	Yes 5 points	Partial yes 4 points	Partial no 3 points	No 2 points	
Procedure is written in a step-by-step format and is repeatable.	Yes 5 points	Partial yes 4 points	Partial no 3 points	No 2 points	
Group collected data and recorded observations.	Yes 5 points	Partial yes 4 points	Partial no 3 points	No 2 points	
Data analyzed with graphs, charts, or calculations.	Yes 5 points	Partial yes 4 points	Partial no 3 points	No 2 points	
Conclusion reflects the results of the experiment.	Yes 5 points	Partial yes 4 points	Partial no 3 points	No 2 points	
				Total Points Earned	

Designing an Experiment

Have student groups design experiments to test these hypotheses.

- Eighth grade students' favorite after school activity is playing video games.
- Using a detergent with water is more effective for removing grease from clothing than water alone.
- Movie tickets in San Francisco are more expensive than in San Diego.
- The most popular book bag color is black.
- Temperature does not affect the rate that seeds develop into plants.
- The average eighth grader is 65 inches tall.

State the research question
▼
Gather information
▼
Develop a testable hypothesis
▼
Identify the variables
▼
Control all variables except the experimental variable
▼
Conduct multiple trials
▼
Keep a record of experimental techniques and procedure
▼
Collect and analyze data
▼
Draw conclusions
▼
Publish results

Figure 2.3: Steps for designing and conducting an experiment.

Investigation 2B: Making Predictions from a Graph

In the last investigation you learned about systems and variables. What is a system? What is a variable?

A system is a group of objects that are related. Each of the objects (or parts of the system) is a variable. Variables affect how a system functions.

Today you are going to work with variables again. This time, you will use a model to investigate the relationship among variables. Who can tell me what a model is?

Students may initially respond that a model is a toy, such as a toy car or airplane.

Those are examples of models but they do not tell us what a model is. A model can be many different things and take many different forms from a toy car to a blueprint of a home. A model is a simple way to represent a relationship between two or more variables. Your goal today is to predict how long it will take for the energy car to reach a specific time on the track. In order to do this, you will need a model that will help to visualize data. You can use this data to make a prediction about the relationship between two variables: time and distance. Who can suggest one type of model that may be helpful to use?

A graph is an appropriate suggestion.

A graph is an appropriate suggestion because it will allow you to compare time and distance.

1 Setting up

Carefully read each step of the procedure on part one. Double-check your setup to be sure that is it correct. Once you start the experiment, do not change the adjustment of the starting block screw. Record your data in Table 1.

Students read each step of the procedure. Display a car and track setup before the class. Use the graphic shown in part one to help students with their setup. Assist as needed.

2 Analyze the data

Before you begin graphing, you must set up the graph properly in order to make sense of the data. Does it matter which variable you put on the *x*-axis?

Yes, the independent variable should be placed on the *x*-axis. In this example, distance is the independent variable. Remind students that the unit (centimeters) must also be included.

Time is the dependent variable. It goes on the *y*-axis.

Use the *Blank Graph* Teaching Illustration in your Teacher Tool Kit to model creating the graph with students.

Review the data you collected in Table 1 and use it to decide on a scale for your graph. You can better understand trends on your graph if it is spread out.

The scale will depend on the data that was collected and the size of the graph paper. Help students set up the increments for the *x*-axis so that the majority of the graph paper will be used. Model this on the transparency. Follow the same method to model the scale for the *y*-axis.

2B Making Predictions From a Graph

Why are graphs useful?

In science, a model is a relationship between variables. For example, the car takes a certain amount of time to roll a certain distance down the track. A model for the cars motion would allow you to predict how long it would take the car to reach any distance down the track. Graphs are a good way to visualize a model that involves two variables, such as time and distance. In this Investigation you will create a model (graph) that predicts how long it takes the car to reach anywhere on the track.

Materials

- Energy car and track
- Timer and photogates
- Meter stick
- Mass balance

1 Setting up

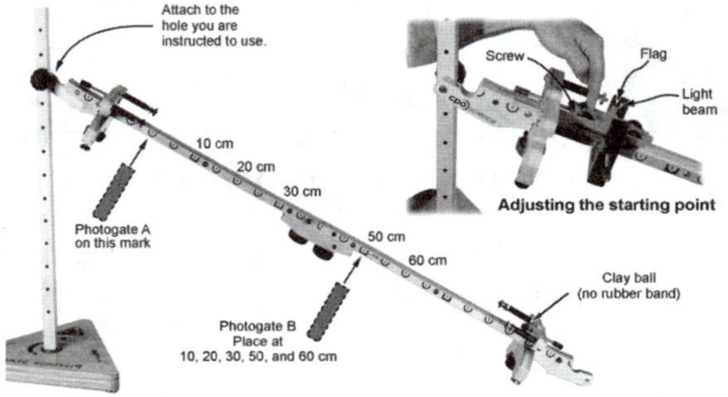

1. Set up the track with a long straight section. Your teacher will tell you which hole in the stand to attach the track. Each group will have a different angle.
2. Put a clay ball on the stop at the bottom. Put the other stop at the top to provide a repeatable starting point for the car.
3. Place photogate A on the mark shown in the diagram. Adjust the screw so the car breaks the light beam just after you release it. Once you start the experiment, do not change the adjustment of the screw.
4. The marks on the track are 5 cm apart. Place photogate B two marks (10 cm) lower than photogate A. Roll the car and record the time from A to B.
5. Move photogate B to 20, 30, 50 and 60 cm from photogate A. Measure and record the time from A to B for each position. You will not be able to measure at 40 cm because of the split in the track. DO NOT measure at 70 cm (yet).

8

s there anything else that needs to be added to the graph?

The graph needs a title. Remind students to select a title that is descriptive of the data that was collected. Solicit suggestions for a title from students.

:reate a graph of the data.

Have one volunteer group share their graph with the class. Use the group's data to demonstrate the correct way to create the graph. Once the points are plotted, use a best-fit curve to extrapolate to the 80 cm mark. Show students how to make a smooth line rather than connecting the dots. Assist students as needed.

3 Test the model

Jse your graph to find the time for the 80 cm distance and write this down as your prediction in he data table. Once you have your prediction, roll the car three times down the track. Measure he time it takes for the car to travel the 80 centimeter distance.

Students follow the procedure.

Add up all three times for the 80 centimeter distance and divide by three to get the average ime. Record this information in the data table.

Students find the average measured time in seconds.

How good do you think your prediction was? You can find out by determining the percent error.

Write the equation on the board.

The percent error can be determined by substituting values in place of each variable represented in the equation. Refer to your data table to locate your predicted and actual times.

Students set up the calculation. Tell students to subtract the measured time from the predicted time first. Then divide that answer by the predicted time. This answer must be multiplied by 100 in order to determine the percent error. Encourage students to include units (seconds) as they complete the calculation. Show them that the units cancel leaving only the percent error in the end.

4 Stop and think

How large was the difference between the prediction and the actual time it took the car to travel 80 centimeters? Was there a significant difference?

Answers will vary by group. Encourage students to share.

Why might there be a difference between your prediction and the actual time?

Answers will vary. Students may discuss a variety of possible errors, such as computing the average time, drawing the best-fit line and extrapolating to 80 cm, or even obtaining the initial times in part one.

How might the prediction at 80 cm be improved?

Answers will vary. Students may suggest taking more measurements of time or measuring every five centimeters instead of ten. This would make the graph more accurate and lead to a more accurate predicted time. Students complete answers to questions in part four. Review their responses before closing the investigation.

Investigation 2B: Data and Answers

1 Setting up

Distance from A to B (cm)	Time from A to B (sec)
10	.2270
20	.3315
30	.4070
50	.5258
60	.5768

2 Analyze the data

a.

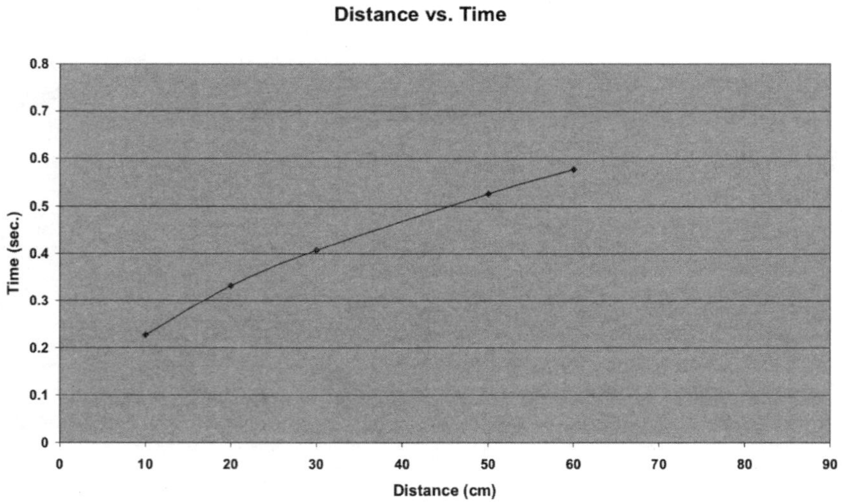

b. Using the graph, our prediction is 0.6750 sec.

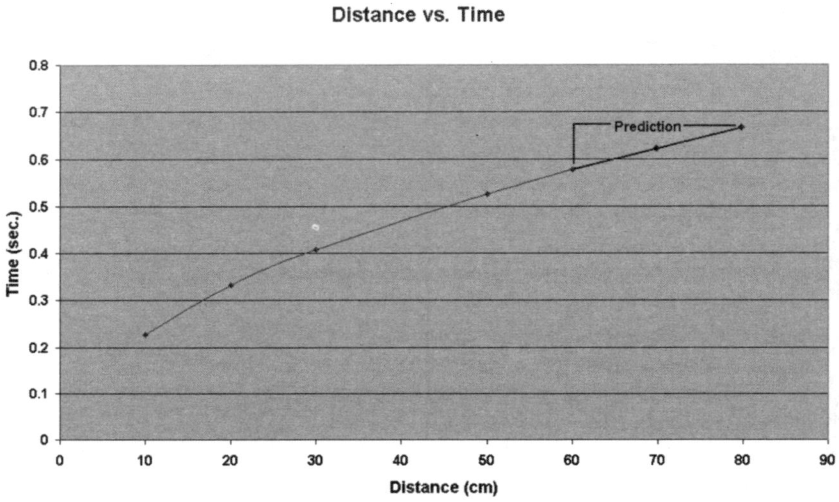

3 Test the model

Predicted time (s)	.6750
Measured times: 3 trials at 80 cm (s)	.6712
	.6716
	.6699
Average measured time (s)	.6709

$$\% \text{ Error} = \frac{\text{predicted - measured}}{\text{predicted}} \times 100$$

$$.6\% = \frac{.6750 - .6709}{.6750} \times 100$$

4 Stop and think

a. The difference was .6%. The difference was not significant.

b. Friction may be slowing the car down a small amount. Reading the lines off the graph is an estimation. The photogates placement could have been off slightly. The release technique could have varied slightly.

c. The time was placed on the y-axis because it was the dependent variable.

d. Use a very large graph to get a more exact prediction. Use a spreadsheet program on a computer to make a graph and prediction.

Notes:

Chapter 2 Answers

2.1 Section Review

1. Any five of the following answers would be correct:
    a. the direction of the applied force
    b. the length of the beam
    c. the width of the beam
    d. the height of the beam
    e. the strength of the wood
    f. where the force is applied
2. c
3. C shows a direct relationship
   B shows an inverse relationship
   D shows a complex relationship
   A shows no relationship
4. independent variable: amount of salt
   dependent variable: temperature of water freezing
5. 1.3 meters
6. $d = v \times t$
7. b and c
8. $c = ab - f$

2.2 Section Review

1. c
2. b
3. If you change more than one variable at once, you won't be able to tell which one caused your results. If only one is kept constant, you can tell that what happened was caused by the variable you changed.
4. No, because all measurements contain some amount of error.
5. The average for group 1 is 0.344. The error for group 1 is 0.005. The average for group 2 is 0.340. The error for group 2 is also 0.005. The average for group 3 is 0.340 and the error is 0.004. The measurements for the three groups are considered the same because they differ by less than the error of 0.005.
6. b
7. When the values are different, but the difference is less than the amount of error.

Connection

1. In a double-blind study, neither the subjects being tested nor the individual conducting the test is aware that a placebo is being used in the test.
2. The placebo effect occurs when a subject's symptoms are improved by using an ineffective treatment. This happens because the individual expects or believes that it will work.
3. A hypothesis is an educated guess or an assumption. It may be accepted as true and used to explain certain facts. A hypothesis can be tested. An experiment will either prove or disprove a hypothesis.
4. Because a doctor's belief that a drug will work also influences a patient's reactions, neither the doctor nor the patient knows if the patient is receiving a real drug of a placebo. Double-blind studies are designed to prevent the placebo effect.

Activity

Applying your knowledge: sample answers

a. Yes, I became better at estimating as time went on. For example, after a while, I had a good feel for the mass of a penny and could use multiple pennies to estimate the larger masses.
b. error on first try = 4.7 g
   error on second try = 0.1 g
c. textbook: 3000 g
   unsharpened pencil: 10 g
   sheet of notebook paper: 5 g
   compact disc: 15 g
d. 300,000 grams; 300 kg
e. I would use kilograms because it is the unit closest to the mass I'm trying to estimate.

Mass (g)	Object	Mass (g) 1st try	Object	Mass (g) 2nd try
1	1 penny	2.6	1 paperclip	0.9
5	2 pennies	5.0	--	--
10	1 pen	5.3	2 pennies + 1 nickel	10.1
25	1 floppy disk	19.7	1 floppy disk + 1 pen	24.3
50	binder clip	32.6	binder clip + 2 nickels + 3 pennies	50.3

Mass (g)	Object	Mass (g) 1st try	Object	Mass (g) 2nd try
100	5 CDs	79.0	5 CDs + 8 pennies	99.8
250	small bottle of water	290	1 apple	247

Chapter 2 Assessment

Vocabulary

Section 2.1

1. graph
2. independent variable
3. variable
4. dependent variable

Section 2.2

5. experiment
6. conclusion
7. procedure
8. experimental variable
9. control
10. average

Concepts

Section 2.1

1. Sample answers:
   a. mass-kilograms-balance
   b. time-second-stopwatch
   c. distance-meter-meterstick
   d. weight-newton-spring scale
   e. angle-degree-protractor
2. Answers are:
   a. A-Inverse
   b. B-No relationship
   c. C-Direct
3. Answers are:
   a. stretch of spring
   b. distance the marble travels
   c. stretch of spring
   d. distance the marble travels

Section 2.2

4. Answers are:
   a. X - could be tested by experiment
   b. cannot be tested by experiment
   c. cannot be tested by experiment
   d. cannot be tested by experiment
5. Answers are:
   a. force
   b. masses added to the cart
   c. The speed of the cart will decrease as the mass of the cart increases.
   d. Masses added and speed values for the cart
   e. The speed of the cart is inversely proportional to the mass of the cart.

Problems

Section 2.1

1. The y-axis should have a scale of 1 box = 1 meter
   The x-axis should have a scale of 1 box = 2 minutes
2. $d = m/V$
   Density and volume have an inverse relationship.
3. Graph B shows an inverse relationship.
4. Answers are:
   a. 0.75 kilometers
   b. The graph shows a direct relationship.

Section 2.2

5. a and c
6. Answers are:
   a. 2.5 g; 2.3 g
   b. Jaden's measurements were more accurate because Kylie's measurements have a higher estimated error.
   c. We cannot know the "true" mass of the penny, but a scientist would assume it is the average of all the measurements.

Chapter 3: Key Concepts in Physical Science

Learning Goals	California Content Standards	Activities and Resources	Assessment
3.1: Mass and the Atomic Theory of Matter The student will: • Describe the composition of matter. • Contrast the position of atoms in solids, liquids, and gases. • Define atom and element. • Identify units of mass.	PS 3b PS 3d PS 3e	**Laboratory Investigation 3A:** Energy **Teaching Illustrations:** Converting Grams to Kilograms, Atoms in Solid, Liquid, and Gas **Graphic Organizer:** Atoms in a Solid, Liquid and Gas **Investigation Skill Sheets:** Measuring Temperature **Reading Skill Sheets:** Science Vocabulary	**Exam View Test Bank** Chapter Pre-Test **Section 3.1 Review**
3.2: Temperature and Energy The student will: • Define energy. • Compare and contrast potential and kinetic energy. • Define temperature. • Apply a rule to convert between temperature scales. • Explain the relationship among heat, temperature, and thermal energy. • State and apply the law of conservation of energy.	PS 9f	**Laboratory Investigation 3B:** Mass and Indirect Measurement **Teaching Illustrations:** Energy Use in a Typical Car, Celsius and Fahrenheit Temperature Scales, Converting Between Fahrenheit and Celsius, Specific Heat Values of Common Materials, Converting Energy **Skill and Practice Sheets:** Temperature Scales, Specific Heat **Connection:** A Mighty Energizing Wind **Chapter Activity:** Your Own Science Experiment	**Section 3.2 Review** **Chapter Assessment** **Exam View Test Bank**

Estimated Instruction Time	Language Tools	Differentiated Strategies	Materials
Six 45-minute class periods	**Literary Picks** **Vocabulary:** kilogram, gram, atom, element	**ELL Strategies:** Listed in the ELL Ancillary **Learning Strategies:** • Cooperative Learning - Investigation 3A • Paired Learning - Motivate • Graphic Organizers - Extend **Addressing Misconceptions:** Atoms and Molecules	**Investigation 3A:** energy car and track, timer and photogates, meter stick, mass balance, Celsius thermometer, foam cups, hot and cold tap water
Six 45-minute class periods	**Vocabulary:** system, energy, force, newton, joule, temperature, thermometer, Fahrenheit scale, Celsius scale, absolute zero, thermal energy, heat, calorie, specific heat **Word Origins:** energy **Closure Cards**	**Learning Strategies:** • Cooperative Learning - Investigation 3B, Explain: It's A-MAZE-ING Indirect Measurement • Active Learning: Motivate: Energy Cards • Graphic Organizers - Extend **Teaching Tip:** Energy Transformation, Temperature and Thermal Energy	**Investigation 3B:** rice (about 1/4 cup per group), cellophane tape, mass balance, scissors, card stock template **Explain - It's A-MAZE-ING:** shoe boxes with lids, cardboard, tape, marbles, **Chapter Activity:** string, assorted sizes of balls, newspaper, meter stick, stopwatch, balance, paper cups, macaroni, rubber bands, paper clips

3.1 Mass and the Atomic Theory of Matter

About the Lesson

In this lesson, students review essential conceps in physical science. Students have been introduced to terms like atoms, elements, and molecules in previous courses. However, it is important to assess students' prior knowledge of these defintions before proceeding with instruction.

Have students choose a partner for this activity. Pose questions to students to assess what they already know about mass, matter, and atomic theory. Encourage students to discuss with their partner before writing an answer to each question.

- What is an atom?
- Name one element. Is water an element or a molecule?
- Would you measure mass in meters or grams?
- Is a rusting nail evidence of a physical change or a chemical change?
- What is energy?
- Name three common states of matter. Give one example of each.
- Are molecules closer together in a solid or a liquid?

Students complete Investigation 3A before reading section 3.1. In this investigation students conduct two experiments designed to help them understand the meaning of energy. In the first experiment students use the car and track to explore the relationship between an object's energy and its position. Students learn about temperature and heat energy in the second experiment.

Investigation 3A: Energy

Setup	Materials	Investigation Vocabulary
1. Allow one class period to complete the investigation. 2. Students work in small groups of three to five. 3. Set up a hot plate to keep the water warm for the duration of the investigation. Maintain a water temperature of at least 50°C. Use ice to keep the water cold. You may bring ice from home in a cooler or obtain it from your school cafeteria.	• energy car and track • timer and photogates • meter stick • mass balance • Celsius thermometer • foam cups • hot and cold tap water	energy temperature kinetic energy potential energy thermal energy

Literary Picks...

Reading Level	Title
3-5	***Counting Frank* by Rod Clement** A well-illustrated book about a boy and his dog Frank. The boy uses Frank as a unit of measurement along with a collection of facts that help introduce the metric system and includes mathematical topics such as counting, size comparison, and a variety of measurement activities.
6-8	***Sun* by Steve Tomecek** This is part of the *Jump into Science* series from National Geographic. The book is a lively story about two kids and a lavender cat, and answers questions about the sun such as its size, temperature, composition, distance from Earth, and the basic reasons why we need the sun to live.
9-12	***Is the Temperature Rising? The Uncertain Science of Global Warming* by George Philander** The author provides background knowledge of how our planet functions in order to really understand the effects of global warming. Includes topics such as the interaction of light and molecules, heat, weather, air, and carbon dioxide.

Students should already know that matter has mass and volume and that it may exist as a solid, liquid, or gas. Use the information in the chart below to review basic facts about matter in each state. Give each student a blank chart. Then have students fill in the chart and draw a picture to explain how they envision the position of molecules in each state.

State of Matter	Shape	Volume	What is the evidence? How can you tell?	Position of Molecules
 Solid	Definite	Definite	• We can see and describe its shape. • We can use a ruler to measure its length, width, and depth and calculate its volume.	
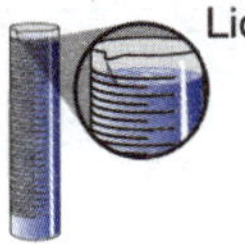 Liquid	No definite shape (takes the shape of its container)	Definite	• We can pour the liquid into containers with different shapes and see that the liquid "changes shape". • We can measure the volume of liquid in the different containers and see that it is the same regardless of its shape.	
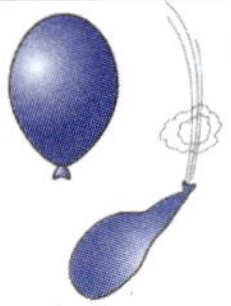 Gas	No definite shape	No definite volume	• We know that gases can expand and contract. • If we blow up a balloon to capacity, we can see changes in its shape and volume. As the air is slowly released, the shape and volume of the balloon also changes.	

1. Have students create a concept map using these terms: atom, element, compounds, matter, molecule, solid, liquid, gas.
2. Students are often curious to know if atoms can be seen with a microscope. Take your students on a field trip to a local college or university to find out. Ask an expert to talk to your students about technological advances in microscopy.
3. Divide the class into small groups and instruct students to use the Internet and other resources to learn about nanotechnology. Have each group make a presentation about one application of nanotechnology.

Students complete section 3.1 review questions.

Teaching Tip . . .Addressing Misconceptions about Atoms and Molecules

A few of students' common misconceptions about atoms and molecules are discussed below.

Atoms move therefore they are alive.

- Atoms are not alive because they do not exhibit the characteristics of living organisms. Ask students to list the qualities of living things (growth, respond to surroundings, made of cells, etc). Then ask students questions like, "Does an atom grow and reproduce?"
- Atoms move because they have energy of motion. The amount of energy in atoms is directly proportional to temperature. As the temperature increases, the energy of atoms also increases.

The structure of an atom and a cell are similar.

- Atoms are significantly smaller than cells. Although atoms (like cells) have a nucleus, they are completely different both in structure and function.
- Ask students to arrange each of the following items in order of decreasing size: the width of a human finger, an atom, a virus, a molecule, the width of a human hair, and a grain of sand. Then have students research to find out if their initial guess is correct. Most students will realize that atoms are much smaller than they originally thought.

Molecules change in size, shape, mass, and texture as matter changes state.

- Size, shape, texture, and mass are physical properties. As matter changes state, these properties of molecules do not change. For instance, students often believe molecules in a solid have greater mass than in a gas. They often say, "ice is heavier than steam" not realizing the difference lies in the number of H_2O molecules per unit of volume. Each water molecule has equal mass (18.01 g/mol).
- Molecules do not expand when they are heated. They only seem to expand because heat causes molecules to move faster and farther apart.

Investigation 3A: Energy

1 Energy in height and motion

In this part of the investigation you will use the energy car and track to make observations about energy. Set up the track with a hill and a level section as shown in part one. Put a rubber band on the bottom and a photogate on the mark near the center of the level section.

Assist students with the setup of the track.

Drop the car from different heights on the hill using the screw on the stop to provide a repeatable starting point. Measure and record the position from the center of the track. Then use the photogate to measure the time it takes for the car to pass through the light beam before and after bouncing off the rubber band. Drop the car from different heights to observe how the time though the photogates changes with position. Record your data in Table 2.

This part of the investigation takes 15 to 20 minutes. Tell students to record all observations.

2 Stop and think

What happened? Did the car ever go back to the height it had when it was released? Why?

The students should observe that the car never reaches the height from which it was dropped after it bounces off the rubber band. Use the experimental results to illustrate the idea of energy and conservation of energy without using mathematics or formulas. Energy can never be created or destroyed, only converted from one form to another. The car has a certain amount of energy at the start. For the car to reach a higher point requires more energy than the car has at the start. If the car is not pushed at the start, then there is no way for the car to acquire this additional energy. This is why the car never gets higher than the height it started from. In fact, since friction slowly drains away some of the car's energy, the car does not even return to the starting height.

How did the speed of the car after bouncing off the rubber band compare to its speed before bouncing off the rubber band? Was the car faster, slower, or about the same after the bounce?

Students observe that the speed of the car after the bounce is always less than it was before. On the level section of track, the energy of the car is kinetic energy. All the energy of speed is absorbed by the rubber band as it stretches and stops the car. The rubber band gives the energy back to the car as it springs back to its original shape and that is why the car bounces off. The key idea is that the energy cannot give more energy back to the car than what it absorbed from the car. That means that the car can never have a higher bouncing speed off than it has going in. The rubber band is not perfect and does not give back all the energy it absorbs. This is why speed after bouncing is always less than speed before bouncing.

You can use the rubber band to launch the car up the hill with the same speed it had coming down. Do you think the car will reach the height it was originally dropped from? Try it.

Students will need to adjust the screw at the bottom of the track to change the amount the rubber band is stretched until they can get the car to pass through the photogate with the same time it had going down. They should observe that the car never reaches the original height.

3A Energy

What is energy?

Energy is a very important idea that is hard to define. Energy is a measure of the ability to change or create change. A car at the top of a hill is able to move because it has energy due to its height. The car's increase in speed as it moves is a change that could not occur without energy. Temperature is another way to measure differences in energy. Warm water has more energy than cold water. Energy can appear in many forms, such as heat, motion, height, pressure, electricity, and chemical bonds between atoms. The key to understanding why things change in physical science is to follow the movement of energy.

Materials

- Energy car and track
- Timer and photogates
- Meter stick
- Mass balance
- Thermometer
- Foam cups
- Hot and cold tap water

1 Energy in height and motion

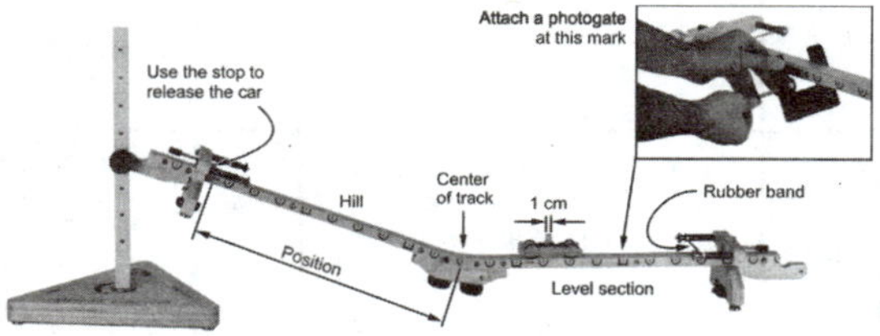

1. Set up the track with the steeper hill and a level section.
2. Put a photogate near the middle of the level section.
3. Drop the car from different heights on the hill using the screw on the stop to provide a repeatable starting point. Measure and record position from the center of the track.
4. Use the photogate to measure the time it takes the car to pass through the light beam before and after bouncing off the rubber band.
5. Drop the car from different heights to get several different speeds.

Table 1: Times before and after bouncing

Drop position (cm from center)	Before rubber band Time through photogate (sec)	After rubber band Time through photogate (sec)

10

Teaching Tip . . .Stop and think (part two)

Students have not yet been given a formal definition of speed in this course. However, it is an important concept in this investigation. Ask students to explain what speed means in their own words. A simple definition like how fast something is moving is sufficient in this case. If you have advanced learners, it is okay to model the speed calculation (distance ÷ time) and have them compare the results.

3 Temperature and energy

Suppose you discover that your neighbor is preparing soup for dinner. You love soup and decide to join her. She places a bowl of soup on the table. Just as you are about to dig in your neighbor tells you, "Wait a few minutes because it is too hot!" How do you determine whether your food is too hot or too cold?

Students may use a number of different methods to detect if their food is too hot or cold. Some examples may include tasting a small amount of the food after stirring it or touching the container to determine if it feels warm.

Words like hot and cold are often used to describe our level of comfort. For example, you may say that your food is too hot or that it is too cold outside on a winter day. Hot and cold are familiar sensations but you may not realize that they are caused by the energy of atoms. In this part of the investigation you will learn more about temperature and energy. Temperature is the measurement used by scientists to quantify sensations of hot and cold. Which would you expect to have a higher temperature: a cup of boiling water or a cup of ice? How can you justify your response?

The cup of boiling water has the higher temperature. You can use a thermometer to measure the temperature of each cup and verify that the temperature is higher in the cup of boiling water.

The cup of boiling water does have a higher temperature than the cup of ice. Let's think about the experiment you will do in this part of the investigation. Two foam cups contain equal masses of water. One cup contains cold water at a temperature of about 0°C. The other contains water at 50°C. What do you think will happen when water from the two cups is mixed?

Students speculate about the resulting temperature of the mixture.

Record the starting temperature of the water from both cups in Table 2. Mix the water, stir well, and measure the final temperature.

Students do the experiment.

4 Stop and think

Before you mixed the water, each cup had the same mass and number of water molecules. The water molecules in each cup are constantly moving, therefore they have energy of motion, or *kinetic energy*. However, not all molecules are moving at the same speed. Some move faster than others. When you add energy to a cup of water, you increase the temperature of the water. You notice a rise in the temperature because more of the water molecules are moving faster and have greater kinetic energy. This is actually how scientists relate temperature and energy. *Temperature* is a measure of the average kinetic energy of molecules.

Write the meanings of temperature and kinetic energy on the board.

Why do yo think scientists define temperature as an object's average kinetic energy?

Each molecule is not moving at the same speed. If a person counted each water molecule in a cup of water and measured its individual speed, he could obtain the sum of the speeds. Dividing the sum by the number of water molecules would give the average kinetic energy.

Potential, Kinetic, and Thermal Energy

In this investigation, students are introduced to three types of energy: potential, kinetic, and thermal.

Potential energy is often defined as energy of position. Students often question what this really means. Potential energy is related to forces. Objects with potential energy are able to exert forces and cause change as energy is transformed. Consider the energy car and track system. The car is in an elevated position relative to Earth. When it is released, gravitational force moves the car from a position of higher to lower potential energy. Other forms of potential energy include chemical, electromagnetic, and elastic.

A moving object can also cause change. Energy of motion is *kinetic energy*, as exhibited by the moving car and the water molecules in the cup. How did the moving water molecules cause change as water in the cups were mixed? The molecules were able to transfer energy.

Temperature is a measure of *thermal (or internal) energy*, which results from the motion of atoms and molecules in matter. Warmer molecules move with greater speed and have more thermal energy than cooler molecules. When a temperature gradient exists between quantities of matter, heat flows. Heat is thermal energy that is transferred from a warmer object to a cooler object. However, a person cannot directly observe thermal energy. Instead one is able to see *evidence* that heat has been transferred, such as a change in temperature.

Investigation 3A: Data and Answers

1 Energy in height and motion

Drop position (cm from center)	Before rubber band Time through photogate (sec)	After rubber band Time through photogate (sec)
49	0.0065	0.0072
44	0.0070	0.0077
35	0.0081	0.0089
25	0.0102	0.0116
15	0.0167	0.0206

2 Stop and think

a. No, it never goes back up the track higher.

b. No, it never has a greater speed after bouncing back.

c. No, it never goes back up the track higher.

d. When the car is released to roll down the track it has a certain amount of energy based on its height. When the car reaches the bottom of the hill the energy from its height is converted to the energy of speed minus any friction that may have been experienced by the car. As the car rolls along the level section of the track it can only slow down as friction continues to subtract from the car's energy. The rubber band stretches as the car hits it, and then pushes back on the car sending it back in the opposite direction. Most of the energy is returned to the car, but not all of it, since it rebounds at a slower speed than it came in. Some of the energy went to friction in the rubber band, maybe heating up the molecules in the rubber as it stretched. As the car rolls back up the hill, all the energy lost to friction on the track and the rubber band have taken away from the original amount it had sitting atop the track, and it can never reach its original height.

3 Temperature and energy

Cold water temperature before mixing (°C)	Hot water temperature before mixing (°C)	Mixture temperature (°C)
1.6	48.6	25.3

4 Stop and think

a. The hot cup has more energy because it has a higher temperature, and temperature is a measure of a material's average thermal energy.

b. Since there are equal amounts of each, the temperature will probably be halfway between the temperature of the cold water and hot water, about 25°C.

c. The energy of the system before mixing will be the same as the energy of the system after mixing.

d. The hot water has more energy than the cold water. When they are mixed together, the overall energy of mixture is the equilibrium point between the high and low energies of the hot and cold water.

Notes:

3.2 Temperature and Energy

About the Lesson

In this lesson, students further explore the relationship between temperature and energy. Some concepts discussed in this lesson include temperature scales and conversions, heat and thermal energy, and energy transformations.

Energy Cards

Have students generate a list of all the different forms of energy they know. It may be necessary to prompt students with ideas. Use the list to create energy cards from white cardstock. Instruct students to draw a picture to represent the form of energy on one side of the card and to write its name on the other side. For example, a flame may be used to represent heat energy.

Assign each student a partner. Read scenarios to students that involve energy transformations. Some examples are shown at right. Once you have completed the reading, ask teams to display cards on their desks for each form of energy described. Walk around the classroom and observe which cards are displayed. Ask volunteers to hold up their cards and discuss each transformation.

Students complete Investigation 3B before reading section 3.2. In this investigation students use indirect measurements to find out the average mass of a single grain of rice. The focus of this investigation is to help students understand how scientists use indirect measurements to determine the size of objects, like atoms, that are too small to be seen.

Investigation 3B: Mass and Indirect Measurement

Setup	Materials	Investigation Vocabulary
1. Allow one class period to complete the investigation. 2. Students work in small groups of three to five.	• rice (about 1/4 cup per group) • cellophane tape • mass balance • scissors • index card	indirect measurement resolution cubic centimeter

Teaching Tip . . .Energy Transformations

Before playing the *Energy Cards* game, be sure that your students understand the meaning of the word *transformation*. Use the word in a sentence and have students use context clues to infer its meaning. For example, "*There are many transformations in the life cycle of a butterfly.*"

When thinking about transformations, the main idea is change. It is important for students to know that energy is not gained or lost; it simply changes form. One way to avoid this misconception is to explain to students that energy is a property of matter. For instance, rather than saying "energy moves" tell students that atoms and molecules move and carry energy.

Energy Cards

Read each scenario and have students identify the energy transformations involved.

- Ian and his friends went on a camping trip. They decided to gather logs and build a fire. What are the energy transformations involved in a burning campfire? (Answer: The burning logs represent chemical energy which produces energy in the form of heat, light, and sound from the crackling of the wood.)
- The gym teacher assigned students to teams for relay races. Describe the energy transformations involved in a student running in the race. (Answer: The student obtains chemical energy from the food she eats, which is converted to mechanical energy as she moves different parts of her body. Heat energy is also given off as she runs.)
- Amelie rang the doorbell when she arrived at her grandmother's house. Does a transformation occur when Amelie rings the doorbell? (Answer: Yes, electrical energy is converted into sound.)

It's A-MAZE-ING: Indirect Measurement

It is often difficult for students to imagine that something exists if they cannot see it. Students also question how characteristics like shape and size can be determined without applying direct measurement. Try this activity to help your students learn more about what they cannot see.

Use shoe boxes (with lids) to construct different mazes. Tape pieces of cardboard to the bottom and sides of the box to create a unique pattern. Place a marble in each box. Be sure that the lid is also taped shut. Label each box with a number or letter.

Divide the class into small groups. Have students in each group conduct a series of tests to figure out the path of the maze. For example, students are allowed to shake the box, swirl it around, invert it, or move it diagonally. As students conduct each test, have them draw a picture of what they believe to be the maze pattern. After five to ten minutes of testing, instruct students to review each preliminary drawing and decide as a group what the maze looks like. Have each group switch mazes with another group and repeat the tests. Be sure that students record the box number for each test. Have students compare drawings. Allow them to open the boxes. Were their drawings correct?

1. Challenge students to work in teams to build a device that demonstrates at least three energy conversions.
2. Divide students into small groups. Have them research to find out about alternative energy sources being studied by scientists. Instruct each group to make a presentation about their findings.
3. Create a concept map with these terms: motion, energy, potential energy, thermal energy, atoms, molecules, kinetic energy, and temperature.
4. Have students use publishing software to create a brochure or pamphlet about energy conservation. Ask students to include helpful tips, such as how to reduce home heating and cooling costs. Distribute the brochure to students throughout the school.

Students complete section 3.2 review questions.

Word Origins...energy (from Greek *energeia* meaning "work")

Energy is often defined as the ability to do work. Physical scientists define work as a measure of the energy transferred as force is applied over a distance.

Ask students to make a list of work or chores they do at home. Go over each item in the list and have students describe how energy is involved. You may even extend this activity to have students describe specific energy transformations related to the work (chores) they do. For example, have a student who does the laundry describe the energy transformations in operating a washing machine.

Teaching Tip . . .Temperature and Thermal Energy

Temperature and thermal energy are not the same. Review the definition of temperature with students. Emphasize that temperature is a measure of the *average* kinetic energy of molecules. Then tell students that thermal energy measures the *total* energy of all molecules within a sample of matter.

Investigation 3B: Mass and Indirect Measurement

What are some ways to measure matter?

Students answers will vary. Sample answers include using a ruler or a balance. Place various objects of different shapes and sizes before students.

Here is a box. Suppose I asked you to determine the length of this box. Can anyone demonstrate how to do it?

Have a student volunteer demonstrate the process. Repeat this process with a few more objects. Choose items with dimensions that can be measured with a ruler or meter stick. Also have students use a balance to determine the mass of objects.

Very good. You can see that these instruments are helpful in obtaining measurements. Do you think it is possible to physically measure all matter?

Students share opinions.

What are some facts that you know about matter?

Students share knowledge about matter. For example, matter has mass and volume or may be solid, liquid, or gas. Accept all reasonable responses.

You have already demonstrated how to determine the mass and dimensions of the objects shown here. What about the forms of matter that you cannot see? For example, how do you think scientists determine the width of a single strand of human hair?

Students offer suggestions. Engage the entire class in discussion.

Let's try to determine the mass of a single hair strand.

Place the strand of hair on a mass balance. Allow students to see that the balance still reads 0.00.

Why is the reading still at zero? Is the hair not matter?

The hair is matter because it has mass and volume. Students should realize that the mass of the single hair strand is too small to be detected by the balance.

There are many situations where scientists need to know the dimensions of something that is too small or large to measure directly. In these situations, scientists use indirect measurement. What does the word indirect mean?

Indirect means in a manner that is not straightforward.

1 Setting up

In today's investigation, you will find out the average mass of a single grain of rice. Would anyone like to guess the mass of a single grain?

It is fun to turn this activity into a guessing contest. Have students write their name and guess on a piece of paper before collecting any data. Collect the pieces of paper. At the close of the investigation review students' initial guesses. Award prizes to students whose guess was closest to the actual mass.

3B Mass and Indirect Measurement

How do scientists measure things too small to put on a balance?

There are many situations where scientists need to measure something that is too small or too large to measure directly. In these situations scientists use indirect measurement. With indirect measurement you calculate the thing you want to know from other measurements.

Materials

- About one quarter cup of rice
- Some cellophane tape.
- A mass balance
- Scissors
- index card

The objective of the experiment is to find the average mass of a single grain of rice. One grain of rice is much too small to weigh accurately with the balance. Instead, you will measure the mass of a three centimeter cube filled with rice. You will then estimate how many grains there are in the cube. By dividing the mass of the cube by the number of grains in the cube, you will get a very good measurement of the average mass of a single grain of rice.

1 Getting started

The first thing we need to do is find out how many grains of rice are in a cubic centimeter.

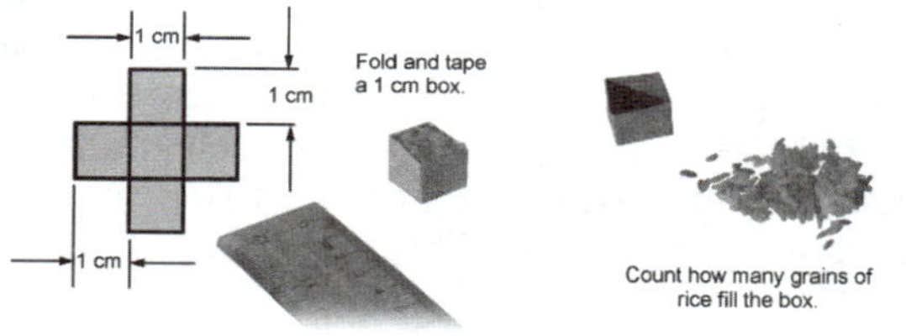

1. Cut out and fold up the small cube (made out of an index card) as shown in the diagram above. Use the cellophane tape to hold the cube together. This cube has a volume of 1 cubic centimeter (1 cc).
2. Fill the cube level with rice.
3. Carefully empty the rice in the cube onto the table. Count how many grains fit into the cube and record the value in Table 1.
4. Calculate the number of grains of rice per cubic centimeter and record this value in Table 1. Use the formula below:

number of grains / volume of cube = grains per cubic centimeter

12

Of course you know that one grain of rice is too small to mass accurately with the balance. Instead you will measure the mass of a cube filled with rice. Then you will estimate how many grains of rice are in the cube. By dividing the mass of the cube by the number of rice grains in the cube, you will be able to get a very good measurement of the average mass of a single grain of rice.

Students create the cube according to the instructions in part one. You may need to review concepts like how volume is measured and calculated. Explain to students that a cubic centimeter is the result of measurement in three dimensions. Display a wooden cube and point out the length, width, and depth of the cube. It may also be necessary to review the procedure for operating the mass balance. Walk around and assist students as needed.

Fill the cube level with rice. Then empty the rice in the cube on the table. Count how many grains fit into the cube and record your answer in Table 1. Calculate the number of grains of rice per cubic centimeter.

Model the calculation. Identify the numerator as the mass of rice in grams and the denominator as the volume of the cube. Advanced learners may recognize this as a calculation of density.

2 Accuracy, precision, and resolution

No instrument in science makes perfect measurements. Scientists must always consider error when reporting measurements.

Review the meaning of accuracy and precision with students. Explain to students that an instrument's resolution influences the precision of measurements.

3 Making a precise measurement

Build a new cube as shown in the diagram in part three. What are the dimensions of the cube? What is its volume?

The cube has dimensions of 3 cm × 3 cm × 3 cm or a volume equal to 27 cubic centimeters.

Record the volume in Table 2. Place the cube on the balance and hit the tare button. The balance should now read zero grams.

Ensure that balances are tared. If the reading is not zero, have students repeat the taring process

Remove the cardboard cube from the balance, fill it with rice, put it back on the balance and record its mass. Use your data to calculate the average mass of one grain of rice.

Students complete the calculations.

4 Stop and think

Review the data you collected with the members of your group. What is the average mass of a grain of rice?

Have students compare their data. Discuss any major discrepancies in masses. Then have students answer the questions in part four. Go over the answers to questions before closing the investigation.

Teaching Tip . . .Indirect Measurement

When learning about atoms and molecules, students often question how scientists know anything about atoms if they cannot be seen. Explain to students that early scientists learned about atoms through indirect measurement. They were able to conduct experiments and make inferences about the structure of atoms. (Students learn more about the work of scientists like Dalton, Bohr, and Rutherford and atomic models in chapter 6.)

Modern technology has provided scientists with many valuable tools in the field of microscopy. For example, scientists are now able to view images of atoms. As a result of milestones like these and repeated investigations, many early hypotheses about atoms have been confirmed.

Investigation 3B: Data and Answers

1 Getting started

Volume of cube: 1

Number of grains of rice: 25

Calculated grains per cubic centimeter: 25

2 Accuracy, precision, and resolution

3 Making a precise measurement

Volume of cube (cm^3)	27	Number of grains of rice in 3 cm cube	675
Mass of rice in cube (g)	10.0	Calculated mass of 1 grain of rice (g)	0.015

4 Stop and think

a. Because removing the paper cube removed the mass of the paper cube. Subtracting that mass from 0.0 g makes the reading be negative.

b. Because we never measured one grain of rice by itself. We measured many grains and divided the total mass by the number of grains.

c. Because recipes usually call for many grains of rice, and not one grain of a particular mass.

Notes:

Chapter 3 Answers

3.1 Section Review

1. b
2. 2,200 grams
3. d
4. Answers may vary. Examples include:
   solid: wood, steel, brick, silver
   liquid: water, mercury, alcohol
   gas: air, helium, nitrogen, oxygen
5. Since a bridge must be able to hold its shape and have the strength to hold up anything that needs to cross it, it would be best to use a solid. Both liquids and gases flow, and you wouldn't want the bridge material to flow away.
6. Water is a compound made up of the elements hydrogen (H) and oxygen (O).
7. a

3.2 Section Review

1. Sample answer: My school's heating and air conditioning system control the temperature of my school. The furnace heats the air, the air conditioner cools the air, and the ventilation system moves the air through the building. They all work together to keep the school at a comfortable temperature.
2. a
3. 30°C
4. b
5. Energy flows from the hot coffee to the air in the room and the table until the coffee, the room, and the table are all at the same temperature. The coffee will cool, and the table and air will very slightly warm.
6. False. Each material has a different specific heat, so each changes temperature at a different rate.
7. In the article, "running out of energy" means that Earth contains a limited amount of energy in the form of oil and gas. These are nonrenewable resources that are converted to other forms when burned, not used up. There is still plenty of energy in other forms.

Connection

1. Wind turns the turbine's blades, causing the shaft to spin. The shaft is connected to the generator, which produces electricity.
2. Wind power costs are much less than those of fossil fuels. Wind energy is clean, not imported from another country, an abundant resource, and requires no mining or drilling.
3. Wind is intermittent and you don't know when it's going to blow. Most wind farms are in remote areas away from the large cities that could benefit from their energy production.
4. Research will help to improve equipment and methods in order to reduce cost and increase efficiency.

Activity

Hypothesis: If a newspaper parachute is attached to a ping-pong ball, the time it takes the ball to fall 3 meters will decrease as the diameter of the parachute increases.

Independent variable: Diameter of parachute

Dependent variable: Time for ball to fall 3 meters

Control variables: Ball used, type of paper for parachute, way the ball is released, length of parachute strings

Procedure:

1. Drop a ping-pong ball from a height of 3 meters and measure the time it takes to fall. Repeat twice, for a total of three trials.
2. Cut out a circular piece of newspaper 3 cm in diameter. Attach 3 pieces of string to the parachute with tape, and tape the strings to the ball to make a parachute. The strings should each be 10 cm long.
3. Hold the parachute so the ball is hanging 3 meters above the ground. Release and measure the time. Do three trials.
4. Repeat with parachutes of 6 cm, 9 cm, 12 cm, and 15 cm. Adjust the string lengths so the ball is always 10 cm below the parachute.

Data:

Parachute diameter (cm)	Time (s)	Time (s)	Time (s)	Average time (s)
0	0.79	0.77	0.80	0.79
3	0.80	0.82	0.81	0.81
6	0.84	0.85	0.87	0.85
9	0.88	0.92	0.94	0.92
12	0.99	0.96	0.99	0.98
15	1.11	1.08	1.07	1.09

Conclusion:

This experiment tested the effect of the size of a parachute on the time a ping-pong ball took to fall three meters. We predicted that the greater the diameter of the parachute, the greater the time the ball would take to fall. Our data clearly supports this hypothesis. As the size of the parachute increased from 0 to 15 cm, the drop time increased from 0.79 seconds to 1.09 seconds.

Applying your knowledge: sample answers

a. Our data supported our hypothesis.
b. We had some trouble making sure the parachutes were equivalent, so we had to adjust the string lengths to make sure the ball was always 10 cm below the parachute. We added that into the procedure as we performed the experiment.

Chapter 3 Assessment

Vocabulary

Section 3.1

1. kilogram
2. element
3. compound
4. atom
5. gram

Section 3.2

6. system
7. energy
8. Calorie
9. newton
10. conservation of energy
11. force
12. thermometer
13. Fahrenheit
14. Celsius
15. heat

Concepts

Section 3.1

1. Quantities are measured in both kilograms and grams because mass occurs in many different sizes. A unit closest to the mass of the object should be used.
2. b
3. A mass balance is most often used to measure mass.
4. Answers are:
   a. compound
   b. element
   c. compound
   d. element
5. The smallest piece of iron would be an atom.
6. Atoms of solids are spaced very close together, moving only by vibrating around a certain position in the solid. Atoms of liquids are free to flow, taking the shape of their container, but limited in volume. Atoms of gases are free to move in all directions, filling their container.
7. a. solid
8. c. gas
9. a

Section 3.2

10. Sample answers: solar system, circulatory system, respiratory system, mass transit system
11. Systems are organized to make the study of complex systems easier.
12. Forces make it possible to transfer energy from one body to another.
13. Potential energy is stored energy. Examples include: gravitational PE, elastic PE, and chemical PE. Kinetic energy is energy in transfer or energy of motion. Examples include: a thrown ball, and a rolling car.
14. Answers are:
    a. KE
    b. PE
    c. PE
    d. KE
15. Steam has the higher kinetic energy. The atoms of a gas (steam) are moving faster than the atoms of a solid (ice), and therefore have higher kinetic energy.
16. At temperatures greater than 10,000°C atoms start to come apart and become a plasma.
17. Thermal energy is the *total* energy stored in the kinetic energy of individual atoms and is measured in in units of joules. Temperature measures the sensations of hot or cold due to the thermal energy in a material and is measured on a temperature scale in °F, °C, (or Kelvins). Heat is thermal energy that is flowing or moving and is measured in joules.
18. high; low
19. the temperature of the human body
20. At absolute zero, -273°C, atoms have the lowest energy they can have. The atoms can be thought of as "frozen" because their energy is so small.
21. a
22. Energy is not used up, but is transformed from forms that are easy to use to forms which are less convenient to use. To conserve energy means to make wise use of the forms of energy that are easily transformed by our current technology.
23. c
24. Power plants transform energy from chemical, solar, nuclear, wind, or falling water into a useful electrical energy.
25. c

Problems

Section 3.1

1. 45,000 grams
2. 0.150 kg
3. 454 grams
4. A

Section 3.2

5. Answers are:
    a. 2 kg of water at 50°C
    b. 1 kg of water at 5°C
    c. 1 kg of steam at 105°C
6. 1 kg of glass
7. The specific heat of water is 4184 J/kg°C and is defined as the amount of energy required to raise 1 kg of water 1°C. Therefore to raise 1 kg of water by 2°C you would need 2 × 4184 joules = 8368 joules.
8. 0.0239 Calories
9. 350 Calories × 4184 C/joule = 1,464,400 joules
   1,464,400 joules ÷ 150 joules/s = 9762 seconds
   9762 seconds or 2.71 hours
10. 80 Calories × 4184 joule/Calorie = 334,720 joules
11. 427°C
12. 12,632 °F
13. Celsius scale; a change of 1°C equals a change of 3.6 °F
14. Answers are:
    a. The stored chemical energy of the pizza is converted to the kinetic energy of moving your body.
    b. Electrical energy is transformed into the heat and light energy from the lamp.
    c. The chemical energy stored in the wood is transferred to heat, light, and sound energy of the fire.

Chapter 4: Density and Buoyancy

Learning Goals	California Content Standards	Activities and Resources	Assessment
4.1: Density The student will: • Apply a rule to define density and solve problems. • Use appropriate units to report density measurements. • Compare the densities of common materials. • Apply various methods to determine the volume of matter.	PS 8a PS 8b PS 9f	**Laboratory Investigation 4A:** Density **Teaching Illustrations:** Density Range of Common Materials, Steps to Solving a Problem, Solving Density Problems **Investigation Skills:** Calculating Volume, Measuring Volume **Skill and Practice Sheets:** Density, Calculating Slope from a Graph	**Exam View Test Bank** Chapter Pre-Test **Section 4.1 Review**
4.2: Buoyancy The student will: • Contrast mass and weight. • Conduct experiments to observe Archimedes' principle. • Define buoyancy. • Explain the relationship between density and buoyancy.	PS 8c PS 8d PS 9f	**Laboratory Investigation 4B:** Sink or Float **Teaching Illustrations:** Example of Buoyant Force, Average Density Explanation **Graphic Organizer:** Learning about Buoyancy **Biographies:** Archimedes **Skill and Practice Sheets:** Buoyancy, Archimedes Principle **Connection:** Airships in the Sky **Chapter Activity:** Will it Float or WIll it Sink?	**Section 4.2 Review** **Chapter Assessment** **Exam View Test Bank**

Estimated Instruction Time	Language Tools	Differentiated Strategies	Materials
Four 45-minute class periods	**Literary Picks** **Vocabulary:** density	**ELL Strategies:** Listed in ELL Ancillary **Learning Strategies:** • Cooperative Learning - Investigation 4A • Active Learning - Motivate • Guided Practice - Explain: Density Calculations • Paired Learning - Extend **Teaching Tip:** Density Stations	**Investigation 4A:** graph paper, ruler, balance, disposable cups, 250-mL beakers, graduated cylinders, steel hex nut (3/8 or 1/2 inch size), glycerin, corn syrup, molasses, water, vegetable oil, food coloring, rubber stoppers **Teaching Tip - Density Stations:** 4 different objects used to determine mass, volume, or density, wooden block, rack of test tubes, different colored water in each test tubes, labels
Four 45-minute class periods	**Vocabulary:** weight, buoyancy, Archimedes principle **Word Origins:** buoyancy **Closure Cards**	**Learning Strategies:** • Cooperative Learning - Investigation 4B • Demonstration - Motivate: Buoyancy Fun **Teaching Tip:** Demonstrating Neutral Buoyancy **Addressing Misconceptions:** Mass and Weight **Multicultural Experience:** John Phillip Holland	**Investigation 4B:** 1/2 stick of clay, displacement tank, beaker, paper cup, graduated cylinder, 15-cm deep plastic container **Motivate - Buoyancy Fun:** deep pan or basin, bath tub toys, balloon, 2-clear and tall drinking glasses, salt **Teaching Tip - Demonstrating Neutral Buoyancy:** large container (like a aquarium), water, resealable sandwich bags **Chapter Activity:** graduated cylinder, balance or scale, plastic cup, assortment of small objects like wood cubes, various plastic items, foam, steel, glass, cork, and aluminum

4.1 Density

About the Lesson

Density is a physical property of all matter. In this lesson, students define density as the ratio of mass per unit volume. Students apply the formula (D = m/v) to calculate the density of materials. They also examine variations in densities of matter in different states.

Display one pound samples of various materials on a table. Some examples may include a wooden block, styrofoam peanuts, a copper or other metal block, cotton balls, play dough, paper clips, wax, and pennies. Ask students which of the displayed samples weighs most. Students often assume that larger size means greater weight. Have student volunteers use a balance to find the mass of each sample. Are students shocked by the masses? How do they explain their observations?

Students complete Investigation 4A before reading section 4.1. In this investigation students conduct experiments designed to help them understand the meaning of density. In the first experiment students infer the relationship between mass and volume. In the second experiment, students create a density column to compare the densities of various liquids.

Investigation 4A: Density

Setup	Materials	Investigation Vocabulary
1. Allow one class period to complete the investigation. 2. Students work in small groups of three to five. 3. Be sure to have enough paper towels and liquid soap on hand to clean messy spills and wash sticky hands. 4. Label and prepare 30-mL quantities of each liquid in plastic cups (part four) before lab to save time and avoid waste. 5. Have at least three different colors of food coloring available. 6. Gather a collection of rubber stoppers that fit easily into graduated cylinders (without touching the sides).	• graph paper and ruler • balance • disposable cups • 250-mL beakers • graduated cylinders • steel hex nuts (3/8" or 1/2" size) • glycerin • corn syrup • molasses • water • vegetable oil • food coloring • rubber stoppers	density slope

Literary Picks...

Reading Level	Title
3-5	***The Magic School Bus Ups and Downs* by Joanna Cole** Join Mrs. Fizzle and her students on a magical underwater tour as she transforms the magic school bus into a submarine. Learn how they finally get the bus to sink by attaching barrels of sand. You can only imagine what they did to make it float!
6-8	***Archimedes and the Door to Science* by Jeanne Bendick** This book is part of the Living History Library series and crosses the content areas of mathematics, history, and science with the captivating story of Archimedes' life. Topics include astronomy, buoyancy, density, specific gravity, and equilibrium.
9-12	***Neutral Buoyancy: Adventures in a Liquid World* by Tim Ecott** A guided tour of the history of underwater exploration and scuba diving as seen through the eyes of journalist and diver Tim Ecott. Discusses the development of diving equipment, hyperbaric physics and physiology, sponge diving, and recreational diving.

Guided Practice: Density Calculations

Write the density formula on the board. Identify the three variables in the formula. Demonstrate how to apply algebra skills to manipulate the density equation to solve for mass or volume. Then lead a discussion to help students complete the example problems listed below.

Example Problems: Use the information provided to solve for the unknown quantity.

a. Find density of an object if its mass is 143 g and its volume is 13.8 mL. (10.4 g/mL)

b. Calculate mass of matter with a density equal to 1.2 g/cm^3 and volume equal to 76 cm^3. (91 g)

c. If the density of a substance is 2.30 g/cm^3 and its mass equals 234 g, what is its volume? (102 cm^3)

d. A block has the following dimensions: length = 3.0 cm, width = 2.0 cm, and height = 2.0 cm. When Eleazar placed the block on a balance, he determined its mass was 14 grams. What is the density of the block? (1.2 g/cm^3)

1. Have students work in pairs to practice solving density problems. As one student solves a problem, the partner verifies the solution.
2. Complete the tasks described at the *Density Stations*. See Teaching Tip at right. Have students record their experimental techniques and procedure at each station. Pool the class data and have students lead a discussion to compare results.
3. **Environmental Impact: Density and Oil Spills**

Have students use the Internet to find out about the Exxon-Valdez oil spill; then answer the following questions. What is the density of petroleum-based oil? How can density be applied to explain the effects of oil spills on marine life? How did the density of oil create challenges for cleaning up the spill?

Students complete section 4.1 review questions.

Teaching Tip . . .Density Stations

Set up various stations in your classroom. Place a different object at each station. Have students work in pairs or small groups to decide how to determine the mass, volume, or density of an object.

Station 1: Apply the displacement method to determine the density of an irregularly shaped object. Use an illustration to explain how you arrived at your answer.

Station 2: The density of a wooden block is 0.60 g/cm^3. Measure the volume of the block. Calculate its mass (without using a balance).

Station 3: Set up a rack of test tubes containing a solution of water tinted with food coloring. You may use different colors and label each test tube with a letter or number. In this manner, students will not know that the solutions are the same. Have students record the letter of the sample they test. Place a graduated cylinder and a balance at this station. Instruct students to find the density of the solution. **Note**: Each group uses only one test tube.

Station 4: Students read the scenario described below, then apply a method to solve the problem. **Note**: Allow students to use Table 4.1 in the student text to identify the unknown material.

Redty found a strange piece of metal while hiking with friends. He was not able to identify it right away but decided to conduct an experiment to determine its identity. Redty used a balance and discovered the mass of the metal to be 7.6 grams. He determined its volume to be 2.8 cm^3. What material did Redty find?

Investigation 4A: Density

Complete the *Motivate* activity before starting the investigation.

Think about each of the samples you examined before. Recall that each sample had equal weight. Is the volume of each sample equal?

Point to each sample. Students should observe that each sample occupies a different amount of space.

Suppose you could compact each of these samples into cubes of equal size, for example one cubic centimeter. Would you predict the masses to be the equal? If not, which sample would be the heaviest?

The mass of each sample will vary. Students speculate about the relative masses of each sample.

Mass is a measure of the amount of matter in an object. The greater the amount of matter in a sample, the greater its mass with be. In this example, you have introduced another variable: volume. Think about comparing one cubic centimeter of cotton balls with one cubic centimeter of copper. Why does one feel heavier than the other?

Emphasize that the volume of each sample is identical but do not disclose any details about density just yet. Steer students toward understanding that the amount of matter that is packed into each cube must somehow be different.

1 The relationship between mass and volume

Do you think there is a relationship between mass and volume? Let's experiment and find out. Each team needs five identical steel nuts. Be sure the nuts are the same size. Zero your balance. Record this mass, in grams, in Table 1. Place one nut on the balance. What is its mass?

Students record the mass of one nut in Table 1. If you have nuts of various sizes, be sure to keep them separated so students have only one size per group.

Continue adding one nut at a time, recording the mass each time.

Students gather data.

Fill your graduated cylinder with water to the 20-mL mark. Record this volume in the "zero nuts" column in Table 1.

This initial volume will vary based on the graduated cylinders you use. What is important is that students have enough water in the cylinder to clearly see the results of adding the nuts. However, students should not place too much water in the cylinder as it may cause too great an increase in the volume and force some of the water out. If necessary, review the proper technique for reading the volume of liquids in a graduated cylinder. Have students tilt the cylinder and allow each nut to slide gently to the bottom of the cylinder.

2 Analyzing your results

What happened to the volume in the cylinder as more nuts were added?

Students observe an increase in volume.

Make a graph to determine if a relationship exists between mass and volume.

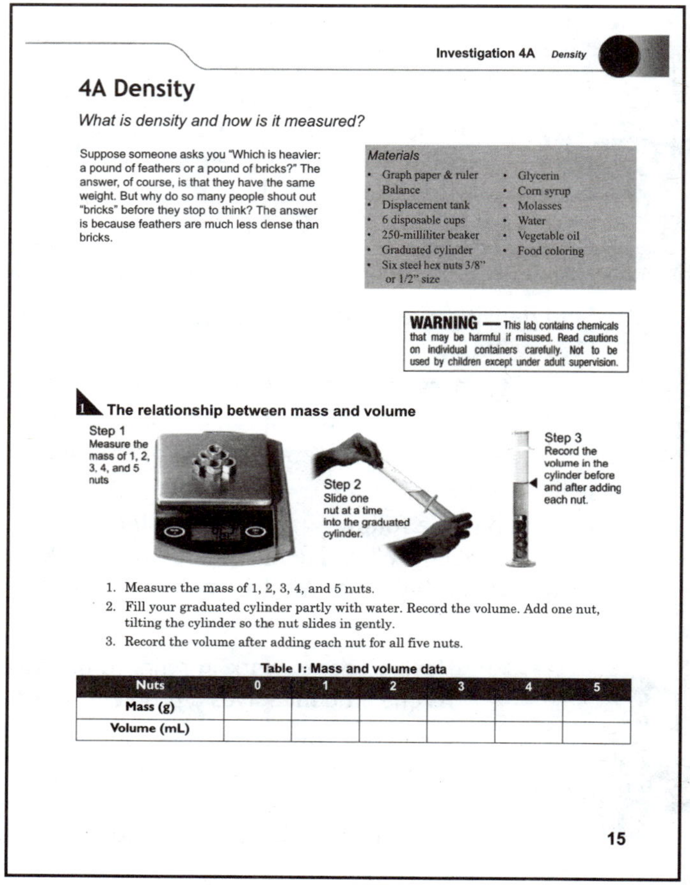

Investigation 4A *Density*

4A Density

What is density and how is it measured?

Suppose someone asks you "Which is heavier: a pound of feathers or a pound of bricks?" The answer, of course, is that they have the same weight. But why do so many people shout out "bricks" before they stop to think? The answer is because feathers are much less dense than bricks.

Materials

- Graph paper & ruler
- Balance
- Displacement tank
- 6 disposable cups
- 250-milliliter beaker
- Graduated cylinder
- Six steel hex nuts 3/8" or 1/2" size
- Glycerin
- Corn syrup
- Molasses
- Water
- Vegetable oil
- Food coloring

WARNING — This lab contains chemicals that may be harmful if misused. Read cautions on individual containers carefully. Not to be used by children except under adult supervision.

1 The relationship between mass and volume

1. Measure the mass of 1, 2, 3, 4, and 5 nuts.
2. Fill your graduated cylinder partly with water. Record the volume. Add one nut, tilting the cylinder so the nut slides in gently.
3. Record the volume after adding each nut for all five nuts.

Table I: Mass and volume data

Nuts	0	1	2	3	4	5
Mass (g)						
Volume (mL)						

15

Students complete questions a-c. Remind students that drawing a line of best fit is not like connecting dots. It is actually a smooth line.

3 Finding the slope

Find the slope of the line on your mass vs. volume graph.

Use Figure 4.1 to guide students to calculating the slope of a line.

The slope of the line tells how many grams of matter are contained in each milliliter of material you tested. The slope is the density of the material.

Write the term, *density*, on the board. Then write, density = grams of matter per mL of material tested.

You have observed that mass and volume are related. This relationship is expressed as the density of matter.

4 The density of liquids

In this part of the investigation you will compare the densities of several liquids. Tare the balance to zero with a clean, dry graduated cylinder. Add about 10 milliliters of molasses. Record the mass and the 10 mL volume in Table 2. Clean the cylinder and repeat the process for each of the remaining liquids. Be sure to record the exact volume of liquid you place in the graduated cylinder each time.

Students collect data.

Density is the mass of a substance divided by its volume. Calculate the density of each liquid using your data in Table 2.

Model the density calculation on the board. Write the formula and label each variable. Point out the units of mass and volume and explain the meaning of *"grams per milliliter."*

5 Deciding how to stack your liquids

Students review the data to help them decide how to stack the liquids; then record the name of each liquid and the order they selected in the space provided.

6 Constructing your density column

Students follow the instructions in part six to construct the density column. Have students add food coloring to the clear liquids so they are easily distinguishable. Use a different color for each clear liquid. Caution students to avoid tilting the graduated cylinder sideways as they pour the liquid as this causes it to run down the sides. Instead, tell students to pour the liquid in the center of the opening as shown in the picture.

7 Stop and think

Which of the liquids in the columns is the most dense? How can you tell? Use your observations to help you answer each question in part seven.

Students complete the answers to the questions (a-c). Discuss students' results. Be sure the rubber stoppers are small enough to fit into the cylinders without getting stuck.

Step 1: Select two points on the line.

The points (3,4) and (2,2) have been selected.

Step 2: Label each coordinate.

Let (3,4) represent (x_1,y_1); and (2,2) represent (x_2y_2).

Step 3: The variable m is used for slope.
Apply the formula.

$$m = \frac{y_2 - y_1}{x_2 - x_1} = \frac{2-4}{2-3} = \frac{-2}{-1} \qquad m = 2$$

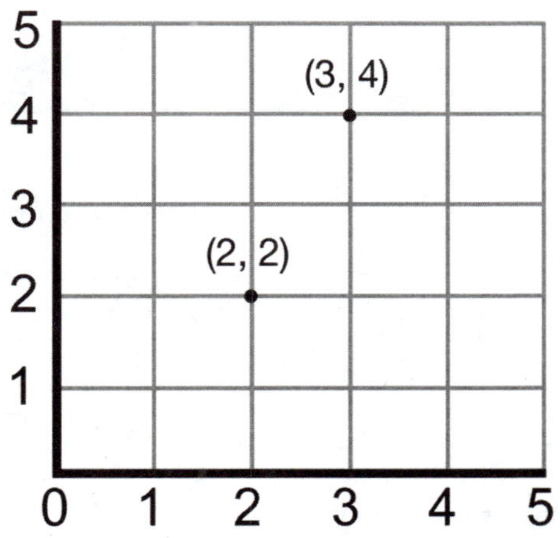

Figure 4.1: Finding the slope of a line.

Investigation 4A: Data and Answers

1 The relationship between mass and volume

Nuts	0	1	2	3	4	5
Mass (g)	0	27.0	54.1	80.7	108.2	135.0
Volume (mL)	0	4.0	8.0	11.8	15.9	19.6

2 Analyzing your results

a.

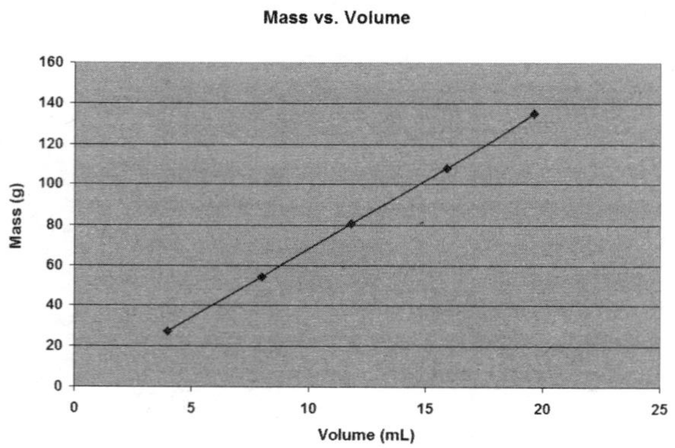

b. The points form a straight line, or direct linear relationship.

c. See student graph above.

3 Finding the slope

a. Our slope was the same as the other groups.

b. A different slope would mean a different density, and in that case, the other group would have measured the density of a different material.

4 The density of liquids

Substances	Mass in grams	Volume in mL	Density in g/mL
molasses	13.7	10 mL	1.37
water	9.9	10 mL	0.99
vegetable oil	9.7	10 mL	0.97
light corn syrup	12.6	10 mL	1.26
glycerin	12.1	10 mL	1.21

5 Deciding how to stack your liquids

Liquid 5 (top)	vegetable oil
Liquid 4	water
Liquid 3	glycerin
Liquid 2	light corn syrup
Liquid 1 (bottom)	molasses

6 Constructing your density column

7 Applying what you have learned

a. Yes, our liquids remained in order. There was a little mixing right when we poured them into the cylinder but they settled out into layers after a few seconds.

b. It would settle just under the water and above the glycerin.

c. It stops sinking when it hits the glycerin. Its density is greater than water and less than glycerin because it sinks through the water but floats in the glycerin. Its density is between 0.99 g/mL and 1.21 g/mL.

Notes:

4.2 Buoyancy

About the Lesson

Buoyancy is an upward force that acts upon objects when they are immersed in a fluid. In this lesson, students learn about buoyancy and its relationship to other properties of matter, such as weight and density.

Buoyancy Fun

Try these activities to introduce students to buoyancy. Observations are explained at right.

- Use a deep pan or a basin to simulate a bath tub or swimming pool. Gather a few bath tub toys. Hold up each toy and ask students whether it will sink or float? Place each toy in the water. Were students' predictions correct? Ask students to explain why they think each toy sank or floated.
- Blow up a balloon. Place it in the tub of water. Ask students to describe what happens as they push down on the balloon. Remove the balloon from the tub. Place it on a table and push down on it. Is the result the same?
- Add water (about half-full) to two clear, tall drinking glasses. Add about 8-10 teaspoons of table salt to one glass. Carefully place an uncooked egg in each glass. Can students explain their observations?

Students complete Investigation 4B before reading section 4.2. In this investigation students apply what they have learned about density to explain principles of buoyancy. Students apply Archimedes' principle to determine whether an object will sink or float.

Investigation 4B: Sink or Float

Setup	Materials	Investigation Vocabulary
1. Allow one class period to complete the investigation. 2. Students work in small groups of three to five. 3. Have plenty of paper towels available to dry the clay and clean spills.	• 1/2 stick of clay • balance • displacement tank • beaker, paper cup • graduated cylinder • 15-cm deep plastic container	buoyancy Archimedes' principle

Word Origins...buoyancy (from Spanish *boyante* meaning "to float")

Bring a lifebuoy to class. A lifebuoy may be thrown to a person in water to keep them afloat and prevent drowning. Ask a student volunteer to identify it and explain how it works. Use student responses to help students infer the meaning of buoyancy.

Buoyancy Fun: *Explanations of Observations*

Students may already know a little about buoyancy. How they respond to these activities will provide hints about their current level of knowledge regarding this subject.

- Be sure to use some objects that will float and others that will sink. An acceptable explanation is that the object sinks because it weighs more than the buoyant force exerted upon it. The toys that sink have a density greater than the density of water. The reverse is also true. An object floats if it weighs less than the buoyant force exerted upon it. A toy floats if its density is less than the density of water.
- Students should observe that the balloon submerged in water "pushed back." When the balloon is placed on a table, there is evidence of pressure pushing against the sides of the balloon in all directions. This is not the same as the buoyant force pushing the balloon back up while it is immersed in the water.
- An object placed in salt water is more buoyant than an object in freshwater because salt increases the density of water. Students should have observed that the egg in saltwater floated higher while the egg placed in freshwater sank.

Addressing Misconceptions: Mass and Weight

Often in conversation, mass and weight are used interchangeably. However, these terms have very different meanings. Ask students a series of questions (see below) to help them distinguish mass and weight.

1. Malik stepped on a bathroom scale and it read 150 pounds. Is 150 pounds a measurement of mass or weight? (Answer: 150 pounds is a measure of weight.)
2. How is the measurement obtained from stepping on a bathroom scale different from the measurement obtained from placing a clump of clay on a balance? (Answer: A balance measures mass in units like grams or kilograms.)
3. How are mass and weight different? (Answer: Mass is a measure of how much matter is in an object. Mass is related to inertia as it is a measure of how an object resists a change in motion. The amount of matter does not vary with location. Weight is dependent on the force of gravity acting on an object. For this reason, weight varies with location. The weight of an object on Earth is a measure of the force of attraction between the object and Earth. In a different location, like Mars or the moon, an object's weight is a result of the force of attraction between itself and Mars (or the moon).

Ask students to consider how their weight may vary on different planets. Have students use the Internet to find out. Entering a phrase like, "weight on other planets" will generate a number of sites where students can convert their weight on Earth.

1. Read *How Does a Ship Float* by Jim Pipe and Catherine Ward. Try some of the projects about sinking and floating.
2. Write a report about the connection between the swim bladder of a fish and buoyancy.
3. Revisit the activities (*Motivate*) on the previous page. Have students use principles of density and buoyancy to explain what they observed.

Students complete section 4.2 review questions.

Teaching Tip . . .Demonstrating Neutral Buoyancy

Fill a large clear container (like an aquarium) with water. Then fill a resealable sandwich bag with water. Squeeze any air out before sealing the bag. Drop the bag into the aquarium. It will neither float on top nor sink to the bottom, but remain suspended in the tank.

Ask the students if the bag of water has weight. Then ask, "What must be counteracting this weight to keep the bag from sinking to the bottom?" Identify the buoyant force involved. Then discuss the idea of neutral buoyancy.

Next, fill another sandwich bag with corn oil and a third bag with molasses. Ask students to predict what will happen when these bags are placed in the tank. Give them an opportunity to test their predictions.

Multicultural Experience: John Phillip Holland

John Phillip Holland (1841-1914) was an Irish-American engineer who is credited with inventing the modern submarine. One of his designs, the *Holland*, became the United States Navy's first submarine.

Encourage students to learn more about J. P. Holland and the science involved in shipbuilding. Many great resources for information about shipbuilding and buoyancy are available through www.pbs.org. Follow the Teacher Source link and conduct a search on these topics.

Investigation 4B: Sink or Float

Complete the *Buoyancy Fun* activities before starting the investigation.

Think about the objects that sank and those that floated. Why do you think some items float while other sink?

Students offer suggestions.

In the last investigation you learned about density. Who can remind the class of what density means?

Density is the ratio of an object's mass per unit of volume. Write the density equation on the board. Point to each variable and identify it as representative of density, mass, or volume. Also, use this time to review the units of density.

Does anyone remember the value for the density of water?

The density of water is 1.00 g/mL.

Very good. As you completed the different activities, you observed that placing an object in water can lead to a number of results. Some items sink and others float. Sometimes an item in water or a different liquid appears to neither sink nor float. Whether an item sinks or floats depends on factors like density and buoyancy. You will learn more about these factors in today's investigation.

Write the term, buoyancy, on the board. Use the *Word Origins* box to help students arrive at a definition for buoyancy.

1 The density of clay

Each group should have one-half stick of modeling clay. Place the clay on the balance. Record its mass. Then, use the displacement tank to find the volume of the clay. Record these measurements in Table 1. Use your data to calculate the density of the modeling clay.

Students complete the measurements. Remind students not to leave their clay in the tank. The clay must be removed as soon as the water has been collected, and then immediately patted dry with a paper towel.

What is the density of modeling clay?

Students share answers. Check for major discrepancies in students' calculations.

Consider the density of the modeling clay. How does it compare with the density of water? Can you use this information to explain why the clay didn't float in the displacement tank?

Students share their explanations. They may relate this experience to the end of investigation 4A, when a rubber stopper was dropped into the density column. When the rubber stopper was denser than the liquid, it sank. Likewise, the clay is denser than water, so it sank in the water.

2 Making the clay float

Steel has greater density than water, yet steel can be fashioned into a shape that floats. Do you think this is a possibility with other materials?

Students suggest possibilities.

Investigation 4B *Sink or Float*

4B Sink or Float

Steel is denser than water so why do steel boats float?

Solid objects float if they are less dense than the liquid in which they are placed. Objects sink if they are more dense than the liquid. You may have noticed that large ships are often made of steel. Steel is much denser than water. So how does a steel boat float? The answer is in the concept of average density. You will soon discover how and why boats can be made of materials that are denser than water.

Materials
- 1/2 stick of modeling clay
- Balance
- Displacement tank
- Disposable cup
- Beakers
- Graduated cylinder
- Container for testing boats (at least 15 cm deep)

1 The density of clay

Find the density of your stick of clay before you change its shape.

1. Measure the clay's mass. Record it in Table 1.
2. Find the volume of your stick of clay using the displacement method:
   - Place a disposable cup under the displacement tank spout.
   - Fill the tank until water begins to run out of the spout (approx. 1,400 mL)
   - When the water stops flowing, remove the cup and replace it with a dry beaker.
   - Gently place your clay into the tank. Collect the water that runs out of the spout.
   - Quickly remove your clay and dry it with a paper towel. Do not allow water to mix with your clay or it will get very slimy.
   - The volume of the water you collected is equal to the volume of your clay. Use the graduated cylinder to measure the volume and record it in Table 1.

Add water until it just starts running out of the spout

Table 1: Density data for clay

Substance	Mass (g)	Volume (mL)	Density (g/mL)
Clay			

3. Calculate the density in g/ml.
4. Did your stick of clay sink or float in the displacement tank? Use the density of your stick of clay and the density of water (1.0 g/mL) to explain why.

19

Now give it a try. Mold all of your clay into a shape that you think will float.
Have students brainstorm with group members about how to mold the clay. Ask students to write two reasons why they chose to mold the clay into the selected shape. Students then test the shape by gently placing their boat in the container of water.

Were you successful in making the clay float? Be sure to record your technique for each test. It is important that you note all details, even if you were unsuccessful. Remember that scientists may learn just as much from an incorrect hypothesis as they do when a prediction is correct.
Once the students have a shape that floats, have them find the mass of their clay boat. The mass should be very close to the mass of their stick of clay.

Teaching Tip . . .Working with the clay

Emphasize the importance of drying the clay quickly. If water is mixed into the clay, it will take on a slimy consistency that will rub off on students' hands, making accurate measurements of mass impossible. Keep a few extra sticks of clay on hand in case a group needs to begin again.

3 Why a boat floats

When a boat floats it displaces a certain volume of water. Do you think your boat will displace the same volume of water as the stick of clay?
Students should record their predictions before measuring the volume.

Now measure the volume of water displaced by the boat.
Students may be surprised to find that the boat displaces more water than the stick of clay.

Use the mass and volume data for your boat to find the average density of your boat.

4 Thinking about what you observed

Look at the data you recorded for the volume of water displaced by the stick of clay and the clay boat. Why are they so different?
The interior of the clay boat is filled with air. Some of that air is below the water line. Therefore, both clay and air were involved in "pushing aside"—or displacing—the water.

What was the average density of your boat? Why do we call it *average* density?
The average density of the boat is very close to the density of water. The average density calculated here reflects all of the boat's mass but ONLY the volume of the portion of the boat that was underwater.

Why does a solid stick of clay sink but a clay boat can be made to float?
Changing the shape of the boat affects the composition of its interior. Rather than solid clay, a large portion of its interior is air. The boat's density reflects the average density of the clay and the air. The boat's average density is less than the water's density.
Archimedes principle states that an object in a fluid is buoyed up by a force equal to the weight of the fluid it displaces. The weight of the object acts downward, and the buoyant force provided by the displaced fluid acts upward. When these two forces are equal, the object floats. Since the boat is less dense than the water, only a fraction of the boat has to be immersed before it has displaced the boat's weight. Therefore, the boat floats.

Investigation 4B: Data and Answers

Notes:

1 The density of clay

Substance	Mass (g)	Volume (mL)	Density (g/mL)
Clay stick	53.9	29.7	1.81

2 Making the clay float

3 Why a boat floats

Mass of boat (g)	Volume of water displaced by the boat (mL)	Average density of the boat (g/mL)
53.9	54.0	0.998

4 Thinking about what you observed

a. The clay boat displaced more water.

b. The volume of water displaced by the clay stick and the clay boat was different because the objects were shaped differently.

c. The boat's average density is less than the density of the clay stick. The boat' average density is slightly less than the density of water.

d. When an object displaces a volume of water with a mass that is the same as its mass, it will float. When an object displaces a volume of water with a mass that is less than its own mass, it will sink.

e. If you add pennies to the boat, you are increasing the boat's mass. You can only add a certain amount of pennies before the boat sinks. As you add pennies, the boat's mass increases. As mass increases but the boat's shape does not change (volume), the average density increases. When the average density is greater than water's, the boat will sink.

Chapter 4 Answers

4.1 Section Review

1. Answers include: mass, volume, density, phase, color, strength
2. milliliter
3. (1) Measure the dimensions of length, width and thickness and find their product; since all dimensions of a cube are the same, measure a side and cube this number. (2) Use the displacement method. Pour water in a graduated cylinder. Read the volume of the water. Add the cube. Read the new volume. The difference between the original volume of the water and the new volume is the volume of the cube.
4. This material would be good for making a cushion. Because its density is low, it has space between molecules leaving room for the material to be compressed. This would make it a good cushioning material for a couch.
5. c
6. physical
7. mass
8. mass and volume (or the dimensions for calculating the volume of a regular solid)
9. $d = m/V$; $V = l \times w \times h = 5\text{ cm} \times 3\text{ cm} \times 2\text{ cm} = 30\text{ cm}^3$
   $d = (18\text{ g})/(30\text{ cm}^3)$; $d = 0.6\text{ g/cm}^3$
   oak (from table 4.1 of densities)
   mass and volume are the two factors that determine a material's density.

4.2 Section Review

1. Volume
2. It increases because more water is displaced.
3. displaces
4. floats or rises in the fluid
5. Answers are:
   a. 1000 cm^3
   b. $d = m/V = 800\text{ g}/1000\text{ cm}^3 = 0.8\text{ g/cm}^3$
   c. The object will float in water since its density is less than the density of water.
6. Iron will float in mercury because it is less dense that mercury.
7. "Heavy" expresses only the physical property of mass. Density, which determines whether or not an object will float in water, is determined by the ratio of mass/volume. An object whose mass/volume ratio is less than one will float in water, no matter what its mass is.
8. The ship is not a solid piece of steel. Because of the way the ship is shaped, most of the ship's hull encloses air. The ratio of the total mass of a ship divided by the total enclosed volume of the ship is less than the density of the water in which it floats. The average density of enclosure we identify as "the ship" is less than the density of water. Therefore, the ship floats.

Connection

1. A blimp is a non-rigid airship. A non-rigid airship is one that does not have a rigid structure to hold gas. Instead, the envelope, made of polyester fabric, takes on its shape when filled with gas. Blimps rely on a lighter-than-air gas for flight. Unlike a hot air balloon, a blimp does not rely on wind for movement. A blimp uses engines, controls, air, and helium for movement.
2. Buoyancy is an upward force that causes an object to float in liquid or gas. For a blimp, buoyancy is caused by the use of helium, which pushes a blimp upward, causing it to rise. Since air is heavier than helium, the pilot inflates or deflates the ballonets to make the blimp fall or rise.
   a. To ascend: Air is released through the air valves. Helium causes the blimp to become positively buoyant and to rise. As helium inside the envelope expands, external air pressure decreases.
   b. To level off: To balance the pressure between helium and air, just enough air is allowed back into the ballonet to create neutral buoyancy. Now the blimp is at a steady state, cruising at a desired altitude.
   c. To descend: To lower the blimp for landing, the density of the blimp must be increased. Air, the denser gas, fills the ballonets to create negative buoyancy. With negative buoyancy, the blimp descends.
3. Blimps are used to provide video coverage of major sporting events and for advertising purposes. Blimps are also used for scientific reasons including monitoring volcanic activity and photographing and identifying whale behavior. In the future, there are hopes to have space blimps capable of carrying humans and equipment into space without the use of rockets.

Activity

Activity Sample data table

Description of object	Prediction: sink or float?	Result:sink or float?	Mass (g)	Volume (cm^3)	Density (g/cm^3)
water	X	X	50	50	1
penny	sink	sink	2.6	0.36	7.2
golf ball	float	sink	45	42	1.1
sm block of wood	float	float	36	75	0.48
tennis ball	float	float	55	145	0.38

Applying your knowledge: sample answers

a. I found the volume using the displacement method. I put a measured amount of water in a graduated cylinder, then added the object and measured the change in water level. The difference is the volume of the object.

b. tennis ball; piece of wood; water; golf ball; penny (the tennis ball and piece of wood floated)

c. The objects with density less than water floated in water.

d. The tennis ball, piece of wood, and golf ball would float in corn syrup because their density is less than that of corn syrup.
The tennis ball, piece of wood, golf ball, and penny would float in mercury because their density is less than that of mercury.
The tennis ball and piece of wood would float in cooking oil because its density is less than that of cooking oil.

Chapter 4 Assessment

Vocabulary

Section 4.1

1. density

Section 4.2

2. buoyancy 3. Archimedes' principle 4. weight

Concepts

Section 4.1

1. They have identical weights.
2. 1.0 g of steel has greater density (7.8 g/cm^3) than 1.0 kg of aluminum (2.7 g/cm^3).
3. 1.0 gram of aluminum has greater volume (0.37 cm^3) than 1.0 gram of iron (0.13 cm^3)
4. Any mass/volume units are acceptable, including: g/cm^3 (or g/cc); g/mL; kg/m^3
5. Answers are:
   a. Solids have the greatest density
   b. Gases have the lowest density.
6. Ice forms on top of ponds and lakes instead of the bottom because solid water is less dense than liquid water and floats on top of it. The result is that the water at the bottom of the pond stays at a fairly steady temperature under a blanket of ice. This protects life in the pond from freezing over the winter.
7. Oak, with a density of 0.60 g/cm^3, could be the material used.
8. It would have a density of 2.7 g/cm^3. The same type of glass in the same phase has the same density no matter how it is formed.
9. c
10. $d = m/V$; rearranging gives $V = m/d$
11. lead; water; cork
12. 18 mL (43 mL - 25 mL)
13. meniscus

Section 4.2

14. They are the same.
15. The ice cube floats in the glass of water because its density is less than the water's and the buoyant force is greater than its weight.

16. The volume of that portion of the object that is submerged.
17. 250,000 N
18. 1.0 g/cm^3.
19. The density of helium is less than air at the same temperature and pressure. The helium floats because its buoyancy, or buoyant force, in air is greater than its weight.

Problems

Section 4.1

1. 20 milliliters
2. 0.123 g/cm^3
3. 27 mL
4. Answers are:
   a. $V = 4/3\pi r^3$; $V = 4/3\pi(15.6\ cm^3)$; $V = 65\ cm^3$
   b. $d = m/V$; $m = dv$; $m = (1.2\ g/cm^3) \times 65\ cm^3$; $m = 77g$
5. ice: $V = m/d = 1000\ g \div 0.92\ g/cm^3 = 1087\ cm^3$
   water: $V = m/d = 1000\ g \div 1\ g/cm^3 = 1000\ cm^3$
6. The density of the steel is $25\ g \div (3.2\ cm^3)$ or $7.8\ g/cm^3$. The volume of the 10-g ball bearing is $10\ g \div (7.8\ g/cm^3)$ or $1.3\ cm^3$.
7. Answers are:
   a. $d = m/V$; $d = 7.8\ g \div 1.728\ cm^3$; $4.5\ g/cm^3$
   b. No, it does not float. The density of the block is greater than the density of water, 1.0 g/cm^3.

Section 4.2

8. 23 N
9. 6.5 N - 5.0 N = 1.5 N
10. 60 mL
11. 100 mL

Chapter 5: States of Matter

Learning Goals	California Content Standards	Activities and Resources	Assessment
5.1: Liquids and Gases The student will: • Compare and contrast the properties of liquids and gases. • Define pressure. • Describe the affect of intermolecular forces on states of matter. • Distinguish melting and boiling points of matter. • Explain how heat is transferred in fluids. • Describe the composition of Earth's atmosphere.	PS 3d PS 3e PS 5d	**Laboratory Investigation 5A:** The Phases of Matter **Teaching Illustrations:** Intermolecular Forces, Temperature and Time Graph, Composition of Earth's Atmosphere	**Exam View Test Bank** Chapter Pre-Test **Section 5.1 Review**
5.2: Solid Matter The student will: • Describe the physical properties of solids. • Describe how heat is transferred in solid matter. • Explain the differences between amorphous and crystalline solids. • Give examples of thermal conductors and insulators.	PS 3c PS 7c	**Laboratory Investigation 5B:** Phase Changes **Teaching Illustrations:** Thermal Conductors and Insulators **Graphic Organizer:** Mechanical Properties **Connection:** Silly Putty: Solid or Liquid? **Chapter Activity:** Make Your Own Viscoelastic Liquid	**Section 5.2 Review** **Chapter Assessment** **Exam View Test Bank**

Estimated Instruction Time	Language Tools	Differentiated Strategies	Materials
Four 45-minute class periods	**Literary Picks** **Vocabulary:** fluid, liquid, gas, pressure, intermolecular forces, melting point, boiling point, evaporation, condensation, convection	**ELL Strategies:** Listed in ELL Ancillary **Learning Strategies:** • Cooperative Learning - Investigation 5A • Discussion - Motivate • Graphic Organizers - Explain: The Kinetic Molecular Theory (KMT) of Matter • Active Research - Extend **Teaching Tip:** Confusing Words	**Investigation 5A:** mass balance, prepared bottle cap with inserted tire valve (See Teaching Tip - Preparing the bottle cap)
Six 45-minute class periods	**Vocabulary:** solid, strength, elasticity, brittleness, ductility, crystalline, amorphous, polymer, heat conduction, thermal equilibrium **Word Origins:** amorphous **Closure Cards**	**Learning Strategies:** • Cooperative Learning - Investigation 5B • Active Learning - Motivate: Pack It In! • Graphic Organizers - Extend **Teaching Tip:** Changing States, Making a Crystal Garden	**Investigation 5B:** **Teaching Tip - Changing States:** dry ice, gloves, tongs, water, glass, scoop, food coloring, empty 16-20 oz plastic soda bottle, balloon, rubber band or twist-tie **Teaching Tip - Making a Crystal Garden:** water, electronic or triple bean balance, spoon or scoop, 50 grams alum (acceptable substitutes include copper sulfate, epsom salts, or borax), stir rod, 250-mL beaker, 2-4 disposable petri dishes, .5-cm diameter dowel **Chapter Activity:** white glue and water solution (1:1 ratio), borax and water solution (5-mL to 60-mL), 8-oz paper cup, stirring stick

5.1 Liquids and Gases

About the Lesson

Liquids and gases are fluids. In this lesson, students learn more about the behavior of atoms and molecules in fluids. Some topics discussed include pressure, intermolecular forces, and the composition of Earth's atmosphere.

Students have already been introduced to states of matter in chapter 3 (and prior to grade eight). Ask students to write defintions of the terms, *liquid* and *gas* in their own words. Then have students discuss how liquids and gases are alike and different. Finally ask students to share examples and nonexamples of liquids and gases.

Students complete Investigation 5A before reading section 5.1. In this investigation students conduct experiments designed to help them compare properties of molecules, such as density and pressure, in solids, liquids, and gases.

Investigation 5A: The Phases of Matter

Setup	Materials	Investigation Vocabulary
1. Allow one class period to complete the investigation. 2. Students work in small groups of three to five. 3. Prepare the bottle caps with inserted tire valve for each group of students in advance. Directions and materials needed for preparing the bottle caps are provided in the Teaching Tip on page 79. 4. Do the experiments ahead of time so you know what to expect.	• mass balance • prepared bottle cap with inserted tire valve • tire pressure gauge • bicycle pump	pressure

Literary Picks...

Reading Level	Title
3-5	***What is the World Made Of? All About Solids, Liquids, and Gases* by Kathleen Weidner Zoehfeld** A fact filled informal text from the *Let's Read and Find Out* series. Provides clear explanations and examples of each state of matter. Includes simple activities that demonstrate the properties of each state.
6-8	***Solids, Liquids, and Gases* by Louise Osborne and Deborah Hodge** This book is sponsored by the Ontario Science Center and features 13 experiments which deal with solids, liquids, and gases. Includes full-color photographs and step-by-step directions that require minimal supervision and allows students to explore each state of matter.
9-12	***Why Things Are the Way They Are* by B. S. Chandrasekhar** An interesting book in narrative format that explain the workings and properties of light and matter. Find out why copper is the color it is, why glass is transparent to light, and why metals conduct heat better than glass.

The Kinetic Molecular Theory (KMT) of Matter

The properties of liquids and gases are explained by the kinetic theory of matter. The *kinetic* component of this theory refers to the energy of particles that make up matter. These particles (atoms and molecules) comprise the *molecular* component. The kinetic theory states that molecules in matter are always moving. As these molecules move, they effect change. The diagram below explains how liquids and gases are described by the KMT.

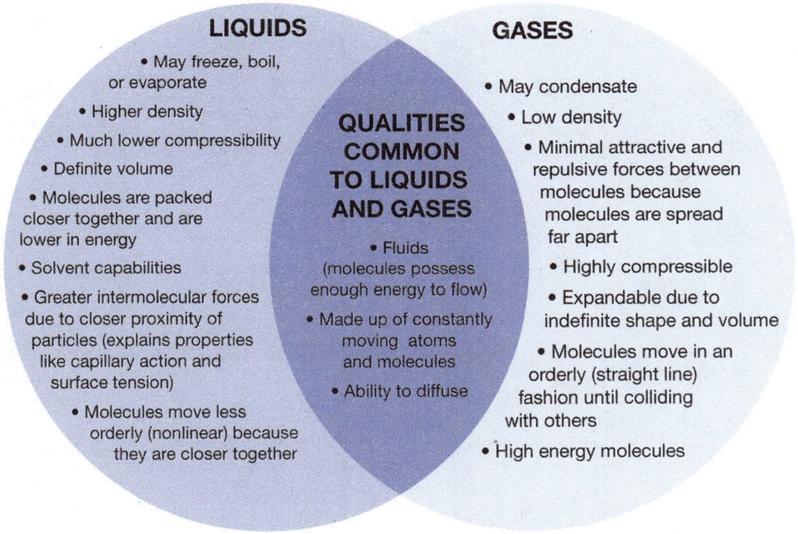

1. What is inside an aerosol can? How does it work? Have students research to find out. Direct students to explain the role of pressure in obtaining the product from these cans.
2. Explore some applications of pressure in daily living. Some examples include barometric pressure and blood pressure. Demonstrate how barometers and manometers work. One option is to invite the school nurse to take students' blood pressure with a sphygmomanometer.

Students complete section 5.1 review questions.

Teaching Tip . . .Confusing words

When speaking about matter, students are often confused by terms like *state* and *gas*. They are most familiar with these words in a different context, like the *state of California* or *fueling the car with gas*. Explain to students that these words have different meanings to scientists.

Physical Properties of Matter

Students learn about many physical properties of matter in chapter 5. A *physical property* can be observed or measured without causing any change to the identity of matter. Physical properties can be further divided into two categories: intensive and extensive.

Intensive physical properties are independent of the amount of matter present. Some examples include density, boiling point, and temperature. Extensive physical properties depend on how much matter is present. Mass and length are extensive physical properties.

As students proceed through chapter five, have them generate a list physical properties they encounter. Tell students to classify each physical property as intensive or extensive.

Investigation 5A: The Phases of Matter

Have you ever heard sayings like, "He is an airhead!" or "She is full of hot air!" What comes to mind when you hear these phrases?

Both imply emptiness, lack of value, or that nothing of worth is there.

Air is a gas. Although it may feel like "nothing", it actually has considerable mass. In this part of the investigation, you will observe how changes in pressure affect the mass of air in a bottle. What do you think of when I say the word pressure?

Students answers may vary. Some common responses include being under stress or pushing against something.

Yes, those are examples of how humans often describe pressure. For instance, you may apply pressure to a wound or a friend may ask you to stop pressuring him. When scientists think of pressure, they have a different way of defining it. Pressure is the result of applying force to fluids. Fluids have the ability to flow. Which states of matter would qualify as fluids?

Liquids and gases are fluids.

Good answer. Therefore air is a fluid. What happens if you apply pressure to a block of steel by pushing down on it?

Push down on a block of steel or some other solid matter.

By pushing down on the block I am applying force to it. Can you see any change to the steel block as I push down on it?

No change is visible.

How about if I push down on this balloon? What do you see?

The walls of the balloon expand outward in all directions.

Pressure is the force applied to matter per unit of area. It acts in all directions, not just the direction in which it is applied (like the steel block). The pressure readings you collect will be in units of psi, which is an abbreviation for pounds per square inch. It is very important that the pressure not exceed 70 psi.

Write this safety note on the board and continue to remind students of this throughout this part of the investigation.

Each investigation team has a prepared cap, with the tire valve inside, and a one liter carbonated soda bottle. Fill the bottle with water to the very top. Transfer the water into a graduated cylinder to measure its volume. Record the volume of the bottle. Replace the cap on the empty bottle and measure its mass. The liquid water has been removed from the bottle, therefore it is full of air.

Students prepare the bottles.

Use your bicycle pump to raise the pressure in your bottle to 10 psi. Use the gauge to check the pressure in the bottle. Use the balance to measure and record the mass of the bottle and air at 10 psi. Repeat the pumping and mass measurement for pressures between 10 psi and 70 psi. Record your observations in Table 1. Remember that the pressure must not exceed 70 psi.

Demonstrate the proper technique. Students can better observe the relationship between pressure and mass with ample measurements, so increase the pressure by 10 psi each time.

5A The Phases of Matter

How do the mass, volume, and densities of solid, liquid, and gas compare?

Air may feel like "nothing" but actually air is matter and has considerable mass! In fact, a cubic meter of air (about the size of an armchair) has a mass of about 1 kilogram. In this Investigation you will measure the amount of matter in solid, liquid, and gas and compare their densities.

Materials
- Thermometer
- Mass balance
- Prepared bottle cap with inserted tire valve
- Tire pressure gauge
- Ice water
- Bicycle pump

1. Safety Note: Be careful with the bottle, and DO NOT exceed 70 pounds per square inch (psi) of pressure.
2. Get the prepared cap with the tire valve inside and a 1 liter carbonated soda bottle.
3. Fill the bottle to the very top with water. Empty the water into a graduated cylinder to measure the volume of the bottle.
4. Put the cap on the empty bottle (full of air) and measure the mass.
5. Use the bicycle pump to raise the pressure in the bottle to 10 psi. Check the pressure with your gauge.
6. Use the balance to measure and record the mass of the bottle (and air) at 10 psi pressure.
7. Repeat the pumping and mass measurement for pressures between 10 psi and 70 psi. DO NOT exceed 70 psi!

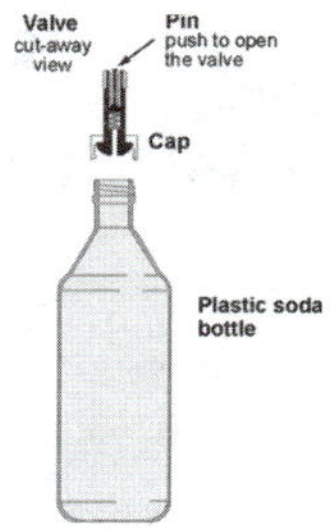

Table 1: Pressure and mass data

Gauge pressure (psi)	Mass (g)	Volume (ml)

Thinking about what you observed

a. Use Table 2 to calculate the mass of air added to the bottle at different pressures.
b. What happens to the mass as you increase the pressure in the bottle?
c. Explain your answer to question b.
d. Make a graph showing the mass of air plotted against the pressure.
e. Is there still air in the bottle when the pressure is zero on the gauge? Use the graph to estimate the mass of air in the bottle at atmospheric pressure (zero on the gauge).

21

1 Thinking about what you observed

Use the data you recorded to help you answer the questions in part two. Your results showed that as pressure in the bottle increased the mass also increased. Why do you think this happened?

Pressure is one factor that affects the density of a gas. Density is the mass per unit of volume. In this part of the investigation, more and more molecules were added to the bottle, increasing the mass each time the gauge pressure was increased. The volume of the bottle remained constant, meaning the air density within the confines of the bottle also increased.

Transfer the gauge pressure readings (in psi) and masses recorded in Table 1 to the first two columns of Table 2. Record the mass of the bottle when the gauge pressure is zero. Subtract the mass of the bottle at zero gauge pressure from the mass of the bottle and air to determine the mass of air added to the bottle. Repeat this calculation for each pressure reading and write your answers in the last column of Table 2.

Model the calculation for the first pressure reading.

Make a graph showing the mass of air (from column 4 in Table 2) plotted against the pressure. Be sure to include an appropriate title, axis labels, and range of data so the trends are easily recognizable. Once you have completed the graph, use it to estimate the mass of air in the bottle at zero gauge pressure.

Explain to students that many tire pressure gauges measure the difference between atmospheric pressure outside and the pressure inside the gauge. Therefore, when gauge pressure reads zero, this is equal to the atmospheric pressure.

2 Liquid and solid phases

Carefully read and follow each step of the procedure in part three. How do you think the volumes of equal masses of solid, liquid, and gas will compare?

Encourage student participation in the discussion. Students should conclude that equal masses of gas, liquid, and solid have very different volumes. They should also notice that the space occupied (volume) by the equal masses of substance varies greatly. For example, the amount of space occupied by the mass of pennies equal to the mass of zero gauge pressure is much less than the space occupied by an equal mass of air.

3 Thinking about what you observed

Mass, like volume, density, or color is a property of matter. Mass describes the quantity of matter present in a material. Atoms make up matter and are arranged differently in solids, liquids, and gases. Consider these concepts as you answer questions in this part of the investigation.

Guide students to the understanding that gases generally are much less dense than liquids or solids because the atoms in a gas have more energy and are spread farther apart. Liquids are usually less dense than solids (water is an exception) but more dense than gases. Atoms in a liquid have more energy than in solids are not as close together. Solids are more dense than liquids or gases because its atoms are very close together in fixed positions and they lack the thermal energy needed to break bonds and move.

> Teaching Tip . . .Preparing the bottle caps
>
> The materials needed are as follows:
>
> - cap from a one liter plastic bottle for each group
> - valve stem for each group (available for purchase at an automotive supply store)
> - electric drill and large drill bits
> - household silicone adhesive/sealant
>
> Drill a hole in the center of the bottle cap large enough to accommodate the black neck and shoulder of the valve stem. Test-fit the valve stem, then remove it. Place a bead of silicone sealant around the hole inside the bottle cap. Replace the valve stem. Allow the silicone to cure 24 hours before using the cap.

Investigation 5A: Data and Answers

Gauge pressure (psi)	Mass (g)	Volume (ml)
0	50.0	1075.0
10	51.2	1075.0
20	51.9	1075.0
30	52.8	1075.0
40	53.6	1075.0
50	54.5	1075.0

1 Thinking about what you observed

a. See Table 2.

b. As the pressure increases, the mass increases.

c. The mass increases because adding pressure is actually adding more material in the form of air. While air seems to be very light, using the pump allows us to put a great deal more air in the bottle than would normally be in there. As pressure is added, air is added, and the mass of the additional air is seen when the mass of the bottle and its contents is measured.

d.

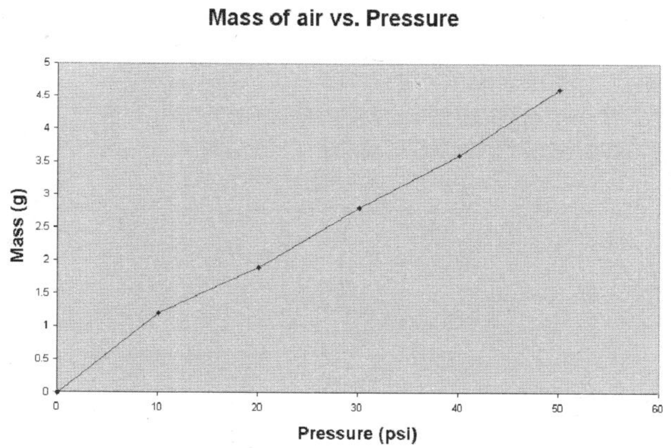

e. There is definitely air in the bottle when the gauge reads zero. The graph shows increases of 0.8 g to 1.0 g of air with each increase of 10 psi, so there is probably around 1.0 g of air in the bottle.

Pressure (psi)	Mass of bottle and air (g)	Mass of bottle at zero gauge pressure (g)		Mass of air added to bottle (g)
0	50.0	50.0		0
10	51.2	50.0		1.2
20	51.9	50.0		1.9
30	52.8	50.0		2.8
40	53.6	50.0		3.6
50	54.5	50.0		4.5

2 Liquid and solid phases

3 Thinking about what you observed

a. These objects all have about the same amount of matter.

b. The density is around 1g/ L. The density is about 1/1000th of water.

c. Answers of the mass will vary. An approximate response is usually around 250 kg. The air in the room has more mass than the mass of a person.

d. Gas is usually the least dense, followed by liquid and then solid.

Notes:

5.2 Solid Matter

About the Lesson

This lesson is about solids. Students define solids in terms of the kinetic molecular theory and examine various properties of solids. Students also differentiate amorphous and crystalline solids and explore how heat is transferred is solids.

Pack It In!

Gather a few small boxes and distribute them to groups of students. Provide students with candies like jelly beans or gummy bears and ask them to pack as many in the box as they can. The only catch is that the integrity of the candy cannot be compromised. Once students have packed in all they can fit, ask them to share their strategy for maximizing the number of candies in the box. Talk about the small amounts of space between candies and tell students that this is similar to how molecules are "closely packed" in solids. Ask questions like, "How much space is between the pieces of candy?" "Are the candies able to move or do they seem locked in their positions?"

Students complete Investigation 5B before reading section 5.2. In this investigation students experiment to find out how much energy is required for matter to change from one phase to another. Exercise caution as students are using hot water in this activity.

Investigation 5B: Phase Changes

Setup	Materials	Investigation Vocabulary
1. Allow one class period to complete the investigation. 2. Students work in small groups of three to five. 3. Use a hot plate to keep water warm. Bring ice from home in a cooler or obtain it from the school cafeteria.	• thermometer • mass balance • foam cups • ice • cold water • hot water	phase change

Word Origins...amorphous (from Greek *amorphos* meaning "without shape")

The root word of amorphous is morph(e), which means to change or transform outward appearance. Amorphous solids do not have a consistent shape or appearance. Instead they can be morphed, or changed to take on various shapes. Some examples of amorphous solids include rubber and glass. Ask students to list five examples of products made from rubber. Then have students compare the shape of the rubber in each product.

Teaching Tip . . .Changing States

Students often ask if matter *always* goes from solid to liquid to gas (and vice versa). Discuss familiar examples of how matter changes state, like melting ice or boiling water. Then talk about examples that may be unfamiliar, like sublimation or condensation.

Sublimation occurs when matter changes from a solid directly into a gas. Dry ice sublimes. Try these fun demos. Do them ahead of time so you know what to expect. **Safety Note:** *Wear safety goggles, gloves, and use tongs to handle dry ice. Use ear plugs (or place cotton in the ears) for the second demo.*

- Show students the dry ice and have them make observations about its physical characteristics. Fill a drinking glass with water and add a scoop of dry ice. What happens? What created the fog?
- Place a scoop of dry ice in an empty 16-20 ounce plastic water bottle. Add colored water (about halfway). Secure a balloon around the next of the bottle. Be sure there are no leaks at the neck. Use a rubber band or a twist-tie to keep the balloon in place. What happens to the balloon? Which states of matter are in the bottle after the balloon pops?

Solids and the Kinetic-Molecular Theory

Like fluids, solids are comprised of moving particles, such as atoms and molecules. However, there are many differences between solids and fluids. Molecules in solids are packed extremely close together. In fact, they are so close together that the intermolecular forces keep particles in a solid essentially in fixed positions. The result is that atoms and molecules are not free to flow; and can only vibrate.

However, these particles are moving and therefore possess kinetic energy. The average kinetic energy of molecules in a solid is significantly less than in fluids. Lower kinetic energy translates into less motion. Some properties of solids are described below.

Shape	Solids have a definite shape. Unlike fluids which conform to the shape of their container, solids are capable of maintaining a defined shape.
Volume	The closely packed arrangement of atoms and molecules in a solid are responsible for its definite volume. Changes in pressure and temperature may have a minimal effect on the volume of solids.
Density	Solids are very dense. In most cases, density is greatest in solids when compared to liquids and gases. An exception of course is water. Liquid water has a greater density than ice.
Compressibility	Solids are practically incompressible. One common misconception is that solid materials, like cork are compressible. Cork has pores which are filled with air. When pressure is applied, these pores are compressed rather than the cork itself.

1. Is it possible to have properties of both solids and liquids? Have your students investigate to find out about liquid crystals and their many applications.
2. Make a compare and contrast chart for methods of heat transfer in solids and fluids.
3. Encourage students to make rock candy at home with their family.

Students complete section 5.2 review questions.

Teaching Tip . . .Make a Crystal Garden

Making a garden is a fun way to study crystals. Students work in pairs.

Materials Needed per group:

- *200 mL water*
- *electronic or triple beam balance*
- *spoon or scoop*
- *50 g alum (acceptable substitutes include copper sulfate, Epsom salts, or borax)*
- *hot plate*
- *glass stir rod*
- *250 mL beaker or small steel pot*
- *2 - 4 disposable petri dishes*
- *10 cm piece of string (cotton or polyester)*
- *.5 cm diameter dowel (or pencil)*

Procedure

1. Pour water into the beaker and place it on the hot plate.
2. Adjust the hot plate temperature to medium heat. Do not allow the water to boil. Have students adjust the heat as needed.
3. Measure 25 g of alum using the spoon and balance.
4. Add the alum to the water. Stir until completely dissolved. Add the remaining alum and stir until it is also thoroughly dissolved. Turn off the heat.
5. Have one student stir the solution, as the partner ties a piece of string to the middle of the dowel.
6. Dip the string into the solution several times.
7. Pour a small amount of solution into each petri dish. Pour various amounts into each petri dish to create different effects.
8. Leave the dowel in the original beaker so it is not touching the sides of the beaker.

Students will notice many crystals growing within an hour. Leave the petri dishes and beaker alone overnight and observe crystals growing on the string and in the petri dishes. Use a microscope to view, draw, and describe the crystals.

Investigation 5B: Phase Changes

Most of the materials we use everyday can be classified as solids, liquids, or gases. What are some characteristics of solids, liquids, and gases?

Students share answers. Sample responses may include: *Solids have a definite shape and volume. Its molecules are very close together. Liquids can flow and have a defined volume. A liquid takes the shape of its container. Molecules are fairly close together in liquids but not as close as in solids. Molecules in gases are very far apart. Gases do not have a defined shape or volume. It is also a fluid.*

Very good. Let us consider water in its solid state. Water molecules in ice are very close together. They are held together by strong connections called *bonds*. As heat energy is applied to an ice cube, the water molecules gain energy. The bonds that hold the molecules together begin to break. Eventually the molecules in ice gain enough energy to overcome the bonds and undergo a *phase change*. Ice changes from solid water to liquid water.

Write *solid water + energy → liquid water* on the board.

Bottled water is a good example of a liquid. A liquid substance can flow, take the shape of its container, and has a constant volume. Suppose you have a 20 ounce bottle of water. As you raise the bottle to your lips to drink it, the water flows. As long as water is in the bottle, it is shaped like the bottle. If you pour the contents of the bottle into a graduated cylinder, its volume is equal to the volume stated on the bottle and its shape changes to the shape of the cylinder.

Demonstrate this to students.

Molecules in a liquid are farther apart than in a solid. Even though atoms and molecules in a liquid are still pretty close together, they have more energy than those in solids and are able to move around. This increase in energy (when compared to solids) helps molecules to make and break bonds with other molecules that are nearby. The gas phase occurs at higher temperatures than the solid and liquid phases. Gases expand because their atoms are free to move. Therefore gases have no defined shape or volume. Who can name an example of matter in the gaseous state?

Answers will vary. Air is a gas.

Changing from one phase to another requires energy. What do you know about the relationship between changes in temperature and energy?

Students recall that heat flow is caused by differences in temperature. Temperature is related to energy and determines whether matter exists as a solid, liquid, or gas.

In this investigation, you will explore how much energy is required for matter to change from one phase to another. Do you think a lot of energy is involved or just a little?

Students share their opinions. Ask students to explain why they think as they do.

You will know whether or not your predictions are correct at the end of the investigation.

Students gather necessary materials. Remind students to use caution when handling the hot water.

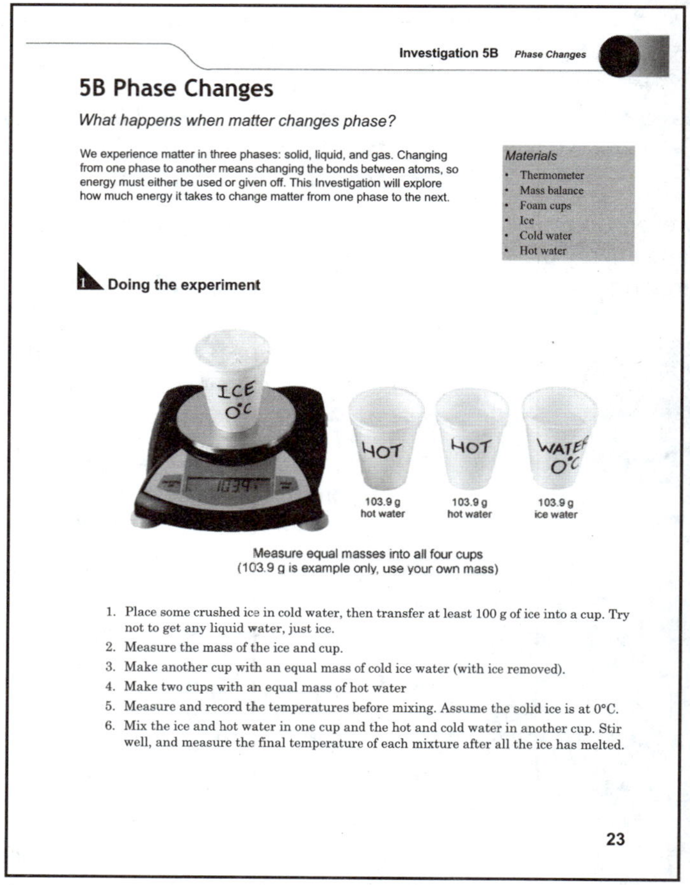
Investigation 5B *Phase Changes*

5B Phase Changes

What happens when matter changes phase?

We experience matter in three phases: solid, liquid, and gas. Changing from one phase to another means changing the bonds between atoms, so energy must either be used or given off. This Investigation will explore how much energy it takes to change matter from one phase to the next.

Materials
- Thermometer
- Mass balance
- Foam cups
- Ice
- Cold water
- Hot water

1 **Doing the experiment**

Measure equal masses into all four cups
(103.9 g is example only, use your own mass)

1. Place some crushed ice in cold water, then transfer at least 100 g of ice into a cup. Try not to get any liquid water, just ice.
2. Measure the mass of the ice and cup.
3. Make another cup with an equal mass of cold ice water (with ice removed).
4. Make two cups with an equal mass of hot water
5. Measure and record the temperatures before mixing. Assume the solid ice is at 0°C.
6. Mix the ice and hot water in one cup and the hot and cold water in another cup. Stir well, and measure the final temperature of each mixture after all the ice has melted.

23

1 Doing the experiment

You will prepare cold liquid water and ice. Each of these will be mixed with hot water. Do you expect the temperature of each mixture to be about the same or very different?

Sketch a diagram on the chalkboard of which substances are to be mixed together. Engage students in a discussion about what they expect to happen.

Now it is time to do the experiment. Place some crushed ice in cold water, then transfer at least 100 grams of ice to a different cup. Try not to get any liquid water, just ice. Measure the mass of the ice and cup. Make another cup with an equal mass of cold ice water, with ice removed. Make two cups with an equal mass of hot water. You should have a total of four cups in all: one with only ice, one with ice cold water, and two with hot water.

Allow time for students to prepare the cups.

Record the temperature of the water in each cup. Write your observations in Table 1. Assume that the solid ice is at 0°C. Mix the ice and hot water together in one cup. Mix the second cup of hot water and the cold water together in another cup. Stir well and measure the final temperature of each mixture after all the ice has melted. Record the temperatures in Table 1.

2 Thinking about what you observed

Look at the data you recorded. Use it to help you answer the questions in part two.

Review students answers to each question. Use the concepts of phase changes and heat of fusion to help students understand their data.

Phase changes require energy. When thermal energy is added to or removed from a material, either the temperature changes or the phase changes. Both of these changes do not happen at the same time. Suppose you heated a block of ice that has a starting temperature of -20°C. The temperature increases as you add thermal energy. Once the ice reaches 0°C, the temperature stops increasing as ice starts to melt and become liquid water. The temperature stays the same. The added energy changes the phase of the water, from solid to liquid. Once all the ice has become liquid water, the temperature begins to rise again as more energy is added.

Have students look at the phase change graph shown on page 101 in the student text.

What do the diagrams in part 2c of the investigation suggest about the position of atoms in different phases of matter and the energy of those atoms?

Atoms in a solid are close together. These atoms tend to stay close together because their thermal energy is too low to break the bonds between them. Matter at low temperatures is often solid. The liquid phase occurs at higher temperatures than the solid phase. Atoms in a liquid are close together, but they can move around and exchange position. Liquids flow because the atoms have enough energy to move around by temporarily breaking and reforming bonds with the atoms nearby. Gas atoms move fast and are separated by fairly large spaces. These atoms have enough energy that the bonds between neighboring atoms and completely broken.

Plasma: A Fourth State of Matter

At temperatures greater than 10,000°C the atoms in a gas start to break apart. In the plasma state, matter becomes ionized as electrons are broken loose from atoms. Because the electrons are free to move independently, plasma can conduct electricity. Lightning is a good example of plasma, and the sun is another. In fact, most of the universe is plasma.

What is the heat of fusion?

The heat of fusion is the amount of energy needed to change one gram of matter from the solid to liquid state without a change in temperature. This transition occurs at the melting point of the substance.

Adding thermal energy to matter causes an increase in temperature. Removing thermal energy causes the temperature to decline. As matter changes from solid to liquid, more energy is needed to disrupt the ordered arrangement of particles in the solid. This additional energy is the heat of fusion, which breaks the bonds between molecules in the solid.

Investigation 5B: Data and Answers

1 Doing the experiment

Liquid cold water plus hot water		
Cold water temperature before mixing (°C)	Hot water temperature before mixing (°C)	Mixture temperature (°C)
.603	46.3	23.8
Solid water (ice) plus hot water		
Ice temperature before mixing (°C)	Hot water temperature before mixing (°C)	Mixture temperature (°C)
0	49.2	0.4

2 Thinking about what you observed

a. The mixture's temperature should be halfway in between 0° and 50°, right around 25°.

b. The temperature of the mixture of ice and warm water was 0.4°, much less than the temperature of the cold water and warm water which was 23.8°.

c. The ice and warm water mixture had a much lower temperature because the energy from the warm water went into changing some of the solid water in the form of ice into a liquid. This process takes a good deal of energy, and the temperature of the water will drop down close to the temperature of the ice until most of the ice has melted. The energy used to change phases from solid to liquid breaks the bonds between the solid water molecules.

Notes:

Chapter 5 Answers

5.1 Section Review

1. Molecules in a gas are free to move around and so gas flows. On average, each gas molecule has enough energy to completely break away from its neighbors. That is why gas expands to fill any container.
2. Molecules in a liquid are free to move around, allowing the liquid to flow. However, the molecules do not have enough energy to break their bonds, so although a liquid can change shape, it cannot expand to fill its container.
3. Pressure acts in all directions in a fluid.
4. The temperature stays the same while ice is melting and heat is added. Once all the ice is melted into water, the temperature will rise. While it is melting, it gains energy.
5. In evaporation, molecules go from liquid to gas at temperatures below the boiling point. This happens because molecules within the liquid have different kinetic energies and the molecules with higher energies sometimes have enough energy to break the bonds of their neighbors causing them to become a gas. This is different from boiling where the energy source is external and it occurs at the substance's boiling point.
6. gas
7. Ice forms on top of ponds and lakes instead of the bottom because solid water is less dense than liquid water and floats on top of it. Frozen water has a very *organized structure* that looks like a honeycomb. This is because each water molecule forms hydrogen bonds with four other water molecules. As water freezes, the *water molecules* separate slightly and cause the volume to increase slightly. The increased volume causes the *density* to decrease, making solid water less dense than liquid water.
8. Convection occurs only in gases and liquids. Convection does not occur in solids because the molecules are locked in a structure. Convection depends on the fluid movement of molecules.
9. Relative humidity tells how much water is in the air compared to how much it can hold in equilibrium. When the relative humidity is high, the air is holding a lot of water vapor already, so any water vapor that evaporates from your skin is condensed right back again and you feel hotter and stickier compared to a dry day.
10. warm; cold
11. Water vapor (gas) condenses to form cloud drops (liquid) which then grow to form precipitation in the form of rain (liquid), hail (solid), snow (solid), and freezing rain (solid/liquid).
12. Florida's Everglades national park is a coastal park, at an elevation just above sea level. Alaska's Denali national park, in contrast, is in a mountainous region. Its highest peak, Mt. McKinley, has an elevation of 6,194 meters (20,320 feet). Because Everglades national park is at a much lower elevation, the atmospheric pressure is greater there.

5.2 Section Review

1. Sample answer: I found a hair scrunchy. If the scrunchy was made of a brittle material, it would break when I tried to it fit around my hair. The elastic in the scrunchy stretches really far so I can put it around my pony tail, then goes back to its original shape to hold the pony tail in place.
2. If the core was made of a brittle material like glass, when the golf ball was hit with the club it would shatter instead of bouncing off the club.
3. ductility
4. While both are quite elastic, rubber has little strength compared to steel. It will break under a relatively light load.
5. Sample answers:
   a. strong rubber bands or hair elastics, waistband elastic, stretchy fabric
   b. glass
   c. diamonds
   d. certain types of taffy candy, "Silly Putty"
6. In sugar crystals, the particles are arranged in an orderly, repeating pattern. In an amorphous solid like plastic, the particles are arranged in a random way.
7. Two properties to consider are brittleness, because a car seat shouldn't shatter easily; and elasticity, because a slightly elastic material could help absorb impact in the event of an accident.
8. c
9. Metals are better conductors of heat. A pan made of metal efficiently conducts heat from the stove's heat source to the food in the pan.
10. Gases, having particles that are spread far apart, are good thermal insulators. Materials with air pockets within them tend to be good thermal insulators. Examples include foam cups, down jackets, and fiberglass insulation.
11. insulator
12. The coolness you feel on your hand when you hold a cold can of soda.

Connection

1. Silly Putty can be used to pick up pet hair, it can be placed underneath a table or chair leg to prevent wobbles, and it can be squeezed to strengthen hand muscles or reduce stress. Encourage students to come up with their own uses.
2. Earl Warrick worked for Dow Corning in the early 1940's. Like James Wright, he was trying to develop a substitute for rubber when he came up with a bouncing compound made from silicone oil and boric acid. While he and James Wright independently developed the compound at about the same time, Wright is usually credited as the inventor because his compound was the one eventually marketed as Silly Putty.
3. The crew of Apollo 8 used Silly Putty to secure their tools so they wouldn't float about the spacecraft.

Activity

Applying your knowledge: sample answers

a. Procedure for bounciness: Drop the putty from 50 cm and have your friend measure how high it bounced.
Procedure for stretchiness: Roll it into a log shape of a certain length and diameter, then pull the putty apart as far you can without separating it and measure the length when it breaks. The ratio of the length when it broke to the original length is the stretchiness.
Not every batch was the same, possibly because of measuring inaccuracies or not mixing completely.

b. We chose to find out how temperature affects bounciness.

c. Hypothesis: As the putty gets warmer, it will bounce higher.

d. Procedure:

   a. Mark off a height of 50 cm.
   b. Make a single batch of putty and divide it into 4 pieces of equal mass.
   c. Put 1 piece of putty in the freezer; 1 piece in the refrigerator; 1 piece at room temperature; and heat up 1 piece by sealing it in a plastic bag and putting it in hot water.
   d. Repeat for each of the four pieces of putty:
   One partner measures its temperature, then drops the piece of putty from the 50 cm height. The other partner measures how high the putty bounces and records the temperature and height in the data table.

e.

Temperature °C	Bounce height when dropped 50 cm
2	3 cm
6	6 cm
20	10 cm
25	0 cm

f. We were surprised that as the temperature got either warmer or cooler than room temperature, our putty got less bouncy. Room temperature was the optimum temperature for bounciness.

Chapter 5 Assessment

Vocabulary

Section 5.1

1. condensation
2. boiling point
3. convection
4. evaporation
5. melting point
6. fluid
7. pressure
8. intermolecular forces

Section 5.2

9. conduction
10. crystalline
11. thermal equilibrium
12. ductility
13. amorphous
14. elasticity
15. brittleness
16. strength

Concepts

Section 5.1

1. Answers are:

a.	L	e.	B
b.	G	f.	G
c.	G	g.	B
d.	L		

2. Pressure comes from collisions between atoms. The pressure acts in all directions on a fluid.
3. Bonding forces occur within molecules and keep the parts of the molecule together. Intermolecular forces occur between free atoms and between molecules.
4. Answers are:
   a. attractive
   b. repulsive
5. At distances greater than the size of the molecule, intermolecular forces are attractive and pull molecules together. Closer than that and the intermolecular forces become repulsive. This is what keeps molecules from overlapping one another.
6. They compete with each other. Intermolecular forces try to bring molecules close and thermal energy causes molecules to vibrate and spread apart.
7. Solid
8.

	Solid	Liquid	Gas
Thermal Energy	low	medium	high
Intermolecular Forces	high	medium	low

9. stronger intermolecular forces
10. The temperature stops increasing until all the ice becomes liquid, at which point, the temperature will increase again.
11. As water freezes, molecules of water separate slightly from each other, which increases the volume a small amount, while the mass stays the same. Thus, the density decreases.
12. As a liquid, oxygen is more dense. More oxygen can be carried in a smaller place.
13. Evaporation takes place when high energy molecules escape from the surface of a liquid. Sweat (liquid molecules) on your skin absorbs the heat from your body giving the water molecules enough energy to leave the skin, and carrying with them the heat energy they have absorbed. This lowers your body temperature.
14. He or she means that evaporation and condensation of water vapor are in equilibrium.
15. Example of natural convection - warm air in a valley rising during the day and cold air on the mountainside sinking into the valley at night.
16. convection
17. carbon dioxide
18. nitrogen and oxygen
19. Over the oceans, warm air nearly saturated with water vapor rises and cools. The water condenses as the ability of the air to hold so much water vapor decreases. Rain is created.

Section 5.2

20. Solids hold their shape because thermal energy is not enough to overcome the intermolecular forces of attraction.
21. Glass is very brittle when it is solid, however, when the glass is heated, the bonds between the atoms are temporarily broken as the heat causes the glass particles to move faster. Thus, the glass can be stretched.
22. Silicon, the regularity of the silicon atoms in the crystal makes the circuits on the computer chips function identically.
23. Good strength, elasticity, low melting point to make molding easier, ease of altering polymer structure to change physical properties.
24. Placing your warm hand on a cold glass of water. The warm hand cools and the cool water warms by conduction until both are in thermal equilibrium at the same temperature.
25. Answers are:
   a. conductor
   b. insulator
   c. insulator
   d. insulator
   e. conductor

Problems

Section 5.1

1. 10 km
2. 25%
3. Answers are:
    a. B-C
    b. C-D
    c. 40°C
    d. more to boil the liquid

Section 5.2

4. You wouldn't hear it because there is no matter to conduct the sound.
5. c
6. solid; the molecules are close together and collide most frequently.

Chapter 6: The Atom

Learning Goals	California Content Standards	Activities and Resources	Assessment
6.1: Fundamental Particles and Forces The student will: • Identify and describe which comprise atoms. • Contrast forces inside atoms. • Describe the effects of radioactivity. • Define isotope. • Describe electric charge as a property of matter.	PS 3a PS 7b	**Laboratory Investigation 6A:** The Atom **Teaching Illustrations:** Units of Elementary Charge, Size and Structure of an Atom, and Electron, Proton, Neutron Table **Biographies:** Ernest Rutherford, Niels Bohr, Marie and Pierre Curie, and Rosalyn Yalow **Skill and Practice Sheets:** Atoms, Isotopes, and Ions	**Exam View Test Bank** Chapter Pre-Test **Section 6.1 Review**
6.2: Electrons in the Atoms The student will: • Compare spectra of elements. • Explain the Bohr atom model. • Describe how elements fill orbitals.	PS 3a	**Laboratory Investigation 6B:** Isotopes and Radioactivity **Teaching Illustrations:** Diagram: Spectral Lines and Energy, Bohr's Model of the Atom, Orbital, Energy Levels **Graphic Organizer:** An Atom **Biographies:** Albert Einstein **Skill and Practice Sheets:** Structure of the Atom **Connection:** Bioluminescence **Chapter Activity:** Atoms and Radioactivity	**Section 6.2 Review** **Chapter Assessment** **Exam View Test Bank**

Estimated Instruction Time	Language Tools	Differentiated Strategies	Materials
Six 45-minute class periods	**Literary Picks** **Vocabulary:** electric charge, elementary charge, neutral, charged, static electricity, electron, proton, neutron, nucleus, atomic number, isotopes, mass number, stable, radioactive **Word Origins:** isotope	**ELL Strategies:** Listed in ELL Ancillary **Learning Strategies:** • Cooperative Learning - Investigation 6A • Active Learning - Motive: It's Shocking! • Guided Practice - Explain: Subatomic Particles • Active Research - Extend	**Investigation 6A:** Atom Building Game
Six 45-minute class periods	**Vocabulary:** spectrum, spectroscope, spectral line **Closure Cards**	**Learning Strategies:** • Cooperative Learning - Investigation 6B • Active Learning - Motivate: Elements of Stars • Guided Practice - Explain: To The Bleachers! • Active Learning - Extend **Addressing Misconceptions:** Electrons and Atoms:	**Investigation 6B:** Atom Building Game **Chapter Activity:** can of pennies, graph paper

6.1 Fundamental Particles and Forces

About the Lesson

Scientists have confirmed through experiments that atoms are not the smallest units of matter. Atoms are made up of smaller particles, like protons, neutrons, and electrons. In this lesson, students learn about these subatomic particles and how scientists discovered their existence. Other topics discussed include static electricity, radioactivity, and periodic properties.

It's Shocking!

Rub your foot against the carpet in the classroom as you walk. Reach out and touch a student volunteer. Did she get a shock? If your classroom is not carpeted, ask students if they have ever had the experience. Most students can relate to this event. Ask students to offer suggestions for why you got "a shock" as you touched their classmate.

Students complete Investigation 6A before reading section 6.1. In this investigation students use the Atom Building Game to create atom models. Students are introduced to particles that make up atoms and learn where in the atom they are located. Read "About the atom model" on the next page before doing the investigation.

Investigation 6A: The Atom

Setup	Materials	Investigation Vocabulary
1. Allow one class period to complete the investigation. 2. Students work in small groups of three to five. 3. Become familiar with the Atom Building Game and its components prior to teaching the investigation. You may choose to play a round of Atomic Challenge with your colleagues before teaching it to your students.	• Atom Building Game	atom proton neutron electron isotope periodic table mass number atomic number

Literary Picks...

Reading Level	Title
3-5	***Atoms* by Melissa Steward** Steward has written more than 70 children's books and specializes in science writing for young readers. This book on atoms is part of the Simple Science Series and explains key scientific concepts in a simple and easy to understand format. Includes colorful photos, glossary, and a fun facts section.
6-8	***Adventures With Atoms and Molecules* by Robert Mebane and Thomas Rybolt** A collection of 30 chemistry experiments about atoms and molecules. Experiments answer questions like: can ions carry electricity? Can sugar be changed to carbon? Includes detailed observations and discussions of the results.
9-12	***Atom in the Family: My Life with Enrico Fermi* by Laura Fermi** Laura Fermi gives a first hand account of her husband's life (an Italian physicist and Nobel Prize recipient). She relives their story of emigrating to the United States and the movement of scientists from Europe. This book is an intellectual biography and history lesson all in one.

Guided Practice: Subatomic Particles

Create a chart like the one shown below on a overhead or transparency. Distribute a periodic table to each student and have them fill in the missing information. Begin this guided lesson by first reviewing important terms: proton, neutron, electron, atomic number, and mass number. Missing answers are shaded and shown in blue.

Element	Symbol	Number of protons	Number of neutrons	Number of electrons	Atomic Number	Mass Number
oxygen	O	8	8	8	8	16
magnesium	Mg	12	12	12	12	24
silicon	Si	14	14	14	14	28
neon	Ne	10	10	10	10	20
argon	Ar	18	22	18	18	40
copper	Cu	29	35	29	29	64
iodine	I	53	74	53	53	127
sodium	Na	11	12	11	11	23
fluorine	F	9	10	9	9	19

1. Find out if gasoline station owners in your area have installed metals pads near each pump. Why do they do this? Have students explain why this simple act is an important safety precaution for motor vehicle operators.
2. Have students work in small groups to research radioactive isotopes used in medicine.

Students complete section 6.1 review questions.

About the atom model...

The model of the atom used in the investigation is based on both the Bohr atom and the more modern quantum description of the atom. The activities are designed to teach the basic ideas behind the structure of the atom. Those ideas include:

- Protons and neutrons are in the nucleus while electrons are outside the nucleus.
- The atomic number is the number of protons in the nucleus and determines which element an atom is.
- The mass number is the total number of particles in the nucleus and determines the isotope.
- Isotopes may be stable or unstable. Unstable isotopes are radioactive.
- Radioactivity occurs because the nucleus changes.
- The electrons in atoms are arranged in well-defined energy levels.

The rows in the periodic table correspond to electrons filling the energy levels in the atom. For example, the first row of the periodic table has two elements (H and He) because the first energy level can only hold two electrons. While the model of the atom is quite effective at illustrating these concepts, there are limits to how realistic the model is with regards to other aspects of atomic structure.

Word Origins...isotope (from Greek *isos* meaning "same" or "equal"; *topos* meaning "place")

Isotopes of an element occupy the same place in the periodic table. Remind students that an element's position is equal to its atomic number and to the number of protons contained within the atom's nucleus. Isotopes of an element therefore have an equal number of protons, which identifies the element's place on the periodic table.

Investigation 6A: The Atom

You already know that matter is made from atoms. Scientists once believed that atoms were the smallest units of matter. We now know that there are even smaller particles inside atoms. The structure of the atom is the foundation for nearly all of the properties of matter we experience. This investigation will lead you through some challenging and fun games that illustrate how atoms are built from three particles: protons, neutrons, and electrons.

1 Modeling an atom

In the atom game colored marbles represent the three kinds of particles. The red marbles are protons, the blue marbles are neutrons, and the yellow marbles are electrons. Take a moment to look at the diagram in the center of the page. You can see that the atom represented in the diagram has three protons, three neutrons, and three electrons. Which particles are shown in the nucleus of the atom?

Tell students that the nucleus is the center of the atom. The protons and neutrons are in the nucleus.

Very good. Where are the electrons located?

The electrons are located in the energy levels. There are two electrons in the first energy level and one in the second energy level.

Build the atom shown in the diagram using the correct marbles. Then fill in the blanks in the empty periodic table box for the atom you constructed.

Have students look at the periodic table (page 28). Ask them locate the element lithium on the periodic table and fill in the missing numbers.

2 Thinking about the atom

You built the element lithium. Its chemical symbol is Li. Let's answer the questions in part two together. Who can tell me which number belongs below the chemical symbol?

The number 3 belongs below the symbol.

What does it tell you about the atom?

It tells you that lithium has 3 protons in its nucleus and 3 electrons. It also tells you that lithium is the third element on the periodic table and its atomic number is three.

The atomic number gives you quite a bit of information about the atom. You can locate any element on the periodic table if you know its atomic number because it is equal to the element number. For example, hydrogen is the first element on the periodic table and it has an atomic number equal to one. What can you tell me about hydrogen based on its atomic number?

Hydrogen has one proton and one electron.

What is the number or numbers above the element symbol called?

These numbers are the mass numbers. This number tells you the total number of protons and neutrons an atom has. In the example, lithium has a mass number equal to six.

Why do you think some elements have more than one mass number above its symbol?

6A The Atom

What is inside an atom?

We once believed that atoms were the smallest units of matter. Then it was discovered that there are even smaller particles inside atoms! The structure of the atom explains why nearly all the properties of matter we experience are what they are. This investigation will lead you through some challenging and fun games that illustrate how atoms are built from protons, neutrons, and electrons.

Materials

- Atom building game

1 Modeling an atom

In the atom game, colored marbles represent the three kinds of particles. Red marbles are protons, blue marbles are neutrons, and yellow marbles are electrons.

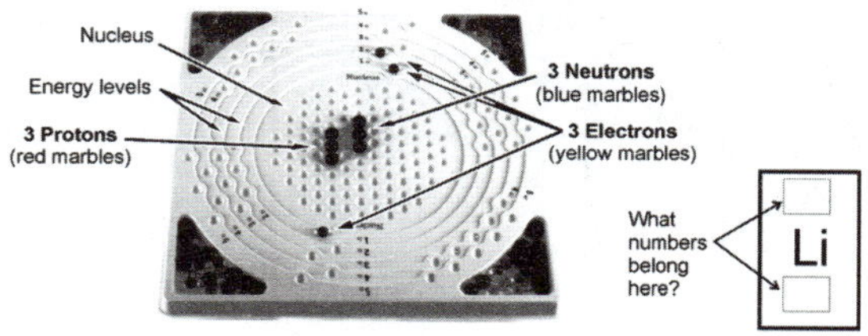

1. Build the atom above using three red, three blue, and three yellow marbles.
2. Fill in the blanks in the empty periodic table box for the atom you constructed.

2 Thinking about the atom

a. What is the number below the element symbol? What does this number tell you about the the atom?

b. What is the number(s) above the element symbol called? What does this number tell you about the atom?

c. Why do some elements have more than one number above the symbol? What are the variations in this number called?

Many atoms have more than one stable form in which it can exist. Lithium is an example. Each number written above an element's chemical symbol represents one of these stable forms, called an *isotope*. One stable isotope of lithium has a mass number equal to six and another stable isotope equal to seven.

An isotope is a form of an atom that has the same number of protons but a different number of neutrons. Consider the isotope of lithium with a mass number equal to seven. It is a neutral atom. How many protons, neutrons, and electrons does it have?

There are 3 protons and 3 electrons. Lithium-7 has four neutrons.

3 The Atomic Challenge

Let's try a game: the atomic challenge. This is a game that simulates the periodic table of elements.

Go over the rules for playing the atomic challenge. Write the three rules on the chalkboard so students can refer to them and ensure they have built a stable atom. Do a few practice rounds with students until they are familiar with the game.

Remember, you must build a stable atom. Check the rules on the board as you go about your turn. The winner of the game is the first player to use all their marbles.

4 Stop and think

Atoms that are not on the periodic table shown may exist in the universe but they are unstable. There are quite a few examples of this. Carbon is a good example. Look at the periodic table provided. Which isotopes of carbon are shown on the periodic table?

Carbon-12 and carbon-13 are shown.

That is correct! Each of these isotopes is stable. Both have 6 protons; carbon-12 has 6 neutrons and carbon-13 has seven neutrons. Carbon has a third isotope, carbon-14. It is unstable and has 8 neutrons. Does anyone know what carbon-14 is used for?

Carbon-14 is used in radioactive dating of fossils and other organic materials.

Your body is made up of many different elements. What four elements make up almost all of the human body?

oxygen (65%), carbon (18%), hydrogen (10%), and nitrogen (3%)

Find the stable isotopes of oxygen. Who can find one element on the periodic table that has no stable isotopes? Which element has 26 protons in its nucleus?

Oxygen has three stable isotopes (O-16, O-17, and O-18). Technetium (Tc) has no stable isotopes. Iron has 26 protons in its nucleus.

On most periodic tables a single atomic mass is listed instead of mass numbers for all stable isotopes. How is this mass related to the different isotopes?

Show students various periodic tables. Point to the number which is the average atomic mass. Students may need help answering this question. See the explanation in the box at right for more details.

Teaching Tip . . .Three Rules for Atomic Challenge

Write these rules on the board.

1. The number of protons matches the atomic number.
2. The total number of protons and neutrons equals a stable mass number.
3. The number of electrons matches the number of protons.

Average Atomic Mass

The average atomic mass is the weighted average of stable isotopes of an element with respect to its abundance. For example, the atomic mass of oxygen is given as 15.9994 atomic mass units (amu). Oxygen-16 is the most abundant isotope, with a natural abundance equal to about 99.7%. Oxygen-17 is only about 0.038% abundant, and oxygen-18 is approximately 0.200% abundant. Explain to students that because oxygen-16 is the most abundant isotope, it makes sense that the atomic mass listed on the periodic table is closest to 16.

Investigation 6A: Data and Answers

1 Modeling an atom

2 Thinking about the atom

a. Below the symbol is the atomic number, which tells you how many protons are in the nucleus of that particular atom.

b. Above the symbol is the mass number, which tells you the total number of both protons and neutrons in the nucleus.

c. Some have more than one number because they can have multiple versions of the same element. Each version has a different amount of neutrons in its nucleus. These different versions are called isotopes. Since isotopes of the same element have the same number of protons, but different numbers of neutrons, they will each have a different mass number. The more neutrons an isotope has, the greater its mass number.

3 The atomic challenge

4 Stop and think

a. Carbon, oxygen, hydrogen, and nitrogen.

b. Oxygen has three isotopes.

c. Tc, called technetium.

d. Fe, called iron.

e. This mass is based on a typical sample of an element. When an element is found in nature, it is almost always made up of a combination of all of its isotopes. Since each isotope has a different atomic mass, the one shown on most periodic tables is sort of like an average of all the possible atomic masses. The relative amount of each isotope in a sample is taken into account, and the atomic mass is calculated from those ratios.

Notes:

6.2 Electrons in the Atom

About the Lesson

The behavior of an atom is mostly due to how its electrons are arranged. In this lesson, students take a closer look at the electron. They also explore principles associated with the Bohr and quantum models of the atom.

Elements from Stars?

Visit www.teachersdomain.org and show the brief video clip, *The Elements: Forged in the Stars* to students. Use this activity to introduce students to the nuclear reactions in stars. Talk about how the strong nuclear force holds the atom together. Then define radioactivity in terms of how atoms "break apart."

When speaking about instability, It is important to remind students that the number of protons in atoms of an element is consistent. Students often mistakenly refer to instability as changing the number of protons. Tell students that when the number of protons changes, the element's identity is changed. Isotopes of an element have the same number of protons, but a different amount of neutrons.

Students complete Investigation 6B before reading section 6.2. In this investigation students use the atom building game to play *Nuclear Reactions*. They learn more about stable and unstable isotopes of atoms while reviewing the arrangement of subatomic particles.

Investigation 6B: Isotopes and Radioactivity

Setup	Materials	Investigation Vocabulary
1. Allow one class period to complete the investigation. 2. Students work in small groups of two to four. 3. Become familiar with the Nuclear Reactions game before the investigation.	• Atom Building Game	radioactivity

Teaching Tip . . .Addressing Misconceptions about Electrons and Atoms

How do students visualize electrons and their arrangement in atoms?

***Misconception #1**: One common tendency is for students to associate a physical structure, like an egg, as a model for an atom.*

This misconception is rooted in using the term *electron shell*. Students envision the yolk as the nucleus and the egg shell as its outer boundary where electrons are housed. Explain to students that "shell" has a very different meaning in this sense.

***Misconception #2**: The electron cloud and its electrons are comparable to a rain cloud and its water droplets. Electrons do not move and are cushioned and held in position by the cloud.*

Electrons are constantly in motion. The electron cloud is comprised of electrons only. Explain the proper context of "cloud" when describing electrons.

***Misconception #3**: Electrons "orbit" the nucleus of atoms like planets around a star.*

Electrons do not move in this manner. The term, *orbital* is most important for understanding the energy of an electron. Drawing perfect circles around the nucleus of an atom is usually the source of students' confusion about this. Read more about *wave-particle* character of light and the Schrodinger model of the atom to learn more.

***Misconception #4**: Electrons belong to specific atoms.*

Students learn that atoms of an element are unique to that element. They often presume that the same is true for electrons. Electrons are not specific to any atom. They can be transferred from one atom to another.

Guided Practice: To The Bleachers!

Take your students to the gymnasium or outdoors to the bleachers for practice filling orbitals and energy levels. Assign the lowest bleacher as the *1s* or first *s* orbital. This orbital can hold electrons representative of hydrogen and helium. Move up to the next level. This is the *2s* or second *s* orbital. It is also filled with two electrons. The third bleacher is designated for the *2p* orbitals. There are three of these (for a total capacity of six electrons). The graphic below shows how to arrange the bleachers through the *4s* orbital and provides an example using the element carbon.

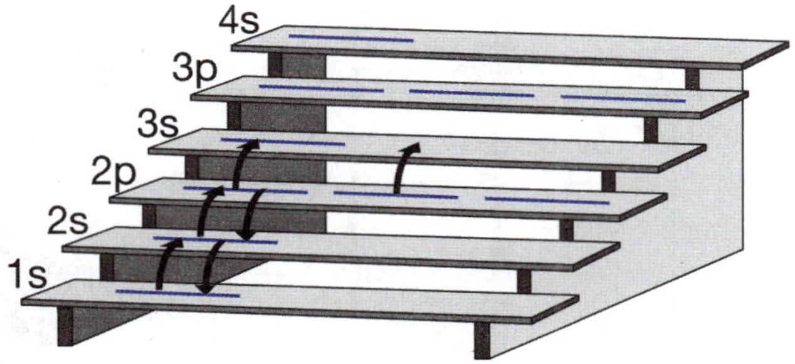

Have your students play the part of electrons. Assign each "electron" a number from 1 to 20. Say the name of an element and have them run the bleachers and fill the orbitals in the right order.

1. Have students build 3-D models of atoms with atomic numbers from 1 to 20.
2. Find out more about the works of scientists whose works have revealed many details about the atom over time. Make a time line to chronicle these important events.

Students complete section 6.2 review questions.

Why Electrons Fill Orbitals as They Do

The manner in which atoms are arranged in atoms is called the electron configuration. All systems in nature want to achieve stability. Therefore, they are usually ordered to have the lowest energy possible. This state of lowest energy is called the ground state. Three rules dictate how and why electrons fill orbitals as they do. Each is briefly addressed below, using the carbon example (left) as a reference.

- Aufbau Principle

The Aufbau principle simply states that electrons fill the orbital with the lowest energy first. The lowest energy orbital is the *1s*.

- Hund's Rule

According to Hund's rule, electrons in a sublevel (like *2p*) fill each orbital with one electron before pairing. Why? Each of these orbitals is equal in energy.

- Pauli Exclusion Principle

The graphic (left) uses arrows to represent the electrons in orbitals. When an orbital is filled with two electrons, the arrows point in opposite direction. This is reflective of the Pauli exclusion principle which states that two electrons in an atom cannot have the same quantum numbers. There are four types of quantum numbers: principle (the main energy levels), orbital (shape), magnetic (spatial orientation), and spin.

Pauli exclusion principle explains that two electrons in the same orbital must have opposite spin (arrows pointing in opposite direction). Think of it this way. Electrons create a magnetic field as they spin. The electrons within an orbital are like magnets. As they are paired, their opposing spin cancels each other out.

Investigation 6B: Isotopes and Radioactivity

In today's investigation you will play a game called *Nuclear Reactions*. As you have seen in the video clip, scientists believe that elements are formed from nuclear reactions in stars. You will make atoms of elements today, but not in the same way they are formed in the stars.

1 Key concepts

Before we begin there are some important concepts we need to review.

Read aloud each of the key concepts. Have students discuss these concepts with classmates. Review each vocabulary term to ensure that students understand their meanings. Define *stable* as an atom whose total number of protons and neutrons equal the mass number of an atom.

2 Playing the game

The objective of the game is to score ten points. The first player to do so wins. To start the game, each player needs five to eight of each colored marble in their pocket of the board. You may choose more marbles at any time without losing a turn. Each team needs a dealer. Who will be the dealer for your team?

Groups choose a dealer.

Your job is to shuffle the deck and give five Nuclear Reactions cards to each player. Look at your cards but don't show them to anyone else.

Students view their cards.

Each player takes a turn. When it is your turn, play one card and follow its instructions. For example, if the card says "add two electrons", the player must add two yellow marbles to the board. Lay the played card face-up on the table. Draw another card to keep five cards in your hand. The person to your right plays next.

Have one person from each group take a turn.

3 Scoring points

Have a student volunteer read the strategies for scoring points aloud while other students follow along. Play a few practice rounds with students to help them become familiar with how to score points.

There are three conditions that when met allow you to score points. The first condition states that the number of protons and the atomic number must be equal. Why do you think this is so?

This is based on the arrangement of elements on the periodic table. Atoms are arranged in order of increasing atomic number, which reflects the increasing number of protons.

The second condition is that the number of protons and neutrons must equal one of the mass numbers shown on the periodic table. This creates a stable nucleus.

Review with students that this is the definition of the mass number, the sum of protons and neutrons.

6B Isotopes and Radioactivity

Are all atoms of the same element exactly alike or are there differences?

This Investigation is a game called Nuclear Reactions. Two to four people may play. As you play, you will try to figure out which nuclear reactions will make real atoms. In order to win, you will need to do this quickly. The game is a little like the way the elements were formed inside stars. At the center of a star, nuclear reactions combine atoms to make new elements.

Materials

- atom game with nuclear reaction cards

1 Key concepts

1. All atoms of the same element have the same number of protons in the nucleus (and the same number of electrons). For example, all lithium atoms have three protons in their nucleus.
2. Atoms of the same element may have different numbers of neutrons in the nucleus. Atoms of the same element with different numbers of neutrons are called **isotopes**.
3. Isotopes are the same element, but they have a different **mass number**. The mass number is the total number of the protons plus neutrons in the nucleus of the atom.
4. Some isotopes are **radioactive**, which means they break apart if left alone. Other isotopes are **stable**, which means they stay together. The periodic table on the next page shows the mass numbers of the stable isotopes for each element.

2 Playing the game

The objective of the game is to score 10 points. The first player to score 10 points wins.

To start the game, each player should have 5-8 of each color marble in their pocket of the board. More marbles may be taken at any time with no loss of turn.

The dealer shuffles the deck and gives five NUCLEAR REACTIONS cards to each player. Each player looks at their own cards but does not show them to the other players.

Players each take a turn. In their turn a player chooses to play one card and does what the card instructs. For example, if the card says "Add 2 Electrons", the player must add two yellow marbles to the board. The played card is laid face-up on the table.

The player then draws a new card to maintain a hand of five and play passes to the next player on the right.

Each player holds 5 cards and may play any one in their turn

Points are scored depending on the atom which is left after a player takes their turn.

Used cards may be shuffled and re-used as needed.

28

The final condition is the number of electrons and protons must be the same. Electrons carry a negative charge and protons carry a positive charge. If these numbers are not the same, the net charge of the atom is not zero. In other words, the atom is not neutral. You want to create neutral atoms as you play the game today.

Students play the game.

1 Stop and think

Hydrogen has three isotopes, hydrogen-1, hydrogen-2, and hydrogen-3.

Make a chart (Figure 6.1) on the board. Guide students through the process of filling in the table.

Isotope	protons	neutrons	electrons	atomic number	mass number
H-1	1	0	1	1	1
H-2	1	1	1	1	2
H-3	1	2	1	1	3

Figure 6.1: Isotopes of hydrogen

What does the number written after the name of the isotope tell you?

It is the mass number, or sum of the protons and neutrons.

You can see that each isotope of hydrogen has one proton. For instance, hydrogen-2 has one proton and one neutron. Hydrogen-3 is unstable, and isn't listed on your periodic table. Suppose you added one proton to its nucleus. Is this element still hydrogen? If not, what is it?

Adding protons to an atom changes its identity to the element with the new number of protons. Adding one proton to hydrogen creates helium. If an atom contains two protons and no neutrons, it is an unstable isotope of helium.

What are the mass numbers of the stable helium isotopes?

three and four

How many protons, neutrons, and electrons are in boron-11?

Boron-11 has five protons, five electrons, and six neutrons.

What element would you have if you added three protons, three neutrons, three electrons to boron-11?

Adding three protons changes the element to oxygen (with 8 protons). The isotope is oxygen-14. This is radioactive. Use this time to emphasize that electrons are not inside the nucleus and do not affect the identity of the isotope.

Boron-10 has five protons and five neutrons. Suppose you wanted to build a fluorine-19 atom. How many protons must be added to the boron-10 atom?

Have students locate the fluorine atom on the periodic table. Fluorine has nine protons, so four more are needed.

How many neutrons are needed to make fluorine-19?

Five more neutrons are needed.

Look at your periodic table. Can you find an element that has four protons and five neutrons?

Beryllium fits the description.

If you combined all the protons and neutrons from boron and beryllium you have created a fluorine-19 atom.

Guide students through the process.

Investigation 6B: Data and Answers

1 **Key concept**

2 **Playing the game**

3 **Scoring points**

4 **Stop and think**

a. You would get helium-3. It would not be radioactive.

b. Oxygen 14. It is radioactive.

c. $^{10}_{5}B + ^{9}_{4}Be \rightarrow ^{19}_{9}F$

Notes:

Chapter 6 Answers

6.1 Section Review

1. b
2. charge is a fundamental property of all atoms.
3. C
4. In order of strongest to weakest: (1) strong nuclear force; (2) electromagnetic force; (3) weak force; (4) gravity
5. neutron
6. b
7. a
8. C
9. b

6.2 Section Review

1. c
2. the one in the middle
3. a
4. the 2nd and the 3rd energy levels
5. 18
6. B
7. c
8. a

Connections

1. A glow stick contains a small glass vial inside a larger flexible plastic tube. The glass vial is filled with an "activator" chemical—usually a hydrogen peroxide solution. The plastic tube contains another chemical, called phenyl oxalate ester, mixed with a fluorescent dye. When the plastic tube is bent, the glass vial breaks. This allows the two solutions to mix, causing a light-producing chemical reaction. The dye determines the color of the light that is given off.
2. No one knows for sure why no bioluminescence has been found in these kinds of organisms, but it may have something to do with the energy needed to make light. More complex living organisms tend to require more energy for life functions; perhaps they can't spare enough energy for bioluminescence.
3. Blue or blue-green is the most common color in marine bioluminescence. Red is sometimes found in deep water jellyfish. All of the colors of the rainbow-red, orange, yellow, green, blue, and violet have been found, but blue is by far the most common. Milky seas are caused by bioluminescent bacteria that glow with a milky-white light. Satellite images have recorded this phenomenon. Sea creatures that swim near the surface often have bioluminescent undersides. This helps them hide from predators below, preventing them from appearing as a dark silhouette against the surface of the water.

Activity

Applying your knowledge: sample answers

a.

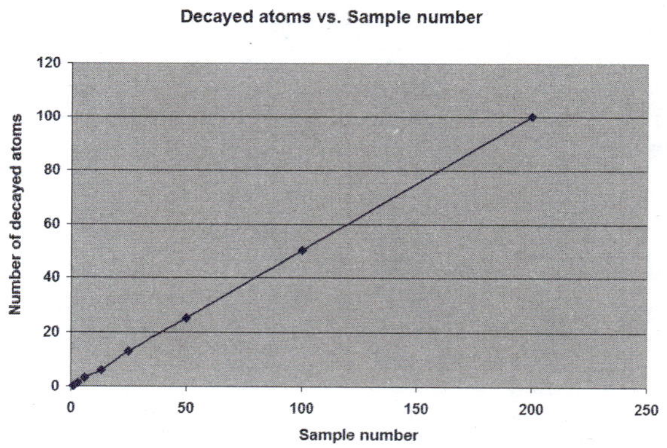

b. The graph is a sloped line. The slope of the line is 1/2.

c. 1 trial

d. 10 trials

e. No, it would look very similar.

f. No, you can't tell exactly when a specific penny decays. You can only tell the probability of how many pennies will decay.

g. About 100 pennies

Trial	Sample #	# of decayed atoms
1	200	101
2	99	50
3	49	26
4	23	11
5	12	5

Trial	Sample #	# of decayed atoms
6	7	3
7	4	2
8	2	1
9	1	1
10	0	0

Chapter 6 Assessment

Vocabulary

Section 6.1

1. neutron
2. elementary charge
3. nucleus
4. isotopes
5. neutral
6. atomic number
7. radioactive

Section 6.2

8. spectral line
9. spectrum
10. spectroscope

Concepts

Section 6.1

1. a and c
2. Electrical forces may be attractive or repulsive while gravitational forces are only attractive.
3. If objects are rubbed together, electrons from one object may be temporarily displaced to the other object, causing one object to become positive, the other negative.
4. A attract; B repel; C repel
5. c
6. c
7. If an atom's nucleus is unstable, it may decay into a more stable atom.
8.

Particle	Place in Atom	Charge	Relative Mass
electron	outside of nucleus	negative -1e	1
proton	in all nuclei	positive +1e	1,835
neutron	in nucleus when present	zero	1,837

9. Strong force; Electrical force; Weak force; Gravitational force
10. Answers are:
    a. Alpha decay occurs when an unstable nucleus ejects two protons and two neutrons.
    b. Beta decay occurs when an unstable nucleus converts a neutron to a proton and an electron and ejects the electron.
    c. Gamma decay occurs when the nucleus emits pure energy in the form of gamma rays.

Section 6.2

11. Examining the light from the hot gas with a spectrometer will produce a bright line spectrum that is unique for the element or elements that comprise the gas.
12. The closer an electron is to the nucleus, the lower its energy.
13. Answers are:
    a. the same as b. lowest c. a specific number of

Problems

Section 6.1

1. Answers are:
    a. 6 b. 12 c. zero
2. They are different elements because the identity is determined by the number of protons.
3. c and d are isotopes because they have the same number of protons.
4. Answers are:
    a. alpha decay
    b. beta decay

Section 6.2

5. electron "Y"

Chapter 7: Elements and the Periodic Table

Learning Goals	California Content Standards	Activities and Resources	Assessment
7.1: The Periodic Table of the Element The student will: • Differentiate physical and chemical properties of elements. • Explain how the periodic table is organized. • Distinguish metals and nonmetals.	PS 3f PS 7a PS 7b	**Laboratory Investigation 7A:** The Periodic Table **Teaching Illustrations:** Periodic Table of Elements, Reading the Periodic Table, Groups in the Periodic Table, Energy Levels in the Periodic Table	**Exam View Test Bank** Chapter Pre-Test **Section 7.1 Review**
7.2: Properties of Elements The student will: • Differentiate the thermal and electrical conductivity of metals and nonmetals. • Explore the unique properties of elements. • Define periodicity. • Identify examples of periodicity.	PS 3f PS 6a PS 7a PS 7c	**Laboratory Investigation 7B:** Periodic Table Group Challenge **Connection:** Naming the Elements **Chapter Activity:** Name that Element	**Section 7.2 Review** **Chapter Assessment** **Exam View Test Bank**

Estimated Instruction Time	Language Tools	Differentiated Strategies	Materials
Six 45-minute class periods	**Literary Picks** **Vocabulary:** physical properties, chemical properties, periodic table, period, group, metal, nonmetal, atomic mass unit, atomic mass, alkali metals, halogens, noble gases **Word Origins:** periodic	**ELL Strategies:** Listed in ELL Ancillary **Learning Strategies:** • Cooperative Learning - Investigation 7A • Active Learning - Motivate: Color By Number, Explain: Chemical Properties vs. Physical Properties • Active Research - Extend **Teaching Tip:** Semimetals	**Investigation 7A:** Periodic Table Tiles
Six 45-minute class periods	**Vocabulary:** periodicity, electrical conductor, thermal conductor, insulator, steel **Closure Cards**	**Learning Strategies:** • Cooperative Learning - Investigation 7B, Explain: Conductors and Insulators • Demonstration - Motivate: Metal Alloys • Graphic Organizers: Extend	**Investigation 7B:** blank periodic table, bingo card sheets, caller's clues and checklist, periodic table, highlighter or marker **Demonstration:** 3 bunsen burners with ring and stand, clay triangle, crucibles, tongs **Explain: Conductors and Insulators:** light bulb in a socket, sandpaper, ruler, electrical tape, 1.5 Volt battery (D-cell), copper wire, samples of metals and nonmetals, goggles **Chapter Activity:** 30 blank 8 x 10 cards, markers

7.1 The Periodic Table of Elements

About the Lesson

The modern periodic table arranges elements in order of increasing atomic number. Elements with similar chemical properties are arranged in columns called groups. In this lesson students learn about periodicity.

Color by Number

Distribute a copy of a periodic table to each students. Have students use crayons or colored pencils to distinguish groups of elements. Do not disclose the names of groups or discuss any periodic trends at this time. Only have students color by number. A suggested color scheme is provided below. Write this information on the board and tell students to copy it onto the back of their periodic table.

Element Numbers	Color	Element Numbers	Color
1	white	5, 14, 32, 33, 51, 52, 84, 85	red
3, 11, 19, 37, 55, 87	yellow	13, 31, 49, 50, 81-83	light green
4, 12, 20, 38, 56, 88	dark blue	6-8, 15, 16, 34	purple
21-30, 39-48, 72-80, 104-112	pink	9, 17, 35, 53	light blue
58-71, 90-103	orange	2, 10, 18, 36, 54, 86	black

Students complete Investigation 7A before reading section 7.1. In this investigation students use the Periodic Table Tiles to build the periodic table. Students learn how the periodic table is arranged and discover trends among elements within specific groups. Students will need textbooks and other reference materials in part four of the investigation.

Investigation 7A: The Periodic Table

Setup	Materials	Investigation Vocabulary
1. Allow one class period to complete the investigation. 2. Students work in small groups of three to five. 3. Become familiar with the Periodic Table Tiles prior to teaching the investigation.	• Periodic Table Tiles	group period

Literary Picks...

Reading Level	Title
3-5	***The Mystery of the Periodic Table* by Benjamin Wiker** A well-written book that will hold the interest of students as they are lead on a journey through the elements of the periodic table. An adventure that includes the scientific likes of Boyle, Stahl and others. Suitable for multiple grade levels.
6-8	***A Guide to the Elements* by Albert Stwertka** This book is an excellent source of information on the history and arrangement of the periodic table. Includes a brief discussion on each element with information like origin of name, important and interesting uses, and chemical and physical properties.
9-12	***The Periodic Kingdom* by P. W. Atkins** Atkins uses a very unique approach (analogy/metaphor) in presenting the basic properties of the elements and the organization of the periodic table. The chemical elements are the kingdom and the chapters are divided into three parts: geography, history, and government and institutions.

Chemical Properties vs. Physical Properties

Students often experience difficulty when asked to distinguish physical and chemical properties of matter. It may also be challenging to have to decide when a change in matter is physical or chemical. Students learned about physical properties of matter in chapters four and five. A physical property can be directly observed. A chemical property is evident when matter undergoes a chemical change.

A chemical change occurs when matter is transformed into a completely different substance with its own unique properties. Chemical changes are the result of chemical reactions. Read aloud a list of changes in matter. Have students classify each change as chemical or physical. Some examples are given below.

Physical Changes	Chemical Changes
tearing paper	combustion of a gas
boiling water	rusting nail
painting your room	tarnishing silverware
stretching a rubber band	digesting food
shattering glass	burning wood
splitting wood	souring milk

1. Do your students think the chemical symbols of elements make sense? Students may think some do, while others do not. Have students research the names and symbols of elements like lead, sodium, copper, iron, mercury, and silver.
2. Have students choose one element to study and present their findings on a poster.
3. Have students work in small groups to create a Power Point presentation about special families of elements, such as the alkali metals or noble gases.

Students complete section 7.1 review questions.

Word Origins...periodic (from Greek *periodos* meaning "way around")

Periodic means to occur at regular intervals. Have students follow along as you demonstrate periodic trends. For example, have each student place a finger on the element hydrogen. Ask students to tell how many electrons are in a hydrogen atom. Then move over to helium and ask the same question.

Remind students that the elements in the first period represent the first energy level of atoms and can only hold two electrons. Arriving at group 18 signals students to make their "way around" to the next period of elements and to begin the process all over again. Repeat this process with elements in the second and third periods.

Ask students to think of other applications of the word, *periodic*. Some examples include birthdays which occur every year or receiving magazines or newspapers (periodicals) monthly or daily.

Teaching Tip . . .Semimetals

Semimetals (or metalloids) have many interesting properties which make them beneficial for a number of industries. For example, silicon is a very important semimetal to the computer industry. Silicon is a semiconductor, which means that it exhibits increasing electrical conductivity as temperature increases. It is one of the main components in integrated circuits. Silicon can be "doped" with other semimetals such as arsenic, boron, and antimony to alter its electrical conductivity.

Take your students on a tour of a silicon wafer fabrication facility or invite an electrical engineer to your classroom to speak to your students about semimetals and integrated circuits.

Investigation 7A: The Periodic Table

Complete the *Color by Number* activity before starting the investigation.

1 Elements and the atomic number

You have already learned that elements are arranged on the periodic table in order of increasing atomic number. What does the atomic number tell you about the atoms of an element?

The atomic number reveals the number of protons in the nucleus and the number of electrons that a stable atom of the element contains. The atomic number also helps one to identify an element's position in the table very quickly because elements are arranged in order of increasing atomic number.

Very good. When you look at the periodic table, you also observe that each element has a chemical symbol. The chemical symbol is like an abbreviation for the element. For instance, rather than writing out the word hydrogen, scientists recognize the letter H as its chemical symbol. The periodic table also displays other facts about elements like its atomic mass, or whether it is a solid, liquid, or gas at room temperature.

2 Building the periodic table

The hard work of many scientists contributed to arranging the modern periodic table. Which techniques do you think these scientists used to determine how to organize the elements?

Students offer suggestions. Steer students toward explaining how the process of science helps humans to make observations and collaborate with other scientists to understand our world. Lead a brief discussion about how scientists came to understand the organization of the periodic table and chemical periodicity. *See Periodic Table Timeline* (Figure 7.1).

For this part of the investigation you are going to build the entire periodic table of elements. You will use the periodic table tiles to do so.

Display a few periodic table tiles. Discuss the information provided on the tiles.

Is the shape of the periodic table shown the same as your colored periodic table?

Students should observe the shapes are different.

The table has a very specific shape which corresponds to the chemical properties of the elements. It is arranged in order of increasing atomic number from left to right. As you build the table be sure to follow the diagram shown to get the correct shape.

Students build the periodic table. Assist as needed.

3 Stop and think

The periodic table you colored in earlier looks like the one shown in the diagram on the right. The table you built has a different shape. Can you think of a reason why the table is often arranged as shown in the right?

Students answers will vary. Students may conclude that the periodic table is easier to read in this arrangement. One important note about this arrangement is that the *s*, *p*, *d*, and *f*-block (based on orbital quantum numbers) are clearly distinguishable.

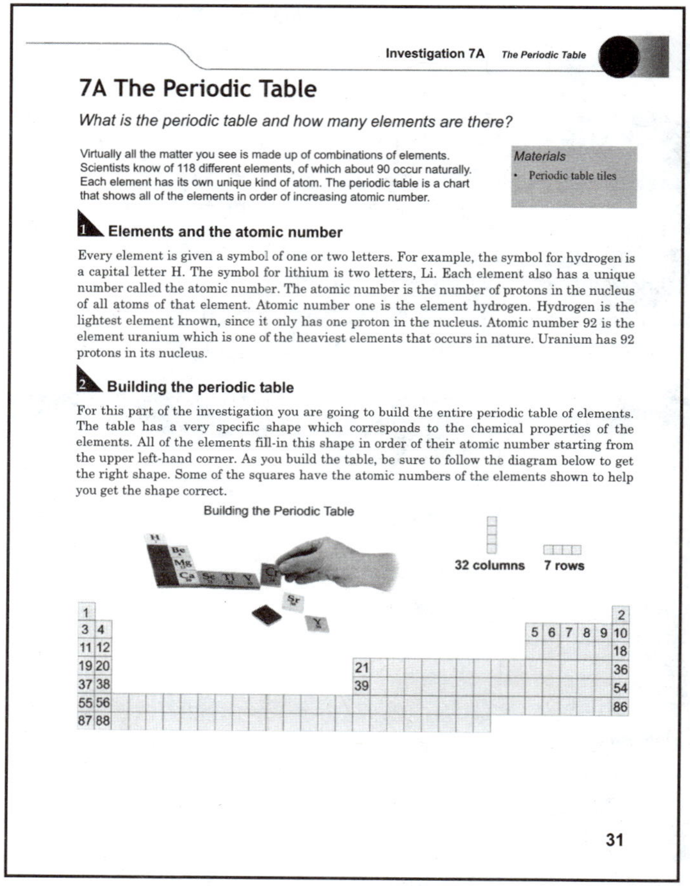
Investigation 7A *The Periodic Table*

7A The Periodic Table

What is the periodic table and how many elements are there?

Virtually all the matter you see is made up of combinations of elements. Scientists know of 118 different elements, of which about 90 occur naturally. Each element has its own unique kind of atom. The periodic table is a chart that shows all of the elements in order of increasing atomic number.

Materials
- Periodic table tiles

1 Elements and the atomic number

Every element is given a symbol of one or two letters. For example, the symbol for hydrogen is a capital letter H. The symbol for lithium is two letters, Li. Each element also has a unique number called the atomic number. The atomic number is the number of protons in the nucleus of all atoms of that element. Atomic number one is the element hydrogen. Hydrogen is the lightest element known, since it only has one proton in the nucleus. Atomic number 92 is the element uranium which is one of the heaviest elements that occurs in nature. Uranium has 92 protons in its nucleus.

2 Building the periodic table

For this part of the investigation you are going to build the entire periodic table of elements. The table has a very specific shape which corresponds to the chemical properties of the elements. All of the elements fill-in this shape in order of their atomic number starting from the upper left-hand corner. As you build the table, be sure to follow the diagram below to get the right shape. Some of the squares have the atomic numbers of the elements shown to help you get the shape correct.

Building the Periodic Table

31

4 Groups of the periodic table

The periodic table is organized into eleven sections. They are shown using the letters a-k. Each section contains elements with similar chemical properties. Most section are single columns. One section includes many columns and two sections are part of entire rows. Locate the element that is the first one in each section, a-k. Write the name of this element in the space provided in Table 1.

Students use the atomic number of each element to locate its name on the periodic table.

Each column of the periodic table is called a *group*. Look at your colored periodic table. How many groups are there?

There are 18 groups on the periodic table.

Which sections in the diagram represent a group?

The letters a, b, f, g, h, i, j and k represent groups of elements.

Elements in a group have similar chemical properties. For this reason, some of these groups have a specific name. Research to find out if these groups have a name.

Students use their textbook to find the names of groups. Those groups with a specific name include a (alkali metals), b (alkaline earth metals), j (halogens), and k (noble gases). Depending on the resources students use, they may also find that section i is also referred to as chalcogens, or the oxygen family.

Do the elements in sections c, d, and e have a specific name?

Be careful not to use the word *group* as you describe these sections. Elements in sections c are the lanthanides. The actinides are section d, and section e is the transition elements.

What are some of the properties common each section of elements?

Have students begin to research the properties of these elements by using their textbook. Students may also use the Internet or other print resources that offer grade-level appropriate description of the elements.

5 Stop and think

The periodic table is arranged by horizontal rows and vertical columns. The rows are called *periods*. Look at your colored periodic table. There are seven periods on the table.

Show students the seven rows. Explain to students that the model of the periodic table they built (using the tiles) has seven periods. Show students how the lanthanides and actinides are pulled away from the table. Even though they are pulled away from the table, these elements are not separate periods. They are part of periods 6 and 7.

Do all the periods have the same number of elements? Why do you think some are different than others? Use your textbook to find out why.

These questions are a great lead-in to discussing periodic trends and properties of elemental groups. It is important that students in grade eight learn to use the periodic table as a resource to locate important information about elements (such as atomic number, atomic weight, etc) and as a tool to learn more about the structure of atoms. In chapter 8, students use the periodic table to better understand how atoms combine to form new substances (chemical bonding).

Aristotle (~330 BC)
Theory of four elements: earth, fire, water, air

Lavoisier (~1770-1789)
Composed the first list of 33 elements. Distinguished metals and non-metals.

Berzelius (1828)
Created a table of atomic weights. Introduced idea of chemical symbols.

Döbereiner (1829)
Developed triads—groups of 3 elements with similar properties.

Newlands (1864)
Arranged known elements in order of increasing atomic weights. Proposed the Law of Octaves.

Meyer (1869)
Compiled a periodic table with elements arranged by atomic weight. Elements with similar properties arranged in columns.

Mendeleev (1869)
Created a periodic table with elements arranged by atomic weight. Elements with similar properties arranged in columns.

Ramsay (1894)
Discovered noble gases.

Moseley (1914)
Determined atomic numbers of each element. Modified the periodic law—an element's properties varies periodically with its atomic numbers.

Seaborg (1940)
Synthesized elements beyond uranium.

Figure 7.1: Periodic Table Timeline

Teaching Tip . . .Color by Number

At the close of Investigation 7A have students identify the colored sections of elements and create a legend for each group or section. Place the periodic table in a page protector and have students secure it in a binder or folder.

Investigation 7A: Data and Answers

1 Elements and the atomic number

2 Building the periodic table

3 Stop and think

4 Sections of the periodic table

Diagram	First element (name)	Group name (if any)	Representative chemical properties
a	lithium	Alkali metals	Highly reactive, especially with oxygen. Combines in ratio of 2 to 1 with oxygen. Does not include hydrogen.
b	beryllium	Alkaline earth metals/ Group 2 metals	Highly reactive with oxygen, but not as reactive as the alkali metals. Combines in ratio of 1 to 1 with oxygen.
c	lanthium	Lanthanides	Many are used in lasers.
d	actinium	Actinides	Only actinium, thorium and uranium occur naturally into Earth's crust, others were synthesized in labs. All are radioactive.
e	titanium	Transition metals	These elements are usually good conductors of heat and electricity.
f	boron	Boron group or family, group 13, or IIIA	This group contains metals and metalloids. Reactivity and properties vary.
g	carbon	Carbon group or family, group 14, or IVA	This group contains metals, non-metals and metalloids, reactivity and properties vary but generally not particularly reactive.
h	nitrogen	Nitrogen group or family, group 15 or VA, or pnicogens	This group contains metals, non-metals and metalloids, reactivity and properties vary.
i	oxygen	Oxygen group or family, group 16, VIA, or chalcogens	This group contains metals, non-metals and metalloids, reactivity and properties vary.
j	fluorine	Halogens	Very reactive, rarely founds in pure form. Combine with alkali metals to form salts.
k	helium	Noble gases	These elements do not usually form chemical bonds with other atoms and are almost always found in their pure state. They are sometimes called inert gases for this reason.

5

a. Each row represents a different energy level. The number of elements in the row indicates how many electrons are in that energy level.

b. The noble gases. These elements do not normally form chemical bonds with other atoms and are almost always found in their pure state.

c. Carbon is in the carbon group, also called group 14 or VA. The elements in this group range from non-reactive (silicon) to reactive, but vary in their properties. All elements in this group have four electrons in their outer shell, allowing for four possible bonds to be made with other elements.

Notes:

7.2 Properties of the Elements

About the Lesson

In this students continue to learn about the properties of elements. Students further explore periodicity and characteristics of elements such as electrical and thermal conductivity. Students also take a closer look at metals, nonmetals, and alloys.

Teacher Demonstration: Metal Alloys

Display materials made from various metal alloys on a table in your classroom. Allow students to make observations about each material. Have students guess the identity of each material. What are some physical qualities of each alloy? For instance, does it appear dull or shiny? Is it heavy or light? Which metals comprise each alloy? A table of metal alloys is shown in Figure 7.2.

Do a demonstration to compare the melting points of tin, copper, and bronze? Which of these do students think will melt first? Set up three bunsen burners, each with a ring and stand, clay triangle, and crucible. Place a small piece of each material in a crucible and heat with a hot flame. Use tongs to handle the crucibles. Allow students to observe the amount of time needed to melt each sample. How does the periodic table explain students' observations?

Students complete Investigation 7B before reading section 7.2. In this investigation students play *Periodic Table Group Challenge*, a bingo-like game which helps students understand the arrangement of elements on the periodic table.

Investigation 7B: Periodic Table Group Challenge

Setup	Materials	Investigation Vocabulary
1. Allow one class period to complete the investigation. 2. Students work in groups of five. 3. Become familiar with the game before students play. 4. Have sufficient copies of the challenge card available.	• blank periodic table bingo card sheets • caller's clues and checklist • copy of the periodic table • highlighter or marker	periodicity

Alloy	Composition	Uses
alnico	Al, Co, Cu, Fe, Ni, Ti	permanent magnets
brass	Cu, Zn	musical instruments, household fixtures
bronze	Cu, Sn	medals, coins, cymbals, screws
pewter	Cu, Pb, Sb, Sn, Bi	tableware
solder	Pb, Sn	electronics, plumbing
stainless steel	Cr, Fe, Ni	surgical tools, appliances, cutlery, hardware, construction
sterling silver	Ag, Cu	flatware, jewelry, serving trays

Figure 7.2: Common alloys.

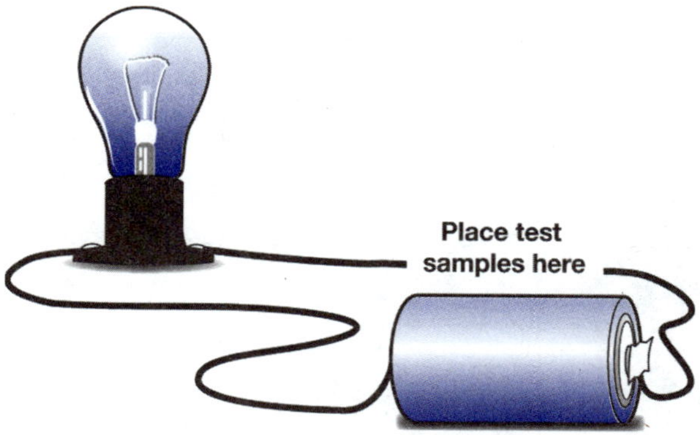

Figure 7.3: A simple conductivity tester

Conductors and Insulators

Students can infer which elements are better conductors or insulators with a conductivity tester. Instruct your students work in small groups to build an electrical conductivity tester using the following materials: light bulb in socket, sandpaper, ruler, electrical tape, 1.5 V battery (D-cell), three pieces of copper wire (each 10 inches long), metal and nonmetal samples for testing, and safety goggles.

Safety Precautions: Be sure that students wear safety goggles throughout the activity. Also, the copper wire ends are sharp. Students must be careful to avoid cutting themselves.

Building the Tester: Have students use the sandpaper to remove an inch of the coating from each end of the copper wires then make a loop at the ends. Use the electrical tape to affix a loop from one wire to the flat end of the battery. The other end of the wire connects to the light bulb socket. Connect a second wire to the other side of the light bulb socket. Use more tape to attach the loop of the third wire to the other end of the battery. The setup is shown on the previous page (Figure 7.3).

Activities: Tell students to touch the two unattached ends of the wire together and record what happens. Then provide samples of metals, nonmetals, and semimetals for students to test. Students place each sample between the two unfastened ends of copper wire and record their observations. Have students infer which types of elements (metals, nonmetals, semimetals) are good conductors or insulators.

1. Find out more about metallurgy, its historical significance, and current applications to materials science.
2. Make a Venn Diagram to compare and contrast properties of metals and nonmetals.
3. Write a report about phosphorus and how glow-in-the-dark materials work.

Students complete section 7.2 review questions.

Allotropes of Carbon

Carbon is a very unique element. As a matter of fact, an entire branch of chemistry is dedicated to it. Organic chemistry is the study of carbon and how it chemically binds to itself and other elements. Carbon compounds are essential to living organisms, but they are also a component of plastics and hydrocarbons, like methane.

Graphite is one allotrope of carbon. The term *allotrope* refers to the different forms of an element. The properties of each allotrope depends on how the carbon atoms are arranged. Graphite is the softest known mineral and is an electrical conductor. It is also an excellent lubricant.

Diamond is another allotrope of carbon. Diamond has many properties that are exactly opposite of graphite. For example, it is the hardest known mineral and is a poor conductor of electricity. It is actually an excellent insulator. Because of its hardness, diamond is very abrasive.

Fullerenes are a more recently identified allotrope of carbon (1985). They are compounds made entirely of carbon. Based on their shape, they are often referred to as *buckyballs* (spherical) or *buckytubes* (cylindrical).

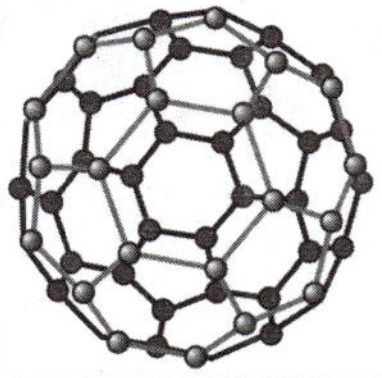

Buckminster Fullerene (Carbon-60)

Scientists are still learning about fullerenes and their possible applications in the fields of nanotechnology and medical research.

Investigation 7B: Periodic Table Group Challenge

You have already learned some facts about elements and the periodic table. Let's review what you already know.

Students volunteer facts they have learned such as atomic number, atomic mass, chemical symbols, arrangement of elements in groups and periods, etc.

The arrangement of elements is not accidental. Elements in each group of the periodic table have properties in common. For example, lithium and sodium have common properties. They are both group one alkali metals and are very reactive. For this reason they are not commonly found in nature as free elements. Each alkali metal is more reactive than the one above it. This means sodium is more reactive than lithium and potassium is more reactive than sodium, and so forth.

Point to each element in the group as you discuss its reactivity.

The most reactive nonmetals are on the opposite side of the periodic table. They are the elements in group 17. What is this group of elements called?

These are halogens.

Chlorine is an example of a halogen. At room temperature it is a toxic gas. However, it can combine with sodium, an alkali metal to create a common substance. Table salt is made when sodium and chlorine atoms combine.

Have students make a few qualitative observations about table salt. Ask students if table salt appears to have derived from the combination of a metal and a gas.

Other halogens also tend to combine with alkali metals. These repeated patterns of behavior within groups is called *periodicity*. You have already learned the meaning of the word periodic. Both periodic and periodicity have the same root word. What is it?

The common root is period.

Good answer. What do you think periodicity means?

Accept all reasonable responses. Periodicity means that properties of elements are a function of their atomic numbers. The reason for periodicity lies in the arrangement of electrons of atoms (particularly valence electrons). Students will learn more about the connection between valence electrons and properties of elements as they study chemical bonding and reactions in the coming chapters. However, it is important to begin laying the foundation so that students understand the significance of electrons and how atoms "behave."

1 The Challenge

Have a student volunteer read the brief description of the challenge aloud.

2 Rules of play

This game is sort of like bingo. Instead of numbers, each square is filled with chemical symbols of elements. One player in each group is the caller. The remaining students are players. Who will be the caller in your group?

Students select the caller for their group.

Investigation 7B *Periodic Table Group Challenge*

7B Periodic Table Group Challenge

How is the periodic table organized?

Each box on the periodic table tells you the element symbol, atomic number, and atomic mass for all the known elements. This is very useful information, but did you know that the arrangement of the elements on the periodic table gives you even more information? Each major column of elements represents a group of elements with similar chemical behavior. Can you see why the arrangement of elements on the periodic table is important?

Materials

- Blank periodic table bingo card sheets; 1 per player
- Caller's clues and checklist; 1 per team
- Copy of the periodic table of elements; 1 per player
- Highlighter or other marker; 1 per player

1 The challenge

Periodic Table Group Challenge is a bingo-like game that helps you understand how the elements on the periodic table are arranged. Each player will fill out their own five-by-five grid with element symbols, and then the caller will read element clues. The players must interpret the clues and highlight any boxes on the grid that fit the clue. The first player that correctly highlights five boxes across, up-and-down, or diagonally in a row is the winner.

2 Rules of play

1. Designate one member of your group to be the caller. The caller will call out clues and keep track of them on the checklist.
2. Each player will have a sheet of 4 blank grids. Fill out one of the grids with random element symbols (other grids can be used for additional games). You may choose elements in the atomic number range of 1 – 54 (hydrogen through xenon). Do not repeat symbols on the card. You will only be able to fit 25 of the possible 54 symbols on the card. You choose which ones to use, and where to place them on the grid.
3. The caller will randomly pick clues from the list, and as a clue is called out, the caller will check off the clue. The answers are only for the caller to check the winner's card!
4. When a clue is called, players check the grid to see if any of the elements fit the clue. Any elements that fit the clue must be highlighted. If no elements fit the clue, then no boxes are highlighted on that turn.
5. When a player has five boxes highlighted in a row up and down, across, or diagonally, play stops. The caller will double check the clue list and answers to see if the clues indeed match the elements. Play continues until a true winner is determined.

33

Each player should have a blank page with boxes. This page is the periodic table challenge card page (part four). How many squares are on the page?

Students count 25 squares.

You will write one chemical symbol for each element you choose in one square. The elements you choose must have atomic numbers in the range of 1 to 54. You may write only one symbol per square. When the called states a clue, check to see if any elements on your grid fit the description. If so, use your highlighter to cross off that element. If the clue does not match any of your elements then no boxes are highlighted on that turn. How can you win?

A winning card matches five clues horizontally, vertically, or diagonally.

Very good. Continue to play until a winner is declared. Take a moment and read over the rules once again.

Students review the rules of play.

3 Caller's clues

Each group has a caller. The caller's job is to call and check off clues. It is also your responsibility to verify the winner after each round. You do not have to call the clues in order. Just be sure to mark off any clues that you call. Only the caller should have the page of caller's clues.

Give the clue page to the caller. If students play more than one round, select a new caller. You may also want to have extra clues available in the event that no winner is declared using the clues provided.

4 Periodic Table Challenge Card

Now you are ready to prepare your periodic table challenge card. Choose 25 elements with atomic numbers between 1 and 54. This includes elements from hydrogen (H) to xenon (Xe). Fill in each square with the chemical symbol for the elements you choose. You may only use an element once.

Students fill in their cards with randomly chosen chemical symbols. Students mark off elements which fit the descriptions read by the caller. A winning card has five symbols marked off horizontally, vertically, or diagonally. Once the caller has verified the winner, students may play the game again. Choose a new caller for each group. A sample challenge card is shown below.

Cl	F	Ga	As	Sr
Mg	Al	O	Li	Zn
P	V	Sr	Na	Be
B	H	Ar	Ti	Ag
Ca	Ne	K	S	Cr

For a fun extension activity, have students use their own periodic tables to create clues for playing the challenge. Allow time for students to play the game with their created clues.

Periodicity

Many trends exist among elements within each period and group of the periodic table. These patterns are directly linked to electron configuration in atoms. A few examples of periodicity are discussed below. Both are important concepts (for your reference) when preparing for chemical bonding, the topic of chapter 8.

- Electronegativity

Electronegativity refers to an atom's ability to attract an electron to itself. In most instances, nonmetals are more electronegative than metals. Halogens are the most electronegative group of elements, due to their almost filled outer energy level of electrons. As one moves down the group of halogens, electronegativity decreases. This trend is consistent with each group on the periodic table (with few exceptions). Electronegativity generally increases across a period.

- Ionization energy

An ion is formed when an atom gains or loses electrons. The energy required to remove one electron from an atom is called the first ionization energy. Removing a second electron is the second ionization energy and so forth.

An alkali metal readily forms a positively charged ion by giving up its one valence electron. Alkali metals have the lowest ionization energy and is the reason why these metals are so highly reactive. Of course, the noble gases have the highest ionization energies because they are relatively nonreactive. Among the main group elements, ionization energy increases across a period.

Investigation 7B: Data and Answers

1 **The challenge**

2 **Rules of play**

3 **Caller's Clues**

4 **Periodic table challenge card**

The assessment for this activity is an inherent part of the periodic table challenge game. As students become more and more familiar with element symbols, and become better at gleaning information from the periodic table, there will be progressively fewer "false bingos." With each new game, winners will be determined more quickly and with greater accuracy. You might want to ask groups to keep track of how many "false bingos" they have for each round of the game, and you might have them measure how much time it takes for a successful winner to be established in each game. Have groups compare their results, and use the data to decide if the class is becoming more familiar with element symbols, names, and the periodic table information.

Notes:

Chapter 7 Answers

7.1 Section Review

1. a and b
2. c
3. magnesium
4. sulfur, selenium, tellurium
5. c
6. chlorine
7. 4th level; it is in the fourth row of the periodic table
8. c
9. sodium, magnesium and aluminum

7.2 Section Review

1. mercury and bromine
2. d
3. The physical or chemical properties are repeated with every row of elements on the periodic table.
4. copper, aluminum and iron
5. copper, aluminum and iron
6. a
7. Steel is an alloy, a mixture of iron and carbon.
8. False, oxygen makes up about 21% of the atmosphere but about 46% of Earth's crust.
9. silicon
10. d
11. a
12. c

Connection

1. Examples of elements named after scientists that students may have studied: Rutherfordium, Bohrium, Meitnerium, Curium, Einsteinium, Fermium, Mendelevium.
2. Cerium and Palladium were named after the asteroids Ceres and Pallas, which were discovered in 1801 and 1802, almost simultaneously with the elements.
3. The three competing groups were the Laboratory for Heavy Ion Research, of Darmstad, Germany; the Joint Institute for Nuclear Research, of Dubna, Russia; and the University of California-Berkeley. The names ratified in 1997 are as follows: 104 Rutherfordium, 105 Dubnium, 106 Seaborgium (after Berkeley researcher Glenn Seaborg, even though IUPAC rules prohibit naming elements after a living person), 107 Bohrium, and 108 Hassium. For more information about the controversy, including a table of other proposed names for these elements, see "Naming the Transferium Elements," http://www.vanderkrogt.net/elements/elem/md.html

Activity

Applying your knowledge: sample answers

a. gold

b. Pb comes from *plumbum*, the latin word for white soft metals, including lead.

c. sodium

d. *carbonis*, or carbon

e. some word examples are: BAlCONY, ReCITe, FLaKY, ClArIFY

Chapter 7 Assessment

Vocabulary

Section 7.1

1. nonmetals
2. atomic mass unit
3. physical properties
4. alkali metals
5. group
6. metals
7. halogens
8. periodic table
9. atomic weight
10. period
11. noble gases

Section 7.2

12. thermal conductor
13. steel
14. insulator
15. periodicity
16. alloy
17. electrical conductor

Concepts

Section 7.1

1. Answers may include: density, phase (state), texture, ductility, specific heat, boiling point, color
2. Physical changes are often reversible while chemical changes are not easily reversible.
3. The table is organized in order of increasing atomic number.
4. Each horizontal row is called a period. Each vertical column is called a group.
5. On the left side of the table.
6. period
7. Answers are:
   a. transition metals
   b. halogens
   c. alkali metals
   d. noble gases

Section 7.2

8. Answers are:
   a. solid: 79
   b. liquid: 2
   c. gas: 11
9. On the extreme right side.
10. Elements with the highest melting points tend to have half-filled energy levels.
11. Answers are:
   a. carbon
   b. silicon and germanium
   c. oxygen
   d. nitrogen
   e. nitrogen or phosphorous
   f. phosphorous
12. The elements appear in the middle and left side of the periodic table.
13. On the right side of the table.
14. Mixing metallic elements may produce materials displaying properties that are superior to those of pure elements for some applications.

Problems

Section 7.1

1. Answers are:
   a. group
   b. period
   c. group
   d. group
   e. period
2. Mg-24, magnesium with 24 amu
3. The period number for an element is the same as the energy level number for the outermost electrons for that element. The first three levels, together, can accommodate 18 electrons (2-8-8). The fourth level can accommodate 18 electrons. The outermost 11 electrons of copper would be found in the 4th energy level. So copper is placed in the 4th period.

Section 7.2

4. Answers may vary but may include:
   a. A. very active metal, solid at room temp
   b. B. good conductor of heat and electricity, solid at room temp
   c. C. insulator, nonmetal, solid at room temp
   d. D. inert, gas at room temp

Chapter 8: Molecules and Compounds

Learning Goals	California Content Standards	Activities and Resources	Assessment
8.1: Compounds and Chemical Bonds The student will: • Distinguish compounds and mixtures. • Write simple chemical formulas. • Explain how the structure of a molecule affects its chemical properties. • Describe how ionic compounds are formed.	PS 3b	**Laboratory Investigation 8A:** Chemical Bonds **Teaching Illustrations:** Chemical Formulas and Structure Diagrams, Chlorine and Sodium Bonds **Skill and Practice Sheets:** Sketching Molecules and Compounds	**Exam View Test Bank** Chapter Pre-Test **Section 8.1 Review**
8.2: Electrons and Chemical Bonds The student will: • Infer the relationship between the number of valence electrons and behavior of atoms. • Predict chemical formulas. • Compare and contrast ionic and covalent bonding.	PS 3b PS 3c	**Laboratory Investigation 8B:** Molecules **Teaching Illustrations:** Periodic Table and Reactivity, Valence Electrons and Bonds, Lewis Dot Diagrams, Example of Oxidation Numbers Adding to Zero, Periodic Table and Electron Acceptors and Donors, Predicting a Chemical Formula **Skill and Practice Sheets:** Dot Diagrams, Chemical Formulas **Connection:** Salt: Common, But Uncommonly Important **Chapter Activity:** Molecular Gumdrop Model	**Section 8.2 Review** **Chapter Assessment** **Exam View Test Bank**

Estimated Instruction Time	Language Tools	Differentiated Strategies	Materials
Six 45-minute class periods	**Literary Picks** **Vocabulary:** compound, mixture, chemical bond, covalent bond, molecule, chemical formula, ion, ionic bond	**ELL Strategies:** Listed in ELL Ancillary **Learning Strategies:** • Cooperative Learning - Investigation 8A • Active Learning - Motivate: Chemistry and Nonperishable Foods, Extend • Guided Practice - Explain: Counting Atoms **Addressing Misconceptions:** Molecules and Compounds	**Investigation 8A:** Atom Building Game **Motivate - Chemistry and Nonperishable Foods:** boxed food, canned food (full or empty)
Six 45-minute class periods	**Vocabulary:** valence electrons, oxidation number **Closure Cards**	**Learning Strategies:** • Cooperative Learning - Investigation 8B • Discussion - Motivate • Active Research - Extend • Active Learning - Assess: Atoms and the Periodic Table Jeopardy Game	**Investigation 8B:** Periodic Table Tiles **Chapter Activity:** toothpicks, colored gum drops

8.1 Compounds and Chemical Bonds

About the Lesson

This lesson is about chemical bonds. Students learn to identify the number and types of atoms involved in chemical bonds. They also learn to predict the behavior of atoms based on an element's position in the periodic table. Finally, students learn to differentiate covalent and ionic compounds.

Chemistry and Nonperishable Foods

Ask each student to bring one nonperishable food item to school. Have students look at the list of ingredients. Ask students to write the names of all the elements that are found in the list of ingredients. Of course, students will most probably not find the name of any individual elements in the ingredients. Ask students to explain why they think this is so?

Note: Many students may not be able to bring an item from home. In this case, collect the boxes and cans of items you use at home and bring in the empty containers. Students can use these to examine the ingredients which comprise these foods.

Extension: Donate students' nonperishable contributions to a local food bank or a homeless shelter.

Students complete Investigation 8A before reading section 8.1. In this investigation students use the Atom Building Game to learn how atoms form chemical bonds. Before beginning the investigation, ensure that students are able to locate elements on the periodic table, know the location of subatomic particles within an atom, and are able to use the periodic table to determine facts like atomic number and the number of electrons in atoms of an element.

Investigation 8A: Chemical Bonds

Setup	Materials	Investigation Vocabulary
1. Allow one class period to complete the investigation. 2. Students work in small groups of three to five.	• Atom Building Game (2 per group)	compound chemical bond chemical formula

Literary Picks...

Reading Level	Title
3-5	***Mixtures, Compounds & Solutions* by Carol Baldwin** This book is part of the *Materials Matter* series which is designed to teach younger readers science concepts. In this book students learn how different materials come together to make new materials. Filled with photographs and easy to understand facts.
6-8	***Carbon* by Jean Blashfield** *Carbon* is part of the *Sparks for Life* series. It is one of eight featured elements in the series. Provides details about the periodic table and the basic chemistry of the element carbon. Includes discussions on how carbon is obtained and used in our daily lives.
9-12	***Mauve: How One Man Invented a Color that Changed the World* by Simon Garfield** A biography about William Perkins, a young chemistry student who accidentally developed a murky residue that became the first aniline dye. The dye became known as mauve and its discovery lead to advancements in medicine and photography.

Guided Practice: Counting Atoms

Write the following terms on the board: subscript, atom, coefficient, molecule, and compound. Review the meaning of each term. Refer students to the graphic shown below (and page 166 of the student text). Ask students questions like, "How many hydrogen atoms are involved in forming this molecule?" "What information does the subscript tell you?"

Reading a chemical formula

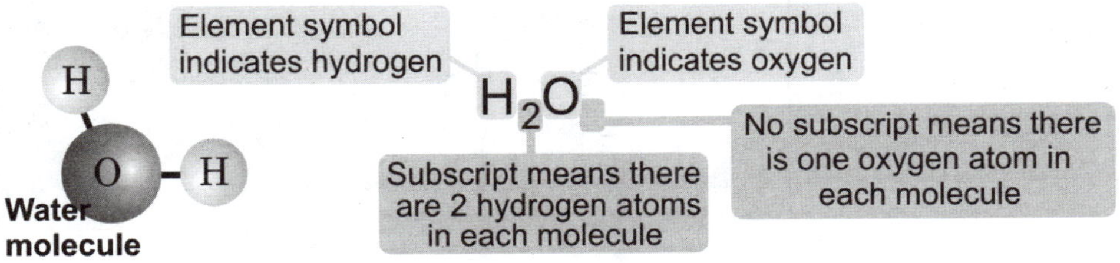

Then have students count and identify the atoms in each of the compounds listed below.

H_2SO_4	2 hydrogen, 1 sulfur, 4 oxygen	C_2H_6	2 carbon, 6 hydrogen
$CaCl_2$	1 calcium, 2 chlorine	Li_2O	2 lithium, 2 oxygen
KOH	1 potassium, 1 oxygen, 1 hydrogen	$KMnO_4$	1 potassium, 1 manganese, 4 oxygen
$AlBr_3$	1 aluminum, 3 bromine	SiO_2	1 silicon, 2 oxygen

1. Use different colors of gumdrops and toothpicks to build compounds. Assign a specific color to each atom. Have students work in pairs. One student builds a gumdrop compound while his partner records how many atoms of each element are present.
2. Make ion flash cards. Write the name and chemical symbol of the element on one side and the ion most likely to form on the other. The most probable ion to form has a noble gas configuration.

Students complete section 8.1 review questions.

Teaching Tip . . .Addressing Misconceptions about Molecules and Compounds

Common misconceptions about molecules and compounds are discussed below.

Misconception #1: Molecules and compounds are the same therefore the terms can be used interchangeably.

Similarities exist between molecules and compounds but they are not the same. Both are pure substances made from atoms that are joined by chemical bonds. However, a compound may be molecular or ionic. An ionic compound is formed when positive and negative ions are attracted to one another and combine to create substances with zero net charge. For example, a positive potassium ion (K^+) combines with a negatively charged chloride ion (Cl^-) and yields potassium chloride (KCl).

A molecular compound (usually called a molecule) is made up of atoms, not ions. Molecules are held together by covalent bonds. An example of this is water (H_2O). The water molecule (shown left) consists of one oxygen atom that is covalently bonded to each of its two hydrogen atoms.

Misconception #2: Molecules are indivisible.

Molecules are comprised of covalently bonded atoms. These atoms may become involved in chemical reactions, which result in atoms being rearranged. Remind students that a chemical bond is not a physical "connector" like a toothpick used to attach gumdrops. Bonds are actually formed by electrons and attractive forces between atoms.

Investigation 8A: Chemical Bonds

By looking at the periodic table you can see that many different elements exist. As a matter of fact, there are more than 100 elements. Think about all the matter you observe daily, like your books, pens and pencils, and the food you had for lunch. Are any of these substances found on the periodic table?

No, they are not.

You are correct, but why do you think this is so?

Students offer suggestions.

Most matter found on Earth is not in the form of elements like iron, copper, and oxygen. Most matter is actually in the form of compounds. Does anyone know what a compound is?

Students should have some prior knowledge in this area. However, this is an effective way for you to pre-assess.

A compound is made up of two or more atoms. These atoms are joined together by a chemical bond.

Write the chemical formula for water (H_2O) on the board.

Water is an example of a compound. Who can identify the atoms that are found in water?

Water is comprised of hydrogen and oxygen.

That is correct. Water is made up of two types of atoms: hydrogen atoms and an oxygen atom. When you see a compound written using the chemical symbols of the atoms that make it up and numbers, this is called a chemical formula. You can see the symbols for hydrogen (H) and oxygen (O) represented by the formula. What else do you observe that is part of the formula?

Students should notice the number two written between H and O.

The number two tells you how many atoms of hydrogen are involved in the chemical bond between hydrogen and oxygen. How many oxygen atoms do you think are involved in this bond?

Students may say, "There is no number shown." Others may guess the answer is one.

There is only one oxygen atom involved in the forming the water. When no number is written after the chemical symbol for an atom involved in a bond, it is understood to be a one.

If students are enrolled in pre-algebra or algebra, they may be familiar with the term coefficient. If so, talk to students about how a variable with no coefficient placed before it is understood to be one.

Today's investigation is about chemical bonds, which hold atoms like hydrogen and oxygen together in compounds.

1 Electrons in an atom

You have already learned that elements are made up of atoms. For instance, the element hydrogen is made of hydrogen atoms. A neutral hydrogen atom has one proton and one electron. Think for a moment only about the electron in a hydrogen atom. Where is it located?

8A Chemical Bonds

How do atoms form chemical bonds with other atoms?

Scientists have identified over 100 elements, but most matter found on Earth is not in the form of pure elements. Most matter is made up of compounds. Even if an element is pure at first, it will combine with other elements sooner or later. For example, if you leave an iron nail out in the rain, it will rust. The iron in the nail will combine with oxygen. It will form a compound called iron oxide, also known as rust. In this investigation you will learn how electrons are involved in forming chemical bonds.

Materials
- 2 atom building games

1 Electrons in an atom

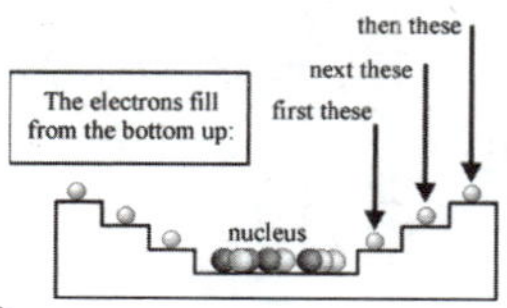

- A neutral atom has the same number of electrons and protons.
- The electrons move around in energy levels that surround the nucleus.
- Since electrons are attracted to the nucleus, they fill the levels closest to the nucleus first.
- Once a lower level is full, electrons start to fill the next level.

2 How many electrons are in the outermost level?

Use the atom board to make a model of each element in the table. For each element, write the number of electrons found in the outermost level. Then write the number of unoccupied or empty spaces in the outermost level.

Table I: Electrons in the Outermost Levels of Atoms

Element	Atomic number	Electrons in outermost level	Unoccupied spaces in outermost level
hydrogen			
helium			
lithium			
fluorine			
neon			
sodium			
chlorine			
oxygen			
carbon			

a. What do lithium and sodium have in common?

b. What do fluorine and chlorine have in common?

c. What do neon and helium have in common?

36

Students have already learned that electrons are in energy levels. Remind students of the "To the Bleachers" activity from chapter six (teacher's guide) in order to review how electrons fill orbitals.

2 How many electrons are in the outermost level?

In chapter seven we discussed periodicity regarding the arrangement of the periodic table. As a result of an element's position on the periodic table, we can make many predictions about how atoms will behave. For example, you know that the noble gases are practically inert because they are in group 18. What is it about elements in group 18 that make them nonreactive?

These elements have a filled outer energy level.

Yes they do. Remember that periodic means to "make your way around" or repeat a pattern. Take out your copy of the periodic table. Start with hydrogen and tell me how many electrons it has in its outer energy level?

Hydrogen has one. Have students place their finger on each element in order and continue to ask the same question. Go all the way to neon and then stop.

Do you notice a pattern? What do you think the pattern is? Make your way around to period three. Continue all the way until you get to the end of the period.

Some observations may include that the number of valence (outer) electrons increases by one across a period. The number of valence electrons corresponds to the group number. For example, group one elements have one valence electron and group 18 elements have eight. Students complete the data table in part two and answer questions a-c.

3 Modeling a chemical bond

Why do you think atoms form chemical bonds? You will find out in this part of the investigation.

Have a student volunteer read each bulleted point aloud. Discuss each point one at a time.

Use two atom boards to build models of chlorine and sodium. Look at the energy level (with electrons) that is farthest away from the nucleus. You can see that sodium has only one electron in the outermost energy level? How many are in the outermost level of chlorine?

Chlorine has seven valence electrons.

The chlorine atom only needs one more electron to fill its outermost energy level. If the sodium atom transfers its lone outermost electron to the chlorine atom, then the chlorine atom has a filled outer energy level. This is when an atom is at its lowest energy. What happens to the sodium atom as it transfers its electron to the chlorine atom?

The sodium atom has no electrons in its third energy level. It is also at its lowest energy.

As sodium transfers an electron to chlorine a bond is created. The new substance formed is a compound called sodium chloride. Do you think you have ever seen sodium chloride before? If so, where did you see it? What are some of its physical properties?

Have some table salt handy to demonstrate the result of this reaction between sodium and chlorine. Guide students as they complete the remaining questions in part three. Use the atom boards to build additional compounds.

Teaching Tip . . .Why bond?

Students often inquire about why atoms bond. They may ask questions like, "Do they have to bond?" or "What is the benefit of forming chemical bonds?" The goal of chemical bonding is to achieve a noble gas configuration, or to fill the outer main energy level with electrons. The result is that an atom is at its lowest energy.

Consider the ionic bond formed as sodium chloride is created. Sodium has one valence electron. Chlorine has seven valence electrons. Chlorine is highly electronegative and is capable of easily attracting the one valence electron from sodium. As the electron from the sodium atom is transferred to the chlorine atom, two ions are created: the positively charged sodium ion and the negative chloride ion. Both of these ions have the same electron configuration as neon.

How do the energies of the sodium and chlorine atoms compare to the ions formed? The ions have much lower energy than the atoms form which they were formed. As a result they are more stable. Remind students that substances in nature "prefer" to exist at the lowest possible potential energy. Chemical bonding helps most atoms achieve this goal.

Investigation 8A: Data and Answers

1

2

Element	Atomic number	Electrons in outermost level	Unoccupied spaces in outermost level
hydrogen	1	1	1
helium	2	2	0
lithium	3	1	7
fluorine	9	7	1
neon	10	8	0
sodium	11	1	7
chlorine	17	7	1
argon	18	8	0
potassium	19	1	7

a. They both have 1 electron in their outermost levels. They also have 7 unoccupied spaces in their outermost levels.

b. They both have 7 electrons in their outermost levels and 1 unoccupied space in their outermost levels.

c. They both have 8 electrons in their outermost levels and no unoccupied spaces in their outermost levels.

3

a. Sodium will tend to lose its only valence electron. It is easier to lose 1 electron than to gain 7. When sodium loses its valence electron, it will become more stable because its outermost level will be full.

b. Chlorine has 7 valence electrons. By gaining 1 electron, chlorine will have 8 valence electrons and will become more stable.

c. When sodium loses 1 electron, it will become positively charged, because it will have more positively-charged protons than negatively-charged electrons. When chlorine gains 1 electron, it will become negatively charged, because it will have more negatively-charged electrons than positively-charged protons. Sodium and chlorine will bond together because opposite electrical charges cause an attraction.

d. Two hydrogen atoms can bond with one oxygen atom. This is called water.

e. Carbon atoms can form four bonds.

Notes:

8.2 Electrons and Chemical Bonds

About the Lesson

In this lesson students learn more about the role of valence electrons in chemical bonding. Students continue to use the periodic table to predict the types of bonds created by atoms and other properties of elements such as oxidation numbers and electron dot structures.

Ask students, "how do you relieve a headache?" Taking aspirin is a common response. Tell students that aspirin is made from three different elements: carbon, hydrogen, and oxygen. Then ask students, "what is your favorite type of candy?" Accept all reasonable student responses. Tell students sugar is one main ingredient in candy and that it is also made from carbon, hydrogen, and oxygen. Display a piece of candy and an aspirin. Ask students to suggest why these two substances have different properties if they are made from the same elements.

Students complete Investigation 8B before reading section 8.2. In this investigation students use the Periodic Table Tiles to learn how atoms combine to create molecules. Students also learn to use the periodic table to predict how atoms in molecules will combine.

Investigation 8B: Molecules

Setup	Materials	Investigation Vocabulary
1. Allow one class period to complete the investigation. 2. Students work in groups of three to five.	• Periodic Table Tiles	oxidation number

Predicting Chemical Bonds

Ionic bonds commonly form between metals and nonmetals; while bonds between two nonmetals tend to be covalent. While this is not true in every instance, it is a simple rule of thumb that students may apply at this grade level. Have students use this rule when predicting which type of bonds are most probable between atoms.

Covalent Bonding

Covalent bonds form when atoms share electrons. If the atoms are equally attracted to the shared electrons then a *nonpolar covalent bond* is formed. Nonpolar covalent bonds are evident in diatomic molecules, like H_2 and O_2.

Polar covalent bonds

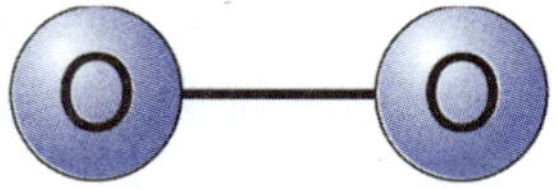

Each oxygen atom has equal attraction for the shared electrons

The term *polar* is often used to describe persons who are on opposing sides of a discussion or those with completely different preferences. This is a useful analogy when thinking about polar covalent bonds, which join atoms that are not equally attracted to a shared electron pair. Polar covalent bonds are common when atoms of very different electronegativity combine. However, the electronegativity difference is not great enough to create an ionic bond. One example is the bond formed between hydrogen and sulfur atoms.

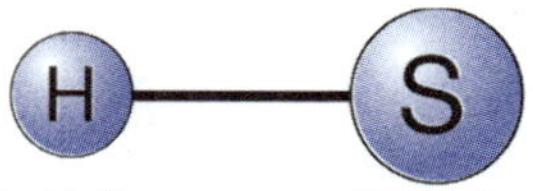

Partially positive   Partially negative

The sulfur atom has a greater attraction for electrons than the hydrogen atom, which means more electrons are closer to the sulfur side of the bond than to the hydrogen side. Since there are more electrons nearer the sulfur atom, it has a partially negative charge and the hydrogen atom (with fewer electrons) is partially positive.

1. Take your students to a local college or university to learn how chemists use computer programs to draw and study models of compounds and molecules. Or, you may invite a chemistry professor to your classroom to show students how to build models. Ask the professor to bring ball-and-stick model kits.
2. Research the work of Linus Pauling, a Nobel Prize winner for his work involving chemical bonding.

Students complete section 8.2 review questions. Then divide students into teams and play the *Atoms and the Periodic Table Jeopardy* game using the clues below. Answers to each clue are provided at right.

Atoms and the Periodic Table Jeopardy Game

	The Periodic Table	Elements	The Atom	Chemical Bonds	Write the Dot Structure*	Potpourri
$	Russian chemist known as the "Father of the Periodic Table"	The lightest element	The positively charged particle in an atom	These electrons are involved in chemical bonds	Magnesium	This is formed when an atom gains or loses electrons
$$	The rows on the periodic table	What elements are made up of	Protons and neutrons are located here	Bond formed when electrons are transferred between atoms	Helium	Metals are good conductors; nonmetals are called this
$$$	How elements on the periodic table are arranged	The only metal that is a liquid at room temperature	Negatively charged particle	Type of bond formed when atoms share electrons	Phosphorus	Carbon-12 and Carbon-13 are examples of this
$$$$	Average mass of all isotopes of an element	This group of elements are the least reactive	An atom at its lowest energy is in this state	Type of bond formed between Ca and F	Aluminum	Pattern of colors produced by an element
$$$$$	The columns on the periodic table	These elements have properties of metals and nonmetals	An atom with an unstable nucleus is called this	A group of atoms joined by covalent bonds	Oxygen	A solid mixture of metals

*Students give the name of the element and write the correct dot structure on paper. Each group shows their structure at the same time.

Atoms and the Periodic Table Jeopardy Answers

The Periodic Table 1. Mendeleev 2. periods 3. increasing atomic number 4. atomic weight 5. groups	**Elements** 1. hydrogen 2. atoms 3. mercury 4. noble gases 5. semimetals or metalloids
The Atom 1. proton 2. in the nucleus 3. electron 4. ground state 5. radioactive	**Chemical Bonds** 1. valence electrons 2. ionic 3. covalent 4. ionic 5. molecule
Write the Dot Structure (correct number of electrons) 1. 2 (unpaired) 2. 2 (paired) 3. 5 4. 3 5. 6	**Potpourri** 1. ion 2. insulators 3. isotopes 4. spectrum 5. alloy

Investigation 8B: Molecules

Complete motivational activity before beginning the investigation.

Think about the types of atoms found in both the candy and aspirin. Are the physical properties of aspirin and candy the same?

No they are not.

What about the chemical properties? Do you think taking candy will relieve your headache?

Most students will agree that candy is not a pain reliever.

When you want a tasty snack, do you eat an aspirin?

Most students will agree that aspirin is not what comes to mind when they think of a snack.

Both sugar and aspirin are molecules made from the same elements. However, the atoms of carbon, hydrogen, and oxygen are arranged differently in each. For this reason, each substance has different chemical properties. Think for a moment about these three elements. What do they have in common.

The targeted response is that these are all nonmetals.

Yes they are. Who can tell me what type of chemical bond commonly joins atoms of nonmetals?

Nonmetals are commonly joined by covalent bonds.

The bonds that join atoms in aspirin and sugar are covalent. A group of atoms held together by covalent bonds is called a molecule.

Write the term *molecule* and its meaning on the board.

Molecules are the subject of today's investigation. The properties of a molecule depend greatly on *how* the atoms in a material are bonded to one another rather than simply identifying the atoms involved in the bonding.

1 Electrons and oxidation numbers

You already know that molecules have covalently bonded atoms. Which part of the atom is involved in covalent bonding?

Valence electrons are involved in chemical bonding.

Molecules are formed as electrons are shared between atoms. Water is a molecule formed by covalent bonds between one oxygen atom and two hydrogen atoms. How many valence electrons does each hydrogen atom have?

Hydrogen has one valence electron.

Good answer. The oxygen atom has six valence electrons which can be represented by an electron dot structure. The electron dot structure shows the number of electrons that are in the outermost main energy level for each atom.

Draw an electron dot structure for an oxygen atom on the board. Have students refer to their periodic tables as you explain why oxygen has six valence electrons.

Look at the electron dot structure for oxygen. Where do you think the hydrogen atoms will bond to the oxygen atom?

8B Molecules and Compounds

What are some molecules and compounds and what atoms are in them?

The properties of a material depend much more on the molecule than on the elements the molecule is made of. For example, aspirin is made from carbon, hydrogen, and oxygen. By themselves, these elements do not have the property of reducing pain. Other molecules formed from the same elements have different properties than aspirin. For example, polyethylene plastic wrap and formaldehyde (a toxic preservative) are also made from carbon, oxygen, and hydrogen. The beneficial properties of aspirin come from the way the atoms bond together in the particular shape of the aspirin molecule.

Materials
- Periodic table tiles

1 Electrons and oxidation numbers

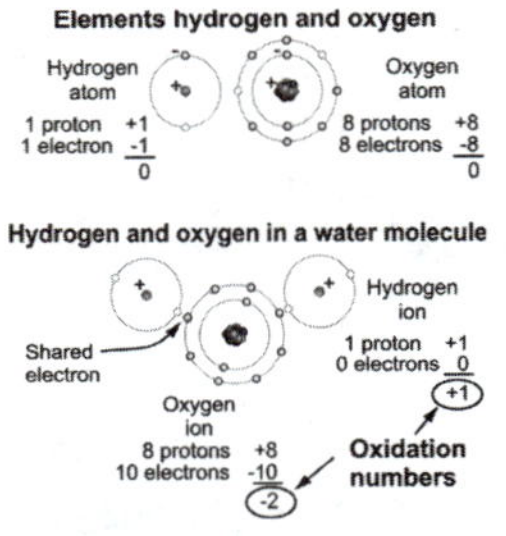

- A molecule (molecular compound) is formed when two or more atoms □□r□ electrons by forming covalent chemical bonds.
- An atom with equal numbers of protons and electrons is neutral (charge of 0). An atom with more protons or electrons is called an ion.
- An ionic compound is formed when electrons are □□□f□rr□□ from one atom to another.
- A positive oxidation number means the element forms chemical bonds to LOSE electrons. Atoms of sodium form chemical bonds that take away an electron leaving a positive charge of +1. Chemists say sodium has an oxidation number of +1 because atoms of sodium lose an electron when forming chemical bonds.
- A negative oxidation number means the element forms chemical bonds to GAIN electrons. Atoms of chlorine form chemical bonds that add an electron, making a negative charge of -1. Chemists say chlorine has an oxidation number of -1 because chlorine atoms gain an electron when forming chemical bonds.

38

Students offer suggestions. They should observe that the oxygen atom has two unpaired electrons. Each hydrogen has one electron and will bond to one of the unpaired electrons on the oxygen atom.

As each hydrogen atom shared its electron with the oxygen atom, a stable molecule (water) is created. By sharing the electrons, oxygen is now surrounded by eight electrons and each hydrogen atom has two. The stability of each atom is equal to the stability of a noble gas. Recall that atoms are most stable when they are at the lowest possible energy.

Model the covalent bonding between these atoms.

You have learned that the number of electrons involved in chemical bonding is linked to its position on the periodic table. The *oxidation number* of an atom is also related to its periodic properties. For example, hydrogen may have an oxidation number of +1 or -1. This means that hydrogen may accept one electron or donate one electron in order to form chemical bonds.

Read aloud the bulleted points in part one. Use the examples to define the oxidation number.

2 Oxidation numbers on the periodic table

The oxidation number of an element predicts how atoms will combine to form molecules and compounds.

Discuss the examples shown.

3 Making a molecule

This part of the investigation is a game. You will use the periodic table tiles to make molecules. The atomic number of each element is equal to its points. For example, each carbon atom used will give you six points. The goal is to make molecules that will give you the most points. Suppose you made methane, a molecule with one carbon atom and four hydrogen atoms. How many points would you earn?

Methane (CH_4) yields ten points. Have a student volunteer read the rules of the game aloud.

4 Rules for building a molecule

The oxidation numbers represent gaining or losing electrons between each atom in the molecule. Since every electron gained by one atom must be lost by another, the total molecule has a charge equal to zero.

Students play the game. Walk around and verify that the molecules students build have zero net charge.

5 Stop and think

Can you apply the oxidation numbers of atoms to determine a chemical formula? Consider a molecule made from hydrogen and nitrogen. What chemical formula might this molecule have?

Students should determine the formula to be H_3N. See the diagram at right for ideas to reinforce oxidation numbers and naming compounds. Discuss the answers to both questions before concluding the investigation. Try additional molecules for extra practice.

Using Oxidation Numbers to Write Chemical Formulas

Step 1: Write the atoms with the proper oxidation number.

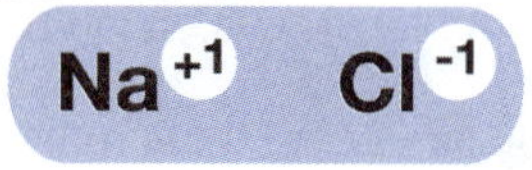

Step 2: The charge of the first atom becomes the subscript of the second.

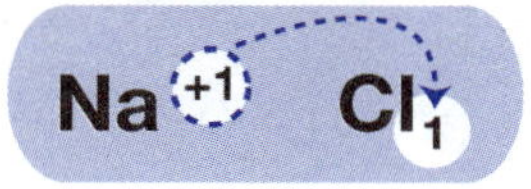

Step 3: Cross the charge of the second atom to determine the subscript of the first.

Step 4: Then write the formula.

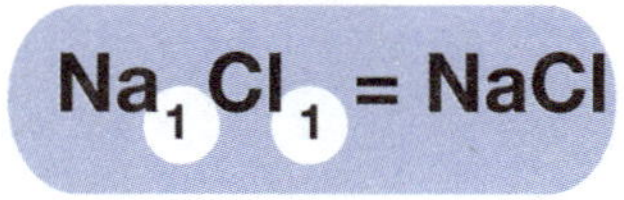

A second example:

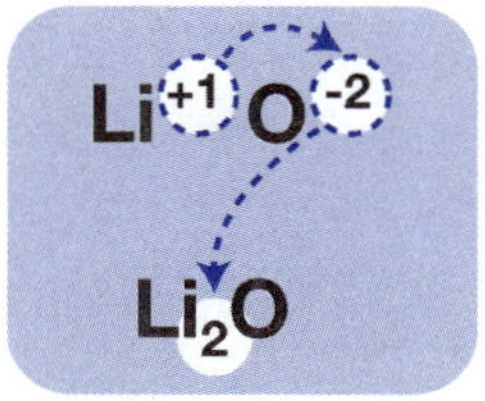

Investigation 8B: Data and Answers

1 Doing the experiment

2 Thinking about what you observed

a. Three hydrogens can bond with one nitrogen. This is called ammonia.

b. There are two iron atoms and three oxygen atoms, Fe_2O_3.

Notes:

Chapter 8 Answers

8.1 Section Review

1. A compound contains two or more elements that are chemically bonded together. A mixture contains two or more elements or compounds that are not chemically bonded together.
2. Compounds: salt, sugar, gasoline, plastic
   Mixtures: seawater, Earth's atmosphere, lemonade
3. 4
4. top left; carbon in center and 4 chlorine atoms around it.
5. Fe_3O_2
6. $C_3H_7O_2N$ the amino acid, alanine
7. 4
8. third from the left
9. a
10. b

8.2 Section Review

1. b
2. bottom left with three electrons in outer level; boron
3. c
4. Correct answer includes any two of the following:
   nitrogen; phosphorus; arsenic; antimony; bismuth
5. b
6. Correct answer includes any three of the following:
   boron; aluminum; gallium; indium; thallium
7. -1
8. a

Connection

1. Salt can be mined from deep below Earth's surface or produced through an evaporation process.
2. Salt was as good as money and used in trading just like gold. Salt was used to keep food edible for long periods of time during early civil at ion.
3. Sodium is on the left side of the periodic table, is a metal, and is soft, light and silvery-white in color. sodium is very reactive and never found in pure form in nature. Chlorine is on the right side of the periodic table, is a halogen, and is found in nature as a gas. Chlorine is a yellowish green, poisonous halogen.
4. The crystal structure of sodium chloride is cubic. Each sodium ion is surrounded by six chlorine ions. Each chlorine ion is surrounded by six sodium ions.

Activity

Applying your knowledge: sample answers

a. Sample sketches of molecules

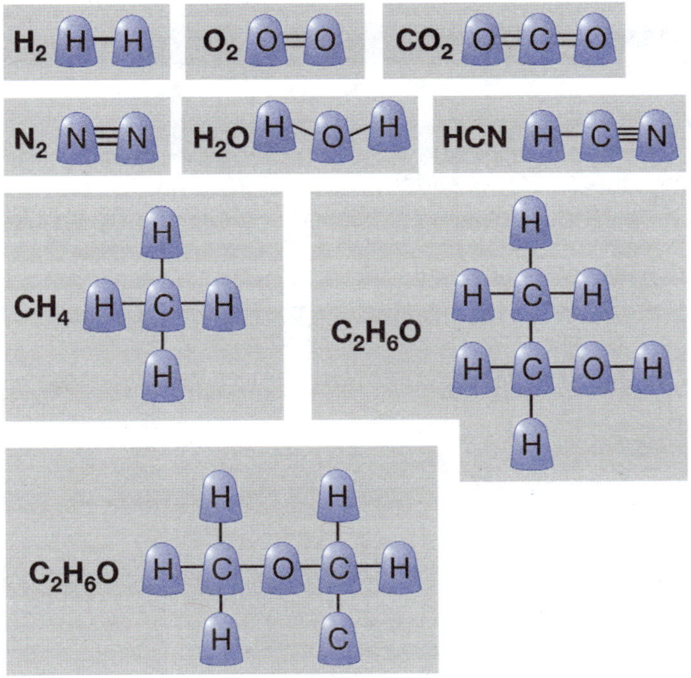

b. There is a single bond in H_2 so two electrons are being shared between the two atoms.

c. There is a double bond in O_2 so four electrons are being shared between the two atoms.

d. Because the bonds, which are shared pairs of electrons, are negatively charged. The negative charged bonds repel each other, pushing the bonds as far apart from other bonds as possible. This will evenly space bonds, and therefore the atoms attached to them, around the molecule.

e. Isomers have different characteristics because physical and chemical characteristics are dependent upon the structure of the molecule, and isomers have different structures. Even though they are made of the same exact atoms, those atoms are attached in a different order and shape. Molecules behave according to their shape, thus different shaped molecules will behave differently.

f. $C_2H_4O_2$; 8 toothpicks, 2 red gumdrops, 2 green gumdrops, 4 green gumdrops.

Chapter 8 Assessment

Vocabulary

Section 8.1

1. chemical bond
2. ion
3. chemical formula
4. compound
5. covalent
6. mixture
7. ionic bond
8. molecule

Section 8.2

9. valence electrons
10. oxidation number

Concepts

Section 8.1

1. H_2O; hydrogen and oxygen
2. The elements in a compound are chemically bonded. A mixture may contain elements or compounds but the particles are not bonded together.
3. Answers will vary but may include:
   Compound: salt, sugar, ammonia
   Mixture: salt water, air, bowl of cereal and milk
4. Atoms form compounds because the atoms have lower energy and are more stable when they are in molecules.
5. covalent
6. molecule
7. Answers will vary but may include water, ammonia, oxygen, methane, and nitrogen gas.
8. There are two hydrogen atoms in the molecule; twice as many as oxygen atoms.
9. There are 2 atoms of carbon and 6 atoms of hydrogen.
10. The chemical formula and the structure of the molecule are most important in determining the function of a compound.
11. Covalent compounds form molecules by sharing electrons. Ionic compounds do not form molecules, but form bonds between all neighboring ions.
12. Energy is released because chemically bonded atoms have less total energy than free atoms.
13. noble gases
14. Metals: alkali (one electron away from the noble gases)
    Nonmetals: halogens (one electron away from noble gases)

Section 8.2

15. Only electrons in the highest unfilled energy levels are involved.
16. 8
17. noble gases already have 8 valence electrons and are stable without bonding.
18. Hydrogen only has one electron, so hydrogen tries to bond to get either 0 or 2 valence electrons.
19. one more valence electron
20. valence electrons
21. 6; 2
22. A negative oxidation number indicates an electron will be gained; a positive oxidation number indicates an electron will be lost.
23. Answers are:
    a. 2+
    b. 3+
    c. 1-

24. zero
25. ionic
26. ionic
27. covalent
28. left; right

Problems

Section 8.1

1. Answers are:
   a. separate elements
   b. mixture
   c. compound
2. Answers are:
   a. 6 carbon, 12 hydrogen, 6 oxygen
   b. 1 calcium, 1 carbon, 3 oxygen
   c. 2 aluminum, 3 oxygen
3. CO_2 carbon dioxide
4. c
5. Answers are:
   a. Li
   b. Ca
   c. S
   d. O

Section 8.2

6. 3
7. Answers are:

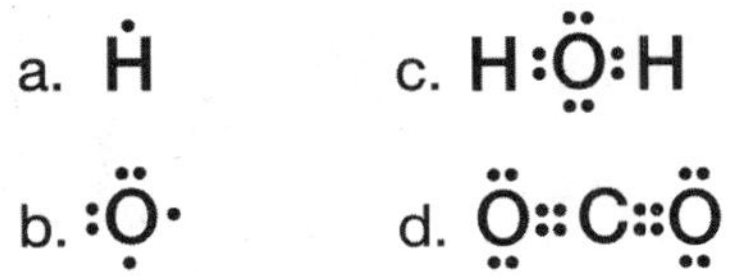

8. Answers are:
   a. Ca^{2+} Cl^{1-}
   b. $CaCl_2$
9. Answers are:
   a. O^{2-}; gains 2 electrons
   b. B^{3+}; loses 3 electrons
   c. Li^{1+}; loses 1 electron
   d. Na^{1+}; loses 1 electron
   e. Mg^{2+}; loses 2 electrons
   f. Al^{3+}; loses 3 electrons
   g. C^{4+}; loses 4 electrons
   h. I^{1-}; gains 1 electron
10. Answers are:
   a. covalent
   b. ionic
   c. covalent
   d. covalent
11. Answers are:
   a. zero
   b. CO_2
   c. covalent
12. Answers are:
   a. zero
   b. CH_4
   c. covalent
13. Answers are:
   a. 6
   b. 12
   c. 6

Chapter 9: Acids, Bases, and Solutions

Learning Goals	California Content Standards	Activities and Resources	Assessment
9.1: Water and Solutions The student will: • Distinguish solute and solvent. • Identify examples of solutions. • Explain the meaning of solubility. • Discuss factors which affect solubility in gases and liquids. • Describe the properties of water that make it a good solvent.	PS 5e	**Laboratory Investigation 9A:** Solubility **Teaching Illustrations:** Solid Dissolving, Water: A Polar Molecule, Supersaturated and Unsaturated Sugar Graph	**Exam View Test Bank** Chapter Pre-Test **Section 9.1 Review**
9.2: Acids, Bases, and pH The student will: • Differentiate acids and bases. • Define pH. • Explain the relevance of acids, bases, and pH to living organisms and the environment.	PS 5a PS 5e	**Laboratory Investigation 9B:** Acids, Bases and pH **Teaching Illustrations:** The pH Scale, pH Ranges in Nature, Neutralization, pH Scale with Acids and Bases **Graphic Organizer:** Acids vs. Bases **Skill and Practice Sheets:** Calculating Concentration **Connection:** Keeping Your (pH) Balance **Chapter Activity:** Acid Rain and Stone Structure	**Section 9.2 Review** **Chapter Assessment** **Exam View Test Bank**

Estimated Instruction Time	Language Tools	Differentiated Strategies	Materials
Five 45-minute class periods	**Literary Picks** **Vocabulary:** solution, alloy, solvent, solute, dissolve, solubility, insoluble, saturated, equilibrium, supersaturated, polar **Word Origins:** saturate	**ELL Strategies:** Listed in ELL Ancillary **Learning Strategies:** • Cooperative Learning - Investigation 9A • Active Learning - Motivate: Milky Matter • Demonstration - Like Dissolves Like • Active Research - Extend **Addressing Misconceptions:** Are Solutions Pure Substances	**Investigation 9A:** graduated cylinder, balance, granulated sugar, hot and cold water, 100-mL beakers, sugar cubes, timer, foam cups, stirring rods, red and blue markers **Motivate - Milky Matter:** glass of milk, microscope, slide **Demonstration:** two large beakers labeled A and B, two foam cups, water, acetone, starch packing peanuts
Five 45-minute class periods	**Vocabulary:** concentration, acid, base, pH, pH scale **Word Origins:** dissociate **Closure Cards**	**Learning Strategies:** • Cooperative Learning - Investigation 9B • Discussion - Motivate • Paired Learning - Explain: Critical Thinking • Active Learning - Extend **Teaching Tip:** Neutralization	**Investigation 9B:** spot plates, pipettes, red cabbage juice, red and blue litmus paper, solutions (See Teachers Guide Setup for preparation), white paper **Motivate:** lemon or lemon juice, vinegar, empty container of ammonia, baking soda **Chapter Activity:** chalk and/or marble chips, beakers, granite chips, white vinegar, Universal Indicator

9.1 Water and Solutions

About the Lesson

Many examples of matter exist as solutions. In this lesson students learn to identify the parts of a solution (solute and solvent) and explore the properties of solid, liquid, and gaseous solutions. Students also learn about the factors that influence solubility and discover properties of water that make it such an effective solvent.

Milky Matter

Pour milk into a glass and place it in an area that is visible to all students. Give students one minute to list all the qualitative descriptions they can think of. Then have students share their list of descriptions. Ask students if they think milk is a pure substance.

Set up a few microscopes and have groups of students prepare a slide with a drop of milk. Ask students to describe how the milk appears when viewed under the microscope. Ask students if they still believe that milk is a pure substance. Explain to students that milk is a homogenized solution and therefore appears to have the same properties throughout when viewed with the unaided eye.

Students complete Investigation 9A before reading section 9.1. In this investigation students learn about solutions and solubility. Students identify the properties of solutions and explore the relationship between temperature and dissolving rate.

Investigation 9A: Solubility

Setup	Materials	Investigation Vocabulary
1. Allow one class period to complete the investigation. 2. Students work in small groups of three to five. 3. Use a hot plate to maintain the water temperature throughout the investigation. 4. Obtain ice from the school cafeteria. 5. Gather sugar cubes (2 per group) for part three of the investigation.	• graduated cylinder • balance • granulated sugar • hot and cold water • 100 mL beakers • sugar cubes • timer • foam cups • stirring rod • red and blue markers	solubility homogeneous solution solute solvent

Literary Picks...

Reading Level	Title
3-5	***Environmental Awareness: Acid Rain* by Mary Ellen Snodgrass** This book is part of the *Environmental Awareness* series. It provides an excellent overview of the effects of acid rain and discusses possible solutions for reducing the damage.
6-8	***Acids and Bases* by Chris Oxlade** Part of the *Chemicals in Action* series, this book describes the properties and uses of acids and bases. Chapter topics include natural acids, testing, pH scale, reactivity, elements, and the periodic table. Also includes several experiments.
9-12	***Acid Rain, Acid Snow* by John Slade** Are you interested in global pollution and how you can help? Then this John Slade book is for you. It is written in layman's terms and provides a look at acid rain, its origin, cause and effects and offers positives options for change.

Addressing Misconceptions: Are Solutions Pure Substances?

As students learn about homogeneous mixtures, they often mistakenly associate the uniformity of the solution with the creation of new substances (chemical changes). For example, students may assume that sugar dissolved in water is evidence of a chemical change because they can no longer see the sugar in its crystalline form. Explain to students that solutions are mixtures, not pure substances, and that each component of a mixture can be recovered.

Try this activity. Prepare test tubes with a mixture of sand, iron filings, water, and food coloring. Do not tell students what is in each test tube, only that there are four components of the mixture. Have students work in small groups to make observations about the materials in the test tube and to discuss possible methods of separating the mixture. Allow class time for each group to test the proposed method(s) of recovering each component of the mixture.

One method for separating the mixture is to pour off the liquid into an evaporating dish. Water and food coloring can be separated by evaporating the water, which has a lower boiling point than food coloring. Students should observe the food coloring remains in the bottom of the dish. Heat the test tube over a flame to remove moisture and pour the sand and iron filings onto a watch glass. Use a magnet to separate the sand and iron filings.

1. Put a spin on the "like dissolves like" demonstration. Create a mixture of starch and foam packing peanuts. Add a starch peanut to the water and have students note their observations. Then add a foam peanut to the water. Of course the foam peanut does not dissolve. Alternate peanuts until students are able to deduce that the peanuts must have different properties.
2. Lean more about Auguste Gaulin and his invention of the "homogenizer."

Students complete section 9.1 review questions.

Teacher Demonstration: "Like Dissolves Like"

Try this quick demonstration to help students infer how polarity affects solubility. Label two large beakers, A and B. Place an empty foam cup into each beaker. Show students that each cup is empty and identical before placing it in the beaker. Then say to students, "I will add some liquid to each cup and I want you to observe what happens." Show students two bottles: one which contains water, the other with acetone. Fill the foam cup in beaker A with water. Wait a few seconds, then fill the cup inside beaker B with acetone. Students should observe that the cup in beaker B breaks down.

Disclose the identity of each solvent to students. Ask students to suggest why water does not dissolve the foam cup while acetone does. Students already know that water is a polar molecule. If the rule of "like dissolves like" is true, then foam must be nonpolar. As a result, students should infer that acetone must also be nonpolar because it dissolved the foam cup.

Disappearing peanuts?

Remove the foam cup with water from beaker A. Be sure the beaker is dry. Display a container of starch packing peanuts before students. Hold up one peanut and ask students to guess how many peanuts can fit into the beaker. Fill the beaker with peanuts, counting aloud as you add each peanut. Once the beaker is full ask students, "do you think I can add any more peanuts to the beaker?"

Slowly pour water over the peanuts. What happens? Ask students, "are the packing peanuts polar or nonpolar?" What is the evidence? Students should infer that starch must have similar properties to water. Ask students, "are the peanuts actually disappearing?" Be sure that students understand that dissolving does not mean that matter disappears.

Investigation 9A: Solubility

The homogenized milk you drink with breakfast or have with your morning cereal is an example of a solution. Can you think of other examples of solutions?

Students offer suggestions. Go through the list of proposed solutions and discuss whether each is an example or a nonexample of a solution.

A solution is a homogeneous mixture. The word homogeneous means to be of the same kind. Therefore a homogenous mixture is a combination of matter that appears to be the same throughout.

Write the term homogeneous on the board. Underline the prefix *homo-* and write the word same beneath it.

Consider a bottle of soda or carbonated beverage. Do you see the individual components in the beverage or does it all appear the same?

It appears the same throughout.

Carbonated beverages are solutions made by combining sugars, water, carbon dioxide gas, and flavorings. Water that you drink from the tap is also a solution. In addition to hydrogen and oxygen, water may also contain dissolved minerals and flavorings. In today's investigation, you will learn more about solutions.

1 Mass and volume in solutions

In part one you will determine the density of sugar, water, and a solution of sugar and water. Which of these do you think is the most dense?

Students speculate about the density of each. Have students write a hypothesis which reflects their prediction.

Let's test your prediction. Place a clean graduated cylinder on the balance and reset it to zero. Then add 80 mL of hot water into the graduated cylinder. Measure its mass in grams and record your answer in Table 1. Pour this water into a foam cup.

Have students label the cup. Remind students to be careful when handling the hot water.

Add sugar to the 40 mL mark of the same graduated cylinder you used to measure the water. Place the cylinder back on to the balance and record its mass. What is the volume of the sugar?

Students added sugar to the 40 mL mark so the volume is also 40 mL. This data is prerecorded in Table 1.

Remove the graduated cylinder with the sugar added from the balance. Pour the hot water from the foam cup into the cylinder. What do you think will happen to the sugar in the cylinder?

The hot water will dissolve the sugar and create a sugar water solution.

Place the cylinder with sugar and water added on the balance. Record the mass and volume of the solution. How does the volume of the solution compare to the volume of the water alone?

Students record the data in Table 1. Then use the density formula to calculate the density of each substance. You may need to remind students of the density formula. Ask a student volunteer to write the formula on the board and to label each variable.

9A Solubility

What does it mean to dissolve?

The water you drink is not pure water. There are many things dissolved in drinking water, such as minerals, oxygen, and sometimes flavorings. A solution is a homogeneous mixture of different compounds, such as water and sugar. Carbonated soda is a solution containing water, sugar, dissolved carbon dioxide gas, and flavorings. Solutions are essential for life because your body's chemical processes happen in solutions. This investigation will look at some of the properties of solutions.

Materials
- Graduated cylinder
- Balance
- Granulated sugar
- hot water
- cold water
- Sugar cubes
- Timer

1 Mass and volume in solutions

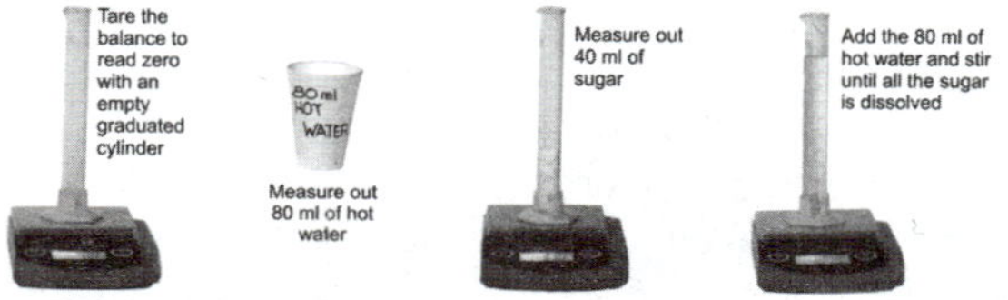

1. Put a clean graduated cylinder on the balance and reset it to zero.
2. Place 80 ml of hot water into the graduated cylinder, measure the mass, then pour the water into a foam cup (you will use this water in step 4).
3. Put 40 ml of sugar into the same cylinder you used for the water. Place the graduated cylinder back on the balance and record the mass and volume of the sugar (you zeroed out the cylinder in step 1, so the mass reading in this step will just be of the sugar).
4. Remove the graduated cylinder. Add the 80 ml of hot water and stir to dissolve all the sugar. Put the graduated cylinder back on the balance and record the mass and volume of the solution.
5. Use your mass and volume data to calculate the density of the sugar, the water, and the sugar and water solution.

Table 1: Mass and volume data for solutions

Substance	Mass in grams	Volume in mL	Density in g/mL
Sugar		40	
Water		80	
Solution of sugar and water			

2 Stop and think

As you prepared the sugar solution, one substance was dissolved into another. Which substance was dissolved?

The sugar was dissolved.

Which substance did the dissolving?

The water dissolved the sugar.

Very good. A solution contains two parts: a solute and a solvent. The solute is the substance that is dissolved and the solvent does the dissolving. In your solution, the sugar is the solute and the water is the solvent.

Write both terms and their meanings on the board.

Whether or not one substance is able to dissolve in another is a measure of its solubility. For example, sugar is able to dissolve in water. Therefore, you can say that sugar is soluble in water. Use your observations from part one to help you answer the questions in part two.

Students answer questions in part two. Discuss the answers to each question before moving on to part three.

3 Temperature and dissolving rate

What are some factors that may affect the rate at which a substance dissolves?

Students answers will vary. Sample responses may include temperature, particle size, stirring or shaking action, and pressure.

Those are all good responses. In this part of the investigation, you will test the effect of temperature on dissolving rate. Each group needs two sugar cubes. Use your red marker to color one face of a sugar cube. Place the sugar cube colored side up in an empty 100 mL beaker. Color one face of a second sugar cube with the blue marker. Place it colored side up into a second empty beaker.

Students prepare each beaker.

Choose one person in your group to act as the time keeper. Two other persons are responsible for pouring water along the sides of each beaker until it covers the sugar cube by about one centimeter. Start adding water to each beaker at the same time. Do not pour the hot or cold water directly on the sugar cubes. Watch the colored face of each cube. Which one dissolves faster?

Students should observe that the hot water dissolves the sugar cube faster.

4 Stop and think

Do you think there is a relationship between temperature and dissolving rate. Consider your results as you answer the question.

Students results should confirm that substances generally tend to dissolve faster when the solvent temperature is increased. Students complete and discuss answers to questions a and b in part four.

Word Origins...saturate (from Latin *satur* meaning "satiated")

Ask students to describe how they feel after eating too much of their favorite food. Common responses include words like "stuffed" or "full." Tell students this feeling of fullness or satisfaction means to be satiated. Use the word in a sentence. For example, "Juanita was satiated after eating an entire pizza."

Then ask students to use a dictionary to compare the meanings of the words *satiate* and *saturate*. Students should observe that the meanings are very similar. Have students explain the meaning of a saturated solution in their own words.

Ask students, "what do you think an unsaturated solution is?" Point out to students that the prefix *un-* means "not", therefore an unsaturated solution does not contain more solute than the solvent can dissolve.

Have students suggest what the relationship between solute and solvent might be in a supersaturated solution. Explain that *super-* means to "go above or beyond." Students should infer that a supersaturated solution has a more solute than the solvent can dissolve.

Investigation 9A: Data and Answers

1 Mass and volume in solutions

Substance	Mass in grams	Volume in mL	Density in g/mL
Sugar	34.8	40	0.87
Water	80.0	80.0	1.0
Sugar and water	114.8	103.5	1.11

2 Stop and think

a. The total mass of the sugar and water before mixing was the same as after mixing, which must mean that is all still there even if it can't be clearly seen.

b. The sugar has a lower calculated density than the water. If this were true the sugar should float, but it sank quickly. The observation does not match the calculation. The sugar must actually have a higher density than water.

c. We expected to have 120 mL when we mixed the 80 mL of water and the 40 mL of sugar, but it only measured 103.5 mL. The sugar really only displaced 23.5 mL of water. There must have been spaces between the sugar crystals that were filled with air.

d. The actual density would be 34.8 g/23.5 mL = 1.48 g/ mL. This matches our observations.

3 Temperature and dissolving rate

4 Stop and think

a. Hot water.

b. Water molecules crash into sugar molecules and knock them loose. The faster they move, the faster the collisions will happen. That's why the hot water made the sugar dissolve faster than the cold water; the water molecules were moving faster in the hot water.

Notes:

9.2 Acids, Bases, and pH

About the Lesson

In this lesson students learn about the properties of acids and bases. They also learn about pH and how the pH scale measures the strength of acids and bases.

Ask students to describe what comes to mind when they hear the word *acid*. Students answers will vary, but common responses include danger or poison. Then show students a lemon or other citrus fruit. Ask students, "do you think a lemon is dangerous or poisonous?" Share with students that while many acids are dangerous and corrosive, others are actually necessary for living organisms. Tell students that lemons contain citric acid, which is important for metabolism (citric acid cycle).

Display other examples of acids, like vinegar and a dilute solution of hydrochloric acid. Do not allow students to touch these. Also exhibit an empty container of ammonia or a box of baking soda. Tell students these are examples of bases. Then have students make qualitative comparisons of acids and bases.

Students complete Investigation 9B before reading section 9.2. In this investigation students use indicators to learn about the pH of common solutions. Students then use their data and observations to determine the identity of two "mystery solutions." Be sure that students wear goggles and a lab apron to protect their eyes and clothing throughout the investigation.

Investigation 9B: Acids, Bases, and pH

Setup	Materials	Investigation Vocabulary
1. Allow one class period to complete the investigation. 2. Students work in groups of three to five. 3. Prepare solutions ahead of time. Solutions for the investigation include: lemon juice, vinegar, seltzer, baking soda, bar soap, ammonia, green tea, antibacterial cleaner, and apple juice. 4. Select two solutions from the list for the mystery solutions (A and B). 5. Prepare the red cabbage indicator the day before the investigation. See page 14.	• spot plates • pipettes • red cabbage juice • red and blue litmus paper • solutions • plain white paper	acid base indicator pH

Acids and Bases from Three Perspectives

The properties of acids and bases described on pages 188 - 189 of the student text fit the Arrhenius definition of acids and bases. Svante Arrhenius (1859-1927) was a Swedish chemist who is well known for his "theory of ionization," which expressed his belief that ions are electrically charged and are produced as molecules are ionized in aqueous solutions. As part of his theory, Arrhenius described acids as compounds that contain hydrogen and ionize in solution to create hydrogen ions (H^+). An Arrhenius base dissociates in aqueous solutions and produces hydroxide ions (OH^-).

Bronsted-Lowry acids and bases

Bronsted acids are proton donors; while bases are proton acceptors. In this model, the proton donor or acceptor may be an ion or a molecule. Consider the reaction between ammonia (NH_3) and hydrochloric acid (HCl) shown below. Ammonia accepts a proton (yielding ammonium ions) and hydrochloric acid donates its proton, forming chloride ions.

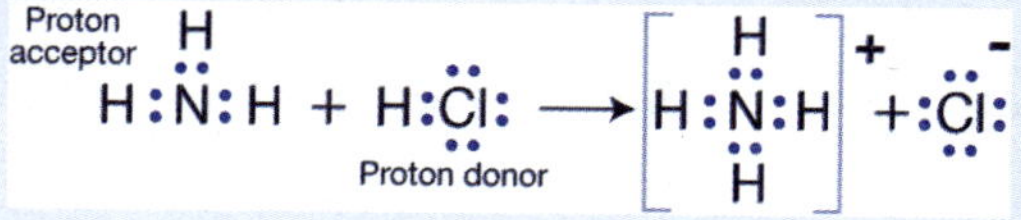

Lewis acids and bases

Lewis acids accept electron pairs. However, Lewis bases are electron pair donors.

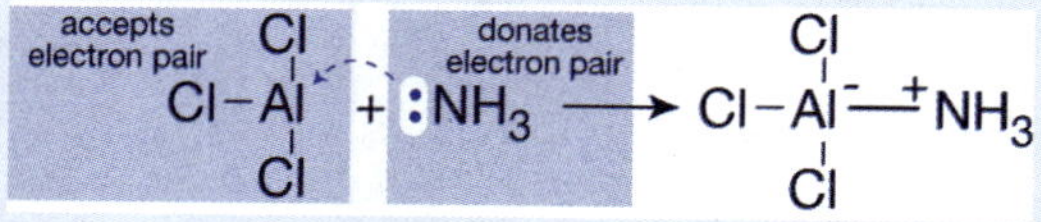

The electron pair from ammonia (a Lewis base) is donated to aluminum chloride (a Lewis acid). The donated electron pair creates a covalent bond between the aluminum atom and nitrogen atom.

Critical Thinking: pH and Living Organisms

Divide students into small groups. Assign each group a problem scenario like the example shown below. Then have each team analyze the problem and write a paragraph to answer the question. Give students the opportunity to present their paragraphs to the class. In the example given below, students will need to draw upon prior knowledge of ecosystems from grades six and seven.

When fossil fuels are burned, gases like sulfur dioxide are released into Earth's atmosphere. These gases combine with water vapor in the air to form acid rain. Any precipitation, such as snow, rain, or fog, with a pH lower than 5.6 is acid rain. The pH of lakes and streams ranges from 6 to 8. The table below describes acceptable pH levels for various organisms within an aquatic ecosystem. Suppose frogs depend on snails as a source of food. Describe how an increase in acid rain might impact the frog population.

Organisms in Aquatic Ecosystems	Lowest acceptable pH levels for survival
Bass	5.5
Clams	6.0
Frogs	4.0
Mayflies	5.5
Snails	6.0

1. Have students work in small groups to create a presentation about acid rain, its causes and effects.
2. Create a poster about acid reflux disease. Have students talk about its symptoms and how the human body responds to protect itself against the effects of the disease. For example, students may discuss how bicarbonate, a base in the salivary glands works to neutralize the acid in the esophagus.
3. Use the Internet to find out how sulfuric acid is made and used.
4. Many household cleansers are bases. Encourage students to share safety tips with their families to protect younger siblings from accidental poisoning.

Students complete section 9.2 review questions.

Word Origins...dissociate (from Latin *dissociare* meaning "to separate from fellowship")

The word *dissociation* has an interesting origin. It actually refers to breaking fellowship or not socializing with a person or object. Tell students that when a molecule like water dissociates, one atom "breaks fellowship" or separates from the water molecule. The atom then finds a new molecule to associate (or bond) with.

Or, you may try an alternative technique to help students with the meaning. Ask students who they are friends with. Then tell them, "you associate with your friends." Have students explain what it means to associate. Sample responses may include to "hang out with" or "spend time together." Then emphasize the prefix *dis-* to help students infer that dissociate has the opposite meaning.

Teaching Tip . . .Neutralization

The subject of the next chapter is chemical reactions. Demonstrate a neutralization reaction to connect concepts from chapters nine and ten. The reaction between sodium bicarbonate and hydrochloric acid shown on page 194 of the student text is a good place to start. Identify the reactants and products involved in the reaction. Then have students tell which reactant is the acid or base.

Investigation 9B: Acids, Bases, and pH

What would you think if a friend told you that she planned to wash her face with acid or use a base to take a bath?

Students express their opinions.

When speaking about acids or bases, it is common to think of them as dangerous. Many acids and bases are dangerous. However, you may use products each day which contain acids and bases without realizing it. Do any of you use facial cleansing products like these?

Display empty containers of common facial cleansers or acne treatments which contain salicylic acid. Many students will recognize these products.

These products contain acids which are formulated for use on skin. The soaps you use for bathing are bases. Would you use battery acid (sulfuric) to wash you face? Why or why not?

Sulfuric acid is a strong acid that would burn and damage skin.

Yes, you care correct. Not all acids and bases are the same strength. Batteries contain sulfuric acid which is too strong to use on skin. Scientists use a scale to determine whether a solution is an acid, base, or neutral. This scale is called the pH scale. It contains values from 1 to 14.

Show students the pH scale.

pH is an abbreviation for the potential of hydrogen. It refers to the activity of hydrogen ions in a solution and determines whether the solution is an acid or a base. Therefore it is a measure of the concentration of ions in a solution.

Students may have previously learned about pH and hydronium ion concentration. It is not important to talk about the specific ions formed in aqueous solution of acids and bases at this point. You may come back to this discussion at the close of the investigation and as students prepare to read the text.

Solutions with a pH value less than seven are acids. Those solutions with pH values greater than 7 are bases.

Point to the pH scale and show students the range for acids and bases.

A solution with a pH equal to seven is a neutral solution. What do you think it means to be "neutral?"

Neutral means not to take sides. A neutral solution is neither acidic nor basic.

In today's investigation you will learn more pH, acids, and bases.

1 Make a pH scale using indicators

In this part of the investigation you will create a pH scale. In order make the scale you will use indicators. An indicator "tells" you whether the solution is an acid or a base. The indicators for this investigation are red and blue litmus paper and red cabbage juice.

Show students which indicators are being used.

Place a piece of white paper under the spot plate. Write the name of the first six solutions from Table 1 on the paper next to one row of wells on the spot plate. Write the pH of each solution next to its name.

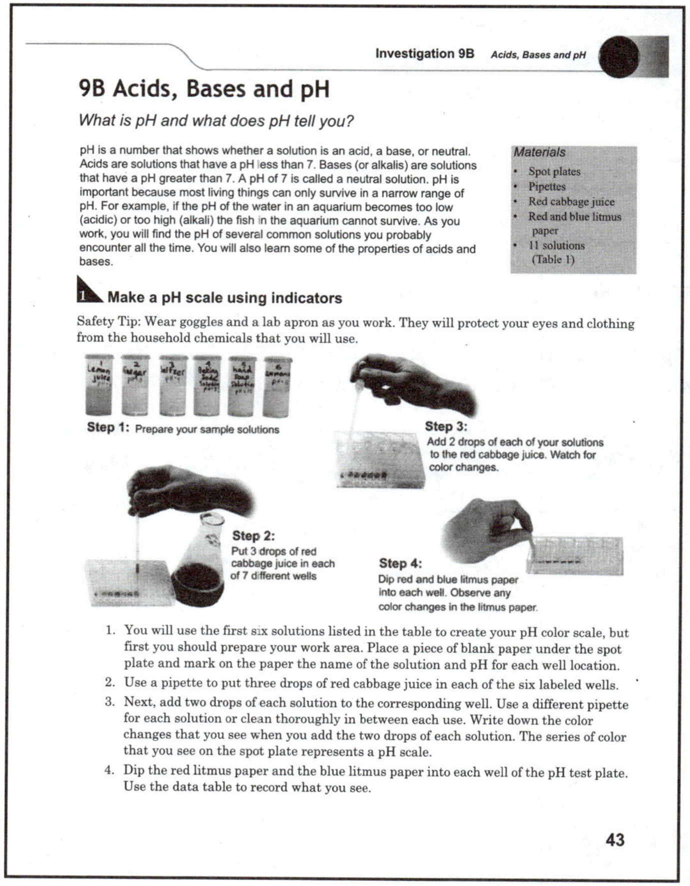

Investigation 9B *Acids, Bases and pH*

9B Acids, Bases and pH

What is pH and what does pH tell you?

pH is a number that shows whether a solution is an acid, a base, or neutral. Acids are solutions that have a pH less than 7. Bases (or alkalis) are solutions that have a pH greater than 7. A pH of 7 is called a neutral solution. pH is important because most living things can only survive in a narrow range of pH. For example, if the pH of the water in an aquarium becomes too low (acidic) or too high (alkali) the fish in the aquarium cannot survive. As you work, you will find the pH of several common solutions you probably encounter all the time. You will also learn some of the properties of acids and bases.

Materials
- Spot plates
- Pipettes
- Red cabbage juice
- Red and blue litmus paper
- 11 solutions (Table 1)

1 Make a pH scale using indicators

Safety Tip: Wear goggles and a lab apron as you work. They will protect your eyes and clothing from the household chemicals that you will use.

Step 1: Prepare your sample solutions

Step 2: Put 3 drops of red cabbage juice in each of 7 different wells

Step 3: Add 2 drops of each of your solutions to the red cabbage juice. Watch for color changes.

Step 4: Dip red and blue litmus paper into each well. Observe any color changes in the litmus paper.

1. You will use the first six solutions listed in the table to create your pH color scale, but first you should prepare your work area. Place a piece of blank paper under the spot plate and mark on the paper the name of the solution and pH for each well location.
2. Use a pipette to put three drops of red cabbage juice in each of the six labeled wells.
3. Next, add two drops of each solution to the corresponding well. Use a different pipette for each solution or clean thoroughly in between each use. Write down the color changes that you see when you add the two drops of each solution. The series of color that you see on the spot plate represents a pH scale.
4. Dip the red litmus paper and the blue litmus paper into each well of the pH test plate. Use the data table to record what you see.

43

Demonstrate this process to students.

Use a pipette to add three drops of red cabbage juice to each labeled well. Then add two drops of each solution to the well you labeled. For example, the first well is for lemon juice so add two drops to the "lemon juice well." Do the same for the five remaining wells. Do any color changes occur? Write what you observe in Table 1.

Students continue to add solutions to the proper wells. Remind students to use a new pipette for each solution.

Use a piece of litmus paper to test each solution. If the red litmus paper turns blue, then write "base" in the table. If the blue litmus paper turns red then write "acid." If nothing happens, then record an "x" in the table.

Students test the solutions with litmus paper.

2 Evaluating the role of the pH test plate

What do your results tell you about the solutions you have tested? Which are acids? Which are bases?

Students answer each question in part two. Have students explain what an indicator reveals about a solution. Discuss the answers to questions in part two before continuing.

3 Using indicators to measure unknown pH

Get a clean spot plate and repeat the experiment (from part one) for solutions 7, 8, and 9. Which of these solutions do you think are acids? Which do you think are bases? Explain why you think as you do?

Students formulate hypotheses about whether green tea, antibacterial cleaner, and apple juice are acids or bases. Students test each solution.

Look at the color changes that happened when you tested these solutions. Compare these results with those from the first six solutions you tested. How can you use the results from part one to infer the pH of these solutions?

Students compare results and note similar color changes.

4 Identifying mystery solutions

Now I will give each group two mystery solutions, A and B. Each mystery solution is the same as one of the solutions you have already tested. Test both solutions. Use your observations to help you identify your mystery solutions.

Students test both solutions and use their data to determine the identity of each solution. Make the investigation interesting by providing each group different mystery solutions.

Were you able to identify each mystery solution? Describe the evidence that helped you decide.

Students discuss their technique for determining the identity of each solution. Close the investigation by reviewing the meanings of important terms like pH, indicator, acid, and base.

Teaching Tip . . .Preparing the Red Cabbage Indicator

Place red cabbage leaves and a cup of water in a blender. Combine the ingredients. Pour the mixture through a strainer to separate the leaves and tinted water. Throw out the processed leaves and store the water in a stoppered flask.

If a blender is not available, you may also boil the cabbage leaves in a pot of water. Collect the colored water and discard the leaves.

Investigation 9B: Data and Answers

1 Make a pH scale using indicators

Name of solution	Color when mixed with red cabbage juice	Red litmus paper	Blue litmus paper	pH
1. lemon	bright pink	x	acid	2
2. vinegar	pink	x	acid	3
3. seltzer	light pink	x	acid	4
4. baking soda solution	aqua	base	x	8.5
5. bar soap solution	blue green	base	x	10
6. ammonia	yellow green	base	x	11
7. green tea	light purple	x	acid	5
8. antibacterial cleaner	bright green	base	x	9
9. apple juice	rose pink	x	acid	3
10. Mystery solution A	Answers will vary			
11. Mystery solution B	Answers will vary			

2 Evaluating the role of the pH test plate

a. A pH indicator is used to gauge the pH of a solution. An indicator is only useful if you understand what the color reaction means. By testing the reaction of solutions of known pH, you can set up a color series that can be used to determine the pH of other solutions. The pH range of red cabbage juice is 1 to 14. The pH range of red litmus paper is 8 to 14. The pH range of blue litmus paper is 1 to 6. Litmus paper only tells you if your solution is an acid or base.

b. Light blue to green to yellow.

c. Reddish pink to purple.

3 Using pH indicators to measure unknown pH

4 Identifying mystery solutions

a. Answers will vary.

b. Answers will vary.

Notes:

Chapter 9 Answers

9.1 Section Review

1. Muddy water is not a solution, it is made up of particles of soil rather than individual molecules.
2. Answers are:
   a. Solvent: water; Solute: salt
   b. Solvent: water; Solute: carbon dioxide gas
   c. Solvent: water; Solute: powdered lemonade mix
3. Steel and Brass are solutions with a solid dissolved in a solid.
   Earth's atmosphere and Natural gas are solutions with gases dissolved in a gas.
   Silver amalgam is a solution with a liquid (mercury) dissolved in a solid (silver).
   Humidity is a liquid (water) dissolved in a gas (air)
   Palladium and Platinum are both solid solvents that will dissolve hydrogen gas.
4. A solute has dissolved when the solute particles are evenly distributed throughout the solvent.
5. Sugar dissolves faster in hot water than in cold water because higher temperature sugar molecules have more energy and are more effective at speeding up the dissolving interactions.
6. More sugar has been added to the milk than can dissolve. The milk is saturated with sugar, and the excess sugar stays undissolved as a solid in the bottom of the bowl.
7. I would dissolve 20 g of salt in 80 g of water to get a 20% salt solution.
   concentration of salt = mass of salt ÷ total mass of saltwater
   0.20 = mass of salt ÷ 100 g
   mass of salt = 20 g
   total mass of saltwater = 100
   mass of water = 80 g
8. Water is called the "universal solvent" because it dissolves many different types of substances. It makes such a good solvent because it is a polar molecule. This gives it the ability to dissolve ionic compounds and other polar substances.

9.2 Section Review

1. Answers may vary. Correct responses include:
   Acids produce hydronium ions; bases produce hydroxide ions.
   Acids have a sour taste; bases have a bitter taste
   Acids turn blue litmus to red; bases turn red litmus to blue
   Acids react with metals to produce hydrogen gas; bases have a slippery feel.
2. Most foods are acidic. Sample answers include
   vinegar, soda water, most fruits, berries, tomatoes, apples
3. Answers are:
   a. Water can act as an both an acid and a base.
   b. The pH of water is 7.
4. b
5. The concentration of hydronium ions the acid produces when dissolved in water.
6. The relative amount of hydroxide ions when the base is mixed with water.
7. litmus, red cabbage juice
8. As more carbon dioxide dissolves in your blood, it produces more hydronium ions and decreases the pH of your blood. As the concentration of carbon dioxide decreases, less hydronium ions are present and your blood pH rises.
9. Students could make a graph of the number of tadpoles in the pond versus year and a graph of pH value for the pond versus year. By overlaying these graphs, it might be evident that as pH increases, the number of tadpoles decreases. This might indicate an inversely proportional relationship between these two variables. As a second step, students could design an experiment to test the effect of solutions of different pH on frogs and tadpoles.

Connection

1. The kidneys and the lungs keep the body's pH in balance.
2. Anxiety is a common cause of hyperventilation.
3. The victim may breathe into a paper bag, which causes more carbon dioxide to be retained.
4. Smoking cigarettes.

Activity

NOTES to Instructor about Activity:

Preparation of the Indicator Solution:

The indicator solution should be made in advance by diluting household vinegar 1:10 with distilled water. This can be done by adding 50 mL vinegar to 450 mL of distilled water. Finally, add Universal Indicator until a red solution is obtained.

Preparation of rock samples:

Some granite may contain some small amounts of limestone. In preparation for the experiment leave the granite chips in vinegar overnight.

For shorter lab blocks:

Depending on time constraints it is possible to speed up the reaction process. If the marble/limestone chips are crushed into a powder it will speed up the reaction and therefore the color change.

Provide the following information to your students:

The indicator solution should change from red (pH 4) to orange to yellow (pH 6.5) to green to blue (pH 9) to violet to red-violet (pH 10) over time.

Time (min)	Color of Solution (marble)	Approx. pH (marble)	Color of Solution (granite)	Approx. pH (granite)
0	orange-red (salmon)	4	orange-red	4
5	orange	5	orange-red	4
10	yellow-orange	6.5	orange-red	4
20	yellow-orange (peach)	6.5	orange-red	4
30	yellow	6.5	orange-red	4
40	yellow-green	7	orange-red	4
overnight	green	8	orange-red	4

Applying your knowledge: sample answers

a. The pH of the solution that contains granite remains unchanged while the pH of the solution that contains limestone changes from 4 to approximately 8.

b. The solution that contains limestone becomes less acidic.

c. The acid rain reacts with the marble and the statues deteriorate.

1. Acid rain penetrates the soil and reacts with the limestone/marble beneath. Over the course of many years, some of the limestone was broken down leaving caves where it was once was.

Chapter 8 Assessment

Vocabulary

Section 9.1

1. solvent
2. solute
3. solution
4. alloy
5. polar
6. equilibrium
7. concentration
8. supersaturated

Section 9.2

9. acid
10. ase
11. pH

Concepts

Section 9.1

1. The solubility increases due to the increased molecular motion of the water.
2. Any of three of the following are acceptable answers: heat the water, shake or stir the container, and increase the surface area of the sugar by crushing the crystals.
3. a
4. c

5. ppm is an abbreviation for "parts per million". Sample answers include: dissolved oxygen in water; the concentration of carbon dioxide or carbon monoxide in air; the concentration of different chemical contaminants in water; the concentration of gases; the concentration of liquids in a gas.
6. If large amounts of hot water are added to a lake, the temperature of the lake would rise. As temperature goes up, the solubility of gases in liquids decreases. Therefore there would be less dissolved oxygen in the water available for fish. Fish may become more exposed to predators as they are forced to move to areas of higher oxygen content. The amount of dissolved oxygen in the water also has a greater effect on immature fish, and they might not survive to adulthood if the oxygen levels of the water are changed significantly. In addition, it's possible that fish may die from lack of oxygen if the temperature of the lake rises too high.
7. When you open the can of soda, you reduce the pressure inside the can. The solution of dissolved carbon dioxide becomes supersaturated, causing the carbon dioxide to bubble out of the water and fizz over the side of the can.
8. When more solute is added to a supersaturated solution, the solute comes out of solution until it reaches the saturation point (rather than the supersaturation point).

Section 9.2

9. The concentration of hydronium ions when the acid is dissolved in water determines the strength of the acid.
10. The strength of a base is determined by the relative amount of hydroxide ions produced when the base is mixed with water.
11. 7
12. Answers are:
    a. A
    b. B
    c. B
    d. A
    e. AB
13. increases, or becomes more basic
14. The mucus lining of your stomach prevents damage occurring to the stomach during digestion. As food leaves your stomach, the liver and pancreas produce a base that neutralizes the acids and protects your intestines.
15. Acid
16. b
17. When ammonia is added to water, it dissociates in water to form hydroxide ions. The ammonia accepts a hydrogen ion from the water to form an ammonia ion and a hydroxide ion.

Problems

Section 9.1

1. concentration = (mass of salt ÷ total mass of solution) × 100%
   concentration = {25 g ÷ (25 g + 75 g)} × 100% = 25%
2. The solubility of baking soda is approximately 10 g per 100 mL H_2O at 25°C
   a. approx. 10 g
   b. The rest of the baking soda does not dissolve and remains solid.
   c. Heat the water
3. Answers are:
   a. Table salt: solubility of 37.7 g/100 mL H_2O
      113.1 g will dissolve in 300 mL of water
   b. Sugar: solubility of 200 g/100 mL H_2O
      600 g will dissolve in 300 mL of water
   c. Chalk: insoluble
      no chalk will dissolve in 300 mL of water
4. concentration = (mass of sugar ÷ total mass of solution) × 100%
   20% = {mass of sugar ÷ (mass of sugar + mass water)} × 100%
   0.2 = mass of sugar ÷ (mass of sugar + 500 g)
   mass of sugar = 125 g

Section 9.2

5. Answers are:
   a. B
   b. A
   c. They would combine to a pH closer to 7 by neutralizing each other.
6. a
7. Adding a base will raise the pH.

Chapter 10: Chemical Reactions

Learning Goals	California Content Standards	Activities and Resources	Assessment
10.1: Understanding Chemical Reactions The student will: • Explain the effects of chemical reactions. • Balance simple chemical equations. • Identify the parts of a chemical equation.	PS 5a PS 5b	**Laboratory Investigation 10A:** Chemical Reactions **Teaching Illustrations:** The Products of the Reaction, Chemical Equation, Balancing Chemical Equations **Graphic Organizer:** A Chemical Reaction **Biographies:** Lise Meitner **Skill and Practice Sheets:** Chemical Equations, Predicting Chemical Equations	**Exam View Test Bank** Chapter Pre-Test **Section 10.1 Review**
10.2: Energy and Chemical Reactions The student will: • Explain how endothermic and exothermic reactions are different. • Identify different types of chemical reactions. • Compare and contrast nuclear and chemical reactions.	PS 5c	**Laboratory Investigation 10B:** Conservation of Mass **Teaching Illustrations:** Exothermic Reactions, Comparing Nuclear and Chemical Reactions **Graphic Organizer:** Exothermic and Endothermic Reactions, Types of Reactions **Skill and Practice Sheet:** Classifying Reactions **Connection:** The Science of Hot and Cold Packs **Chapter Activity:** Explore Hot and Cold Packs	**Section 10.2 Review** **Chapter Assessment** **Exam View Test Bank**

Estimated Instruction Time	Language Tools	Differentiated Strategies	Materials
Five 45-minute class periods	**Literary Picks** **Vocabulary:** chemical reaction, reactants, products, chemical equation	**ELL Strategies:** Listed in ELL Ancillary **Learning Strategies:** • Cooperative Learning - Investigation 10A • Graphic Organizers (K-W-L) - Motivate • Guided Practice - Explain: Balancing Chemical Equations • Active Learning - Extend **Addressing Misconceptions:** Chemical Reactions **Multicultural Experience:** Women and Minorities in Chemistry	**Investigation 10A:** zipper sealed plastic freezer bag, balance, epsom salts, water, ammonia, phenol red, raw potato, hydrogen peroxide, baking soda, vinegar, calcium chloride, graduated cylinders, glow stick
Five 45-minute class periods	**Vocabulary:** exothermic, endothermic, activation energy, addition reaction, polymerization, precipitate, nuclear reaction **Word Origins:** endothermic, exothermic **Closure Cards**	**Learning Strategies:** • Cooperative Learning - Investigation 10B • Discussion - Motivate • Active Learning - Explain: Types of Chemical Reaction • Active Research - Extend **Addressing Misconceptions:** Energy in Chemical Reactions	**Investigation 10B:** periodic table tiles, effervescent tablets, balance, water, small paper cups **Motivate:** candle, matches **Explain - Chemical Reactions:** gumdrops, toothpicks **Chapter Activity:** thermometer, styrofoam cups, hot pack, cold pack, safety goggles, apron

10.1 Understanding Chemical Reactions

About the Lesson

In this lesson students learn about chemical reactions. They learn to identify the reactants and products and balance simple chemical equations. Students also differentiate chemical and physical changes and identify evidence that a chemical reaction has occurred. Finally, students apply the law of conservation of mass to explain that matter is not "lost" but is rearranged as substances undergo chemical changes.

Students have already learned about chemical reactions in previous grades. Find out what students already know about chemical reactions by using a K-W-L chart. Then have students generate a list of ways they can tell if a chemical reaction has occurred. Some examples may include emitting light, changes in temperature or color, forming gas bubbles, pungent odors, or creating an entirely different substance.

Students complete Investigation 10A before reading section 10.1. In this investigation students learn to identify the evidence that a chemical reaction has occurred. Students carry out a series of reactions and record their observations. Be sure that students wear goggles and lab aprons throughout the investigation. See safety precaution on page 8.

Investigation 10A: Chemical Reactions

Setup	Materials	Investigation Vocabulary
1. Allow one class period to complete the investigation. 2. Students work in small groups of three to five. 3. Prepare potato slices and smaller dropper bottles of solutions a day in advance. 4. Caution students not to ingest any chemicals used in the investigation. Discuss the consequences for disregarding lab safety rules before starting the investigation. 5. Prepare "waste" containers in advance.	• zipper sealed plastic freezer bags • balance • epsom salts • water • ammonia • phenol red • raw potato • hydrogen peroxide • baking soda • vinegar • calcium chloride • graduated cylinders • glow sticks	chemical reaction chemical change reactant product chemical equation

Literary Picks...

Reading Level	Title
3-5	***Chemical Chaos*** **by Nick Arnold** Part of the *Horrible Science* series that keeps children amused while learning. This book is a nice introduction to basic chemistry and includes chapters such as "Chaotic Chemical Changes," "It's a Gas," and "Rotten Reactions."
6-8	***Chemical Reactions*** **by Jacqueline Barber** Recommended by NSTA as a book that allows students to do, read, and experience real science. Chemical reactions are mixed in a plastic bag and then explained with terms like reactants, products, exothermic and endothermic reaction.
9-12	***Chemistry in Action: The Molecules of Everyday Life*** **by Nina Morgan** This book is part of the *New Encyclopedia of Science* and features the history of chemistry and how discoveries and processes have shaped our lives. Includes information on elements, bonding, molecules, compounds and mixtures, chemical reactions and combustion.

Guided Practice: Balancing Chemical Equations

Lead a discussion to help students identify the reactants, products, subscripts, coefficients, and atoms involved in chemical reactions. Explain to students that a balanced chemical reaction means that matter is conserved when substances react. Use the simple equations shown below as guided practice. Proper coefficients are shown in blue.

$$2Na + 2H_2O \longrightarrow 2NaOH + H_2$$

$$CL_2 + 2NaBr \longrightarrow 2NaCl + Br_2$$

$$2KClO_3 \longrightarrow 2KCl + 3O_2$$

$$2Ag_2O \longrightarrow 4Ag + O_2$$

1. Have students complete the K-W-L chart they created in the "motivate" activity. Ask students to share what they have learned with the class.
2. Instruct students to keep a "chemical reactions journal." Have students record each example of a chemical reaction they observe for an entire week. Discuss students observations at the end of the week.
3. Use an online tutorial to practice balancing equations. Entering the words "balance equations" should generate many good resources.

Students complete section 10.1 review questions.

Multicultural Experience: *Women and Minorities in Chemistry*

Encourage students to learn more about the contributions of women and minorities in chemistry. Have students use any creative format of their choice to present what they learn to the class. Students may choose to make a Power Point presentation, write a song, rap, or poem, create a poster, make a speech, or interview the chemist if he or she is still living. Students may conduct research to find their own person of interest or choose from the list below.

- Percy L. Julian
- Yuan Tseh Lee
- St. Elmo Brady
- Alejandro Zaffaroni
- Lloyd Augustus Hall
- Ahmed Zewail
- Dorothy Crowfoot Hodgkin
- Retha Clark King
- Kenichi Fukui
- Ruth Mary Rogan Benerito
- Gerty Radnitz Cori
- Gertrude Elion

Teaching Tip . . .Addressing Misconceptions about Chemical Reactions

Students often believe that chemical reactions cause the reactants to "disappear forever," or that only physical changes are reversible. Tell students that equilibrium reactions are reversible as they occur simultaneously in both directions.

Investigation 10A: Chemical Reactions

Can you imagine a world without chemical reactions?

Students share opinions.

Chemical reactions are constantly going on all around you. Who can think of some examples?

Students answers will vary. Sample responses may include reactions within the human body, like digestion, reactions within industrial facilities, combustion reactions, or photosynthesis in plants.

Those are all good examples. When chemical reactions occur, matter changes. So you can say that chemical reactions involve chemical changes. What do you think is happening with atoms when matter experiences a chemical change?

Atoms are rearranged as a result of chemical changes.

Yes, this is true. Consider a chemical reaction that you are familiar with. Water molecules (H_2O) are formed as hydrogen atoms (H_2) combine with oxygen atoms (O_2).

Write the unbalanced reaction on the board.

How are the hydrogen atoms and oxygen atoms rearranged in this reaction?

The atoms combine as water is formed.

The chemical reaction between hydrogen and oxygen is shown here as a *chemical equation*. An equation means that what is written on one side is equal to what is written on the other. Look at the equation on the board. Are the number of each type of atom shown on both sides of the equation the same?

Have students count the number of atoms shown in the reactant and product sides of the equation. Students should observe that they are not equal.

The hydrogen and oxygen atoms shown on the left are called *reactants*. They are the substances that are coming into contact with one another. Water is the *product*. The product is the new substance created as a result of the reaction.

Write these terms on the board.

An important rule in chemistry is the law of conservation of mass. When scientists say that matter is conserved they mean that it is not lost as reactions occur. For this reason a chemical equation must be balanced. In order to balance an equation, the number atoms in the products must equal the number of atoms in the reactants.

Guide students through balancing the equation on the board.

In today's investigation you will carry out six different chemical reactions. Observe the properties of the reactants and products involved in each reaction. Make a note of the clues that provide evidence that a chemical change has occurred.

Discuss the safety rules for this reaction. Be sure that students are aware of the proper disposal techniques for each product. Show students where the "waste" containers are located. Walk around and observe that students are not engaging in unsafe practices.

Investigation 10A *Chemical Reactions*

10A Chemical Reactions

How do you know when a chemical reaction has occurred?

Chemical changes take place all around you and even inside your body. One way to tell if a chemical change has taken place is to carefully observe substances around you. When two substances combine in a chemical reaction, they form a new product that may have very different properties than the original substances. As you work, you will carefully observe some chemical reactions. Make a list of any proof that you see of chemical changes.

Materials

- Sealable plastic freezer bags
- Balance
- Epsom salts
- Water
- Ammonia
- Phenol red
- Raw potato
- Hydrogen peroxide
- Baking soda
- Vinegar
- Calcium chloride

Safety tip: Always wear goggles and an apron while you work with chemical reactions.

WARNING — This lab contains chemicals that may be harmful if misused. Read cautions on individual containers carefully. Not to be used by children except under adult supervision.

- Your teacher will give you everything you need to do each chemical reaction.
- You will do six different chemical reactions.
- Carefully follow all of the directions below.
- For each reaction, write down exactly what you see with lots of details.
- Describe what each substance looks like before and after the chemical change takes place.
- Follow your teacher's instructions for disposal of your reactions when you are done.

1 Reaction #1

1. Put 5 grams of epsom salts into a baggie.
2. Add 10 milliliters of ammonia solution to the baggie. Close it tightly.
3. Feel the baggie with your hands as the reaction is taking place.
4. Let the baggie sit until you are done with all six reactions. Then look at this baggie again.

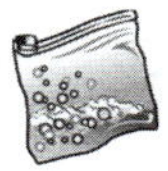

2 Reaction #2

1. Place a potato slice into a baggie.
2. Add 10 milliliters of hydrogen peroxide to the baggie. Close it tightly.
3. Feel the baggie with your hands as the reaction is taking place. You can even hold it up to your ear and listen to the reaction.
4. Let the baggie sit until you are done with all six reactions. Then look at this baggie again.

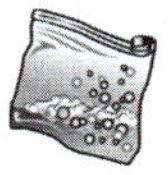

45

Safety Precaution: Phenol red

Phenol red may irritate the eyes or skin. In case of accidental contact, flush the eyes or skin with cold water for 10-15 minutes. Also, avoid inhaling phenol red. If a student accidentally inhales phenol red, allow him to sit or lie calmly in a well-ventilated area.

1 Reaction #1

Place five grams of epsom salts into a plastic baggie. Pour 10 mL of ammonia into a graduated cylinder and transfer it to the baggie. Close the bag. Feel the bag as the reaction is occurring. What do you feel? Write down any observations. Place the bag on your lab table and come back to it later.

Have one student in each group rinse and dry the graduated cylinder.

2 Reaction #2

Place a potato slice into a new baggie. Use the graduated cylinder to add 10 milliliters of hydrogen peroxide to the bag. Close the bag. What do you feel? Are you able to hear anything?

Students record their observations. Clean and dry the graduated cylinder.

3 Reaction #3

Place 5 grams of baking soda into a baggie. Add 10 milliliters of phenol red and 10 milliliters of vinegar to the bag. Feel the bag and observe what happens.

Students record their observations. Clean and dry the graduated cylinder.

4 Reaction #4

Place 10 grams of calcium chloride into a baggie. Add 5 grams of baking soda to the bag. Add 10 milliliters of phenol red to the bag. Close the bag. Feel the bag and observe what happens.

Students record their observations. Clean and dry the graduated cylinder.

5 Reaction #5

Activate a glow stick by bending it slightly. This breaks the inner container allowing the two solutions to mix. Then shake the glowstick several times to help the solutions to mix thoroughly. Record your observations.

Model the process for students.

6 Reaction #6

Activate a heat pack. Feel the heat pack with your hands as the reaction is taking place. How does it feel?

Students note observations.

7 Stop and think

Look at each bag (1-6). Did a reaction occur in each bag?

Students complete questions a-c. Discuss the results of each reaction. Have students describe the evidence of each chemical reaction.

Chemical Reactions in this Investigation

Reaction #1 *(use ammonia with care!)*

Magnesium sulfate (epsom salt) reacts with ammonia to produce ammonium sulfate and magnesium hydroxide. $Mg(OH)_2$ is a precipitate.

$MgSO_4 + 2NH_4OH$ -----> $(NH_4)_2SO_4 + Mg(OH)_2$

Reaction #2

An enzyme from the damaged potato cells (catalase) decomposes the hydrogen peroxide into water and oxygen. The gas bubbles you see are pure oxygen! The same reaction happens when you use hydrogen peroxide on a cut.

$2H_2O_2$ -----> $2H_2O + O_2$

Reaction #3

Acetic acid (vinegar) reacts with sodium bicarbonate (baking soda) to produce sodium acetate (a salt), water, and carbon dioxide. This is an acid/base reaction.

$CH_3COOH + NaHCO_3$ -----> $CH_3COONa + H_2O + CO_2$

Reaction #4

Calcium chloride is sold as an ice melter, usually in pellet form. This is a great reaction that shows gas formation, temperature change, and a color change.

$2NaHCO_3 + CaCl_2$ -----> $CaCO_3 + 2NaCl + H_2O + CO_2$

Reactions #5 and #6

The chapter 6 Connection features bioluminescence, and this chapter's Connection (10) features hot and cold packs.

Investigation 10A: Data and Answers

1 Reaction #1

The contents of the bag turned milky. The bag felt cool to the touch. After the bag sat for awhile, there was a white solid at the bottom.

2 Reaction #2

Bubbling and foaming occurred on the potato slice. There didn't seem to be any temperature change. When I held the bag up to my ear, I could hear snapping and popping sounds! After we waiting awhile, the bubbling stopped and the potato turned brown in spots.

3 Reaction #3

The contents of the bag bubbled and foamed quite a bit. The bag felt slightly cool to the touch. The color started as red, turned orangey, and then turned back to a reddish-orange color and the bubbling stopped.

4 Reaction #4

The contents of the bag foamed and bubbled a lot and turned very warm! The bag expanded quite a bit. The color went from red to bright yellow. After the bag sat for awhile, there was a chalky, white solid sitting on the bottom. The color turned back to a an orangey pink.

5 Reaction #5

The glowstick began to glow very bright. The temperature did not change. After: The stick was still glowing and there was no change in the temperature.

6 Reaction #6

The pack felt very hot. After: the pack cooled down a little.

7 Stop and think

a. see table

b.

observation category	evidence of chemical change
bubbles/gas formation reaction #2, 3, 4	This indicates that a new substance that exists as a gas at room temperature was probably produced.
formation of a solid that settled out of the solution reaction #1, 4	This indicates the formation of a new substance that does not dissolve in water.
color change reaction #3, 4	This indicates the formation of a new substance that has a different pH
change in temperature or other energy produced (light) reaction #4, 6 (heat) and #5 (light)	This means that there is a change in energy and could mean that chemical bonds are being broken and reformed

c. Evidence for Chemical Changes:

- The formation of a gas indicates that a chemical change has occurred because the gas is a new substance formed as a result of breaking and reforming chemical bonds.
- The formation of a solid that settles out of a solution indicates that a chemical change has occurred because a new substance has seem formed that cannot dissolve in water.
- A color change may indicate that a chemical change has occurred because the color change could mean that a new substance with different pH than original substances has been formed.
- If the temperature changes, a chemical change might have occurred because when chemical bonds are broken and reformed, there is sometimes a change in energy (heat) or even light produced.

Notes:

10.2 Energy and Chemical Reactions

About the Lesson

In this lesson students explore different types of chemical reactions and are introduced to nuclear reactions. Students learn that chemical reactions involve energy which is contained in the chemical bonds between atoms.

Light a candle and have students observe it as it burns. After a minute or two ask students to describe what is happening in terms of energy as the candle burns. Students should observe that lighting the candle provides the heat necessary to initiate the reaction. Then ask students, "it is necessary to add more heat in order for the candle to continue to burn?" Once the candle is lit, no additional heat is required. Ask students, "is energy being absorbed or released by the burning candle?" Students should observe that energy is released in the form of light by a burning candle.

Explain to students that energy is involved in chemical reactions. As atoms are rearranged to create new substances, chemical bonds are broken and energy may be absorbed or released. Use the Word Origin box (right) to help student differentiate endothermic and exothermic reactions.

Students complete Investigation 10B before reading section 10.2. In this investigation students carry out a chemical reaction in order to infer the law of conservation of mass. Be sure that students wear safety goggles and a lab apron while completing the experiment.

Investigation 10B: Conservation of Mass

Setup	Materials	Investigation Vocabulary
1. Allow one class period to complete the investigation. 2. Students work in groups of three to five.	• periodic table tiles • effervescent tablets • balance • water • small paper cups	law of conservation

Word Origins...endothermic, exothermic (*endo-* and *exo-* from Greek meaning "in" and "out" respectively; *therme* from Greek meaning "heat")

Endothermic reactions take in or absorb heat; and exothermic reactions give out or release heat. Have students revisit their data from the chemical reactions in Investigation 10A. Ask students to identify which of the reactions were endothermic or exothermic.

Teaching Tip . . .Addressing Misconceptions about Energy in Chemical Reactions

When speaking about endothermic and exothermic reactions, avoid using phrases like, "energy is used up" or "created." These are incorrect ways of describing what happens to energy in chemical reactions. Remind students that energy is actually stored in the bonds between atoms. As atoms are rearranged during a chemical reaction, energy may be released, but it is never used up or created.

Types of Chemical Reactions

Before introducing students to the specific examples of the different types of chemical reactions, try using the model below. Use different colored gumdrops or modeling clay to represent the A, B, C, and D atoms involved in each reaction. Use toothpicks to represent the chemical bonds. Have students rearrange the gumdrops to show the movement of atoms.

Type of Reaction	Basic Form
Addition	A + B → A B
Decomposition	A B → A + B
Single Replacement	A + B C → B + A C
Double Replacement	A B + C D → A D + B C

1. Find out more about Antoine Lavoisier and his contributions to chemistry.
2. Collaborate with a social studies teacher to have students research the historical significance of nuclear reactions. Discuss the Manhattan Project and the scientists who were involved in it.

Students complete section 10.2 review questions.

Precipitation Reactions

A precipitate is an insoluble solid that separates or "falls out" of a solution. Have students think about the meaning of *precipitate*. Most students will agree that rain (or some other form of precipitation) falling to the ground comes to mind.

Precipitation reactions usually involve ionic compounds. For example, aqueous solutions of silver nitrate ($AgNO_3$) and sodium chloride (NaCl) react to produce sodium nitrate ($NaNO_3$) and silver chloride (AgCl). The solid AgCl is a precipitate formed as the positive silver ion from silver nitrate combines with the negative chloride ion from sodium chloride.

$$AgNO_{3\,(aq)} + NaCl_{(aq)} \rightarrow NaNO_{3\,(aq)} + AgCl_{(s)}$$

Net ionic equation:

$$Ag^{+}_{(aq)} + NO^{-}_{3\,(aq)} + Na^{+}_{(aq)} + Cl^{-}_{(aq)} \rightarrow Na^{+}_{(aq)} + NO^{-}_{3\,(aq)} + AgCl_{(s)}$$

Why is silver chloride the precipitate?

Solubility rules govern whether a precipitate is formed in a chemical reaction. In the example shown above, the silver chloride precipitate is formed because specific metal chlorides are insoluble. (Silver is included in the list of insoluble chlorides.) The sodium nitrate does not precipitate out of solution because both sodium and nitrates form soluble compounds.

Spectator ions

In the reaction above, the sodium and nitrate ions do not get involved in creating the AgCl precipitate. When ions in a chemical reaction do not participate in forming a precipitate they are called "spectator ions."

Investigation 10B: Conservation of Mass

As you have already learned, scientists balance chemical equations because matter is not destroyed or created when chemical changes occur. This statement is the law of conservation of mass. Another way of stating this is by saying that the mass of reactants in a chemical reaction is equal to the mass of products.

Write the law of conservation on the board.

The products of a chemical reaction may exist in many forms. For instance, a solid called a *precipitate* may form or a liquid or gaseous substances may be produced. If the product is a solid or liquid, it is easy to see; but what about a gaseous product? How do you know it is there?

Students share opinions.

1 Testing the reaction

In the first part of the investigation, you will conduct an experiment to "test" the conservation of mass. Begin by measuring the mass of an effervescent tablet. Record the mass in Table 1.

ELL students may have difficulty with the pronunciation of this term. Pronounce the term slowly. Write it on the board and repeat it once more.

Place the empty cup on the balance. Zero the balance. Place the cup on the balance and add water until the cup is about half-filled. Record its mass. Total the mass of the effervescent tablet and the water from the first two rows in Table 1. Write this answer in row three. This is your total starting mass.

Be sure that students have not placed the tablet in the water at this point. If students prematurely add the tablet to the water, have them start the procedure over.

Leave the cup on the balance and add the effervescent tablet to the water. What do you see happening?

Students should observe fizzing or bubbling.

After the reaction has stopped, measure the mass of the cup and its contents. Tap the cup gently to release as many bubbles as you can.

Students record the mass once the reaction has ended.

Do you think a chemical reaction occurred as you added the tablet to the water? How can you tell?

Students share their observations.

What were the reactants in this reaction? What were the products? Is there a the difference between the mass you started with (from row three of Table 1) and the mass of the products?

Students subtract the initial mass (in step three) from the final mass in step five. The mass of the products should be less than the starting mass of reactants.

2 Stop and think

Was mass conserved in this reaction?

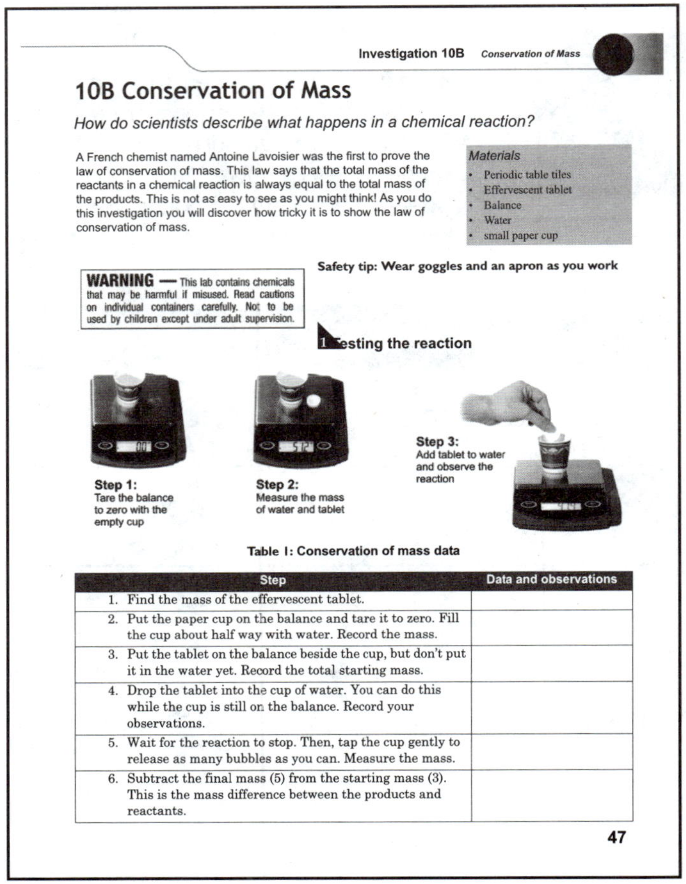

Investigation 10B *Conservation of Mass*

10B Conservation of Mass

How do scientists describe what happens in a chemical reaction?

A French chemist named Antoine Lavoisier was the first to prove the law of conservation of mass. This law says that the total mass of the reactants in a chemical reaction is always equal to the total mass of the products. This is not as easy to see as you might think! As you do this investigation you will discover how tricky it is to show the law of conservation of mass.

Materials
- Periodic table tiles
- Effervescent tablet
- Balance
- Water
- small paper cup

Safety tip: Wear goggles and an apron as you work

WARNING — This lab contains chemicals that may be harmful if misused. Read cautions on individual containers carefully. Not to be used by children except under adult supervision.

1 Testing the reaction

Table 1: Conservation of mass data

Step	Data and observations
1. Find the mass of the effervescent tablet.	
2. Put the paper cup on the balance and tare it to zero. Fill the cup about half way with water. Record the mass.	
3. Put the tablet on the balance beside the cup, but don't put it in the water yet. Record the total starting mass.	
4. Drop the tablet into the cup of water. You can do this while the cup is still on the balance. Record your observations.	
5. Wait for the reaction to stop. Then, tap the cup gently to release as many bubbles as you can. Measure the mass.	
6. Subtract the final mass (5) from the starting mass (3). This is the mass difference between the products and reactants.	

47

Students share their opinions. Based on the results students may believe that mass was not conserved, since the final mass is less that the mass of the reactants.

Why was the mass of products less than the mass of reactants?

Students offer explanations. Explain that mass was conserved in this reaction. One product of this reaction is carbon dioxide gas. As the reaction proceeds, the bubbling or fizzing indicates that a gas is formed. Because the container is open, the carbon dioxide escapes. This creates the appearance that mass is somehow "lost." Be sure that students understand that the mass is neither missing nor does it disappear.

3 Modeling the reaction

You can use the periodic table tiles to model the reaction between the effervescent tablet and water. The tablet contains sodium bicarbonate, a base with the chemical formula $NaHCO_3$. When it reacts with water, three products are formed. The reaction is shown on your investigation handout in part three. Use the periodic table tiles to build the reaction.

Students build the reactants and products with the periodic table tiles.

4 Stop and think

Count the number of each type of atom involved in the reaction. Record your data in Table 2. Be sure to keep the number of atoms in the reactants separate from the number of atoms in the products.

Students count and record the numbers of atoms.

Is the number of atoms equal on both sides of the equation?

The equation is balanced.

What does a balanced equation have to do with the conservation of mass?

A balanced equation means that mass is conserved.

Can you identify the phases of the reactants and products in this reaction?

Solid sodium bicarbonate reacts with liquid water. The products of this reaction are an aqueous solution of sodium hydroxide, carbon dioxide gas and liquid water.

You know that mass was conserved in the reaction even though the masses of product and reactant were not equal when placed on the balance. How could you improve this investigation to better demonstrate the law of mass conservation.

Students offer ideas. A sample response may include to complete the reaction in a covered or stoppered beaker or flask. If the carbon dioxide is not permitted to escape, then the masses of reactant and products should be equal.

Effervescent Tablets

Effervescent tablets are commonly used to relieve symptoms of acid indigestion. The sodium bicarbonate in the tablets undergoes a neutralization reaction with stomach acid. Relate this reaction to what students learned in the previous chapter on acids and bases.

Investigation 10B: Data and Answers

1 Testing the reaction

Step	Data and observations
1. Find the mass of the effervescent tablet.	3.2 g
2. Put the paper cup on the balance and tare it to zero. Fill the cup about half way with water. Record the mass.	51.2 g
3. Put the tablet on the balance beside the cup, but don't put it in the water yet. Record the total starting mass.	54.4 g
4. Drop the tablet into the cup of water. You can do this while the cup is still on the balance. Record your observations.	It bubbled violently for about two minutes
5. Wait for the reaction to stop. Then, tap the cup gently to release as many bubbles as you can. Measure the mass.	52.9 g
6. Subtract the final mass (5) from the starting mass (3). This is the mass difference between the products and reactants.	1.5 g

2 Stop and think

a. The measurable mass does not agree.

b. The violent bubbling means that some gas was produced by the reaction. As the gas was released, the mass went down. The missing mass is the mass of the gas that was produced.

3 Modeling the reaction

4 Stop and think

a.

Element	Reactant	Products
Hydrogen	3	3
Carbon	1	1
Oxygen	4	4
Sodium	1	1

b. The numbers are the same on both sides of the equation. This means that no mass is lost, and that what is there before the reaction is there after the reaction.

c. Answers are:

Reactants - H_2O/liquid, $NaHCO_3$/solid

Products - NaOH/liquid, CO_2/gas, H_20/liquid

d. The reaction could be made to take place inside an airtight container like a plastic baggie with a seal. This way, any gas that is produced will be retained inside the baggie and not lost during the reaction.

Notes:

Chapter 10 Answers

10.1 Section Review

1. Physical changes result in a change of size or shape. Chemical changes result in different substances being formed. Atoms are rearranged, and often energy is released or absorbed in chemical changes.
2. Answers are:
   a. physical change
   b. chemical change
   c. physical change
   d. chemical change
   e. physical change
3. Answers are:
   a. Reactant side: 1 carbon atom; 4 hydrogen atoms; 4 oxygen atoms
   b. Product side: 1 carbon atom; 4 hydrogen atoms; 4 oxygen atoms
   c. Yes; the number of atoms on the product side equals the number of atoms on the reactant side.
   d. 2 molecules of H_2O
4. d
5. It means that no mass is lost or gained during the reaction, the atoms are only rearranged into different groups of molecules. The same number of atoms are in the reactant side and the product side.
6. A chemical formula tells you which types of atoms and how many of each are present in a compound. It tells you which types of compounds (ions or molecules) and how many of each are present. It also shows the conservation of mass in a chemical reaction.

10.2 Section Review

1. At the start of a reaction, bonds between atoms must be broken; Energy is released when new bonds form as atoms recombine into the products.
2. Endothermic: forming new bonds releases less energy than it takes to break the original bonds: Exothermic: forming new bonds releases more energy than it takes to break the original bonds.
3. Answers are:
   a. (s) means solid and (g) means gas
   b. Exothermic
   c. addition reaction
4. Activation energy is the energy needed to start a reaction and break chemical bonds in the reactants. Examples include: supplying a spark of energy for methane gas to initiate and then continue the chemical reaction of burning.
5. A polymer is a large molecule made up of repeating segments. Polymerization is an addition reaction.
6. Answers are:
   a. decomposition reaction
   b. displacement reaction
   c. addition reaction
   d. combustion reaction
7. Nuclear reactions can change an element into a different element; chemical reactions cannot.
   Chemical reactions involve electrical forces while nuclear reactions involve the strong nuclear force.
   Mass and energy are separately conserved in chemical reactions; mass can be converted to energy in nuclear reactions.
8. Refining petroleum means to separate the complex mixture of petroleum into molecules with different numbers of carbon atoms. Raw petroleum out of the ground is not a very useful product, and refining it breaks it down into its useful components.

Connection

1. An exothermic reaction refers to a chemical reaction in which energy is released as heat. When an exothermic reaction occurs, bonds are made between atoms. During this bond making, energy is lost to the surroundings. Therefore, exothermic reactions produce hot temperatures or may be explosive. An endothermic reaction refers to a chemical reaction in which heat is absorbed from the environment. In endothermic reactions, bonds are broken between atoms. Here, the energy required for bond breaking is absorbed from the surroundings. Therefore, endothermic reactions produce cold temperatures.
2. Hot and cold packs consist of large pouches made of thick, strong plastic. These pouches contain a dry chemical in the form of a powder. Within the large pouch is an inner pouch made of thin, weak plastic. This inner pouch contains water. The packs are activated when the seal of the inner water pouch is broken and the contents are vigorously shaken. The released water mixes with the chemical in the outer pouch to create a chemical reaction.

3. Hot packs are made of chemicals that produce an exothermic reaction when mixed with water. They usually contain calcium chloride or magnesium sulfate. These chemicals release heat energy when mixed with water and raise the temperature of the pack. Cold packs are made of chemicals that produce an endothermic reaction when mixed with water. Often, the chemical used in cold packs is ammonium nitrate. It absorbs heat energy and lowers the temperature of the pack.
4. Physical therapists and sports trainers use heat therapy to reduce muscle spasms, reduce pain, and increase blood flow. Researchers think that heat inactivates nerves fibers that can force muscles to spasm. Heat may also induce our body to release endorphins, which are chemicals that block the transmission of pain by our nerves. Heat applied to body parts also relaxes the walls of blood vessels, resulting in increased blood flow. Cold therapy may also be used to reduce muscle spasms by making muscles fibers less sensitive to being stretched. Cold therapy also alleviates pain by slowing the speed of nerve impulses. In addition, tissue swelling drastically decreases when cold and compression are applied to an injured area. The cold temperature constricts the walls of blood vessels, while the compression reduces the blood flow to the injured body part.

Activity

Experiment	Starting Temperature (°C)	Final Temperature (°C)
Hot Pack	23.5	108.5
Cold Pack	23.5	-9.5

Applying your knowledge: sample answers

a. Hot pack: 85.0°C Cold pack: -33.0°C
b. Hot Pack: calcium chloride ($CaCl_2$) Cold Pack: ammonium nitrate (NH_4NO_3)
c. The heat absorbed when the chemical dissolves is removed from your skin causing your skin to feel cool.

Chapter 10 Assessment

Vocabulary

Section 10.1

1. chemical reaction
2. products
3. reactant
4. chemical equation

Section 10.2

5. exothermic
6. activation energy
7. nuclear reactions
8. polymerization
9. endothermic
10. addition reaction
11. decomposition reaction

Concepts

Section 10.1

1. physical change
2. they are broken
3. d
4. b
5. d
6. b
7. A balanced chemical reaction shows the same number of atoms on both the reactant side and the product side.
8. c
9. physical changes: the wax melts
chemical change: the wick burns and the wax burns

Section 10.2

10. Forming new bonds releases more energy than it takes to break the old bonds.
11. endothermic
12. Examples include: heating or cooking with a carbon based fossil fuel (propane, methane, or fuel oil); a vehicle's internal combustion engine; respiration, the process the body uses to burn glucose and turn it into carbon dioxide and water.
13. b

14. In nuclear reactions, Einstein's formula $E = mc^2$, explains that mass can be converted to energy, therefore mass may not be conserved. However, the total of mass and energy in the reaction is conserved.
15. $Li_2CO_3 \rightarrow Li_2O + CO_2$

Problems

Section 10.1

1. Answers are:
   a. 1 calcium atom; 1 sulfur atom; 4 oxygen atoms
   b. 4 sodium atoms; 4 oxygen atoms; 4 chlorine atoms
   c. 1 iron atom; 2 nitrogen atoms; 6 oxygen atoms
   d. 4 aluminum atoms; 6 carbon atoms; 18 oxygen atoms
2. yes
3. Yes, the law of conservation of mass is upheld because we can account for the gain in mass. The iron bolt is being oxidized (rusting). The law of conservation of mass states that the total mass of the reactants is equal to the total mass of the products. In this case, oxygen and iron are the reactants, and they combine to produce rust (Fe_2O_3). Some of the original iron remains unreacted as well.

Section 10.2

4. Answers are:
   a. Displacement
   b. Exothermic
   c. $Al + 3NaOH \longrightarrow Al(OH)_3 + 3Na^+$
      $2Na^+ + H_2O \longrightarrow Na_2O + H_2$

Chapter 11: The Chemistry of Living Systems

Learning Goals	California Content Standards	Activities and Resources	Assessment
11.1: The Chemistry of Carbon The student will: • Distinguish the four main macromolecules found in living organisms. • Explain the role of carbon in processes like photosynthesis and respiration. • Identify the elements which comprise carbohydrates, lipids, nucleic acids, and proteins.	PS 5c PS 6a PS 6b PS 6c	**Laboratory Investigation 11A:** Organic Chemistry **Teaching Illustrations:** Composition of the Human Body, Photosynthesis Reaction, ADP/ATP **Graphic Organizer:** Biological Molecules **Skill and Practice Sheets:** Classifying Reactions	**Exam View Test Bank** Chapter Pre-Test **Section 11.1 Review**
11.2: Proteins, Fats, and Nucleic Acids The student will: • Distinguish saturated and unsaturated fats. • Discuss the functions of macromolecules. • Explain how enzymes function. • Describe how DNA reproduces.	PS 6a PS 6b PS 6c	**Laboratory Investigation 11B:** The Structure of DNA **Teaching Illustrations:** Enzymes **Connection:** Good Fats vs. Bad Fats **Chapter Activity:** The Scoop on Nutrition Labels	**Section 11.2 Review** **Chapter Assessment** **Exam View Test Bank**

Estimated Instruction Time	Language Tools	Differentiated Strategies	Materials
Seven 45-minute class periods	**Literary Picks** **Vocabulary:** organic chemistry, carbohydrates, proteins, fats, nucleic acids, cellular respiration	**ELL Strategies:** Listed in ELL Ancillary **Learning Strategies:** • Cooperative Learning - Investigation 11A • Discussion - Motivate • Graphic Organizer (Venn Diagram) - Explain • Active Learning - Extend **Addressing Misconceptions:** Simple vs. Complex Carbohydrates, Photosynthesis and Cellular Respiration	**Investigation 11A:** periodic table tiles, reference book of compounds
Six 45-minute class periods	**Vocabulary:** saturated fat, unsaturated fat, amino acids, catalyst, enzymes, protein synthesis, mutation **Closure Cards**	**Learning Strategies:** • Cooperative Learning - Investigation 11B • Discussion - Motivate: What's in Your Combo Meal? • Guided Practice - Explain: Biologically Important Molecules • Active Learning - Extend **Teaching Tip:** Enzymes	**Investigation 11B:** licorice sticks, round toothpicks, colored gumdrops **Chapter Activity:** nutritional labels from food packages or from the internet, poster board

11.1 The Chemistry of Carbon

About the Lesson

This lesson is all about carbon. Students discover more details about carbon and its importance to living organisms. Some topics discussed in the lesson include organic structures, biological macromolecules, and life sustaining carbon reactions such as photosynthesis and cellular respiration.

Students learn a great deal about photosynthesis and cellular respiration in grade seven. Find out what students already know about these topics by asking a few basic questions. Examples are given below.

- How do plants make food?
- Describe the conditions in which green plants grow best.
- Why is the sun important to plants?
- What are mitochondria?
- What is ATP and why is it important to living organisms?

Students complete Investigation 11A before reading section 11.1. In this investigation students are introduced to organic chemistry. Students examine the different ways that carbon atoms are arranged and discover the bonding tendencies of organic molecules.

Investigation 11A: Organic Chemistry

Setup	Materials	Investigation Vocabulary
1. Allow one class period to complete the investigation. 2. Students work in small groups of three to five. 3. Gather a variety of books and other references for students to use during the investigation. Helpful resources may include the CRC Handbook of Chemistry. organic chemistry textbooks, or biochemistry references.	• periodic table tiles • reference book of organic compounds	organic chemistry protein amino acid macromolecule

Literary Picks...

Reading Level	Title
3-5	***Fats, Oils, and Sweets* by Carol Parenzan Smalley** Contains large colorful pictures with simple text that explains the importance of fats. This is part of the *Rookie Read-About Health* series that is geared to younger students and lower level readers.
6-8	***Carbohydrates for a Healthy Body* by Hazel King** This book will answer questions about carbohydrates, sugars, starches and more. Part of the *Body Needs* series that includes other books which deal with nutrients and the human body.
9-12	***Nutrition for Dummies* by Carol Ann Rinzler** Rinzler, an award-winning food and health author addresses the misconceptions surrounding fat, cholesterol, vitamins, carbohydrates, proteins, water and nutrients. Includes insights into healthy eating and how your body processes the food you eat.

Addressing Misconceptions: Simple vs. Complex Carbohydrates

Carbohydrates are essential molecules for living organisms. Students may believe that carbohydrates are "bad for them." Explain that not all carbohydrates have the same chemical structure, so the body processes simple and complex carbohydrates in a different manner. Similarities and differences between simple and complex carbohydrates are discussed below.

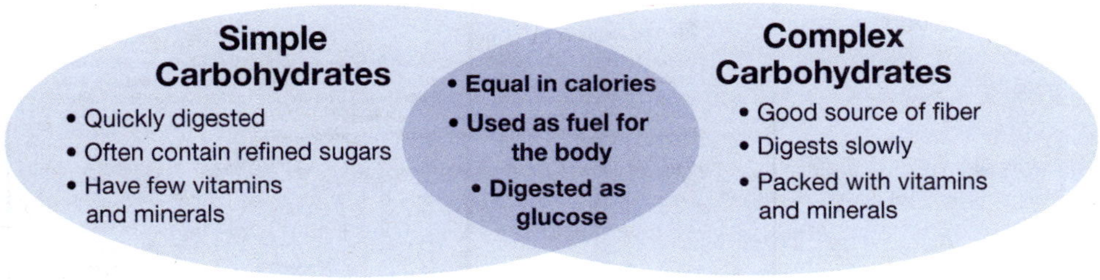

1. What are organic foods? Have students find out what this term means. Are there standards for what is (or is not) organic? Is there a government agency that regulates organic foods? Instruct students to share what they learn with the class.
2. Invite a dietician or nutritionist to speak with your students about the importance of engaging in physical activity and eating healthy foods.
3. Get involved with community health initiatives.

Community Involvement Project

Work with your students to organize a health and wellness fair at your school. Contact your local hospital's wellness center and collaborate with medical professionals to offer screenings for hypertension, diabetes, and cholesterol. You may also choose to have a blood drive. Set up tables with informative pamphlets and brochures about health issues like the dangers of smoking, the importance of prenatal care, and nutrition.

Students complete section 11.1 review questions.

Teaching Tip . . .Misconceptions about Photosynthesis and Cellular Respiration

There are many misconceptions about photosynthesis and cellular respiration. A few of them are addressed below.

- Photosynthesis and cellular respiration are often shown as one-step reactions. There are actually many intermediate steps involved. For example, students often learn that when photosynthesis occurs, carbon dioxide and water react and yield oxygen and glucose. It is important that students realize that there are many other requirements for photosynthesis. Some examples include sunlight, other organic compounds in chloroplasts (not just chlorophyll), and energy storing compounds, like ATP.
- The product of photosynthesis is usually starch or sucrose, rather than free glucose molecules. This is why students are able to use iodine to test for the presence of starch in the leaves of plants. Starch is a very large molecule and sucrose is a combination of glucose and fructose sugar units.
- Cellular respiration in plants does not occur only at night. It is actually a continuous process.
- Plants do use some green light in photosynthesis. While it is true that chlorophyll absorbs very little green light, other leaf pigments do absorb green light. The energy from these pigments is in turn passed on to chlorophyll which enhances the process of photosynthesis.
- Breathing and cellular respiration are not the same but there is a connection between the two processes. Cellular respiration occurs in two phases. The anaerobic phase, glycolysis, does not requires oxygen. The aerobic phase does require oxygen. Most of the ATP that cells need to function is created aerobically via the Krebs cycle and the electron transport chain. Most cells cannot survive with only the ATP produced by glycolysis.

Investigation 11A: Organic Chemistry

1 The chemistry of carbon

Carbon is an important element to living organisms. One major reason is that carbon can bond to itself and other atoms to form many complex molecules. Look at the periodic table. How many bonds can a carbon atom make?

Carbon can make four bonds.

That is correct. Carbon has four unpaired valence electrons. Therefore atoms can bond to carbon in four places. Carbon atoms may be involved in single, double, or even triple bonds.

Have students examine the carbon structures in part one.

Is each carbon atom bonded in four places?

Students find each carbon atom in the three structures shown. Then count the number of bonds each carbon atoms makes with other atoms. Point out the double bonds in the benzene ring.

2 Single carbon molecules

In this part of the investigation, you will look at molecules that contain only one carbon atom. The first molecule shown is carbon dioxide. It has the chemical formula CO_2. If carbon makes four bonds but only two oxygen atoms are connected to it, what does that tell you about the bonds between the carbon and each oxygen atom.

Write the linear arrangement of carbon and oxygen atoms for carbon dioxide. Be sure that the atoms appear in this order: O=C=O. Students should be able to infer that each oxygen atom is double bonded to the carbon atom. Students build the carbon dioxide molecule with the periodic table tiles.

Now build the methanol and methane molecules. How many bonds does carbon make in each?

Students build the molecules. Carbon makes four bonds in each.

How do you think the tiles should be arranged to build a model of carbonic acid?

Students share opinions.

Let' s look at the atoms involved in the bond. It is a good idea to choose carbon as the central atom since it can make the most bonds of all the atoms in the molecule. Which atoms do you expect to bond directly to the carbon?

Students should guess oxygen since the it makes more bonds than hydrogen.

Try a few arrangements. When you think you have the correct order raise your hand and I will check your model.

Allow time for students to infer the proper arrangement of atoms. The proper structure of carbonic acid has carbon as the central atom with one oxygen atom double bonded to it. Carbon make its two other bonds singly with an oxygen atom on each side. One hydrogen atom is bonded to each of these oxygen atoms. See Figure 11.1.

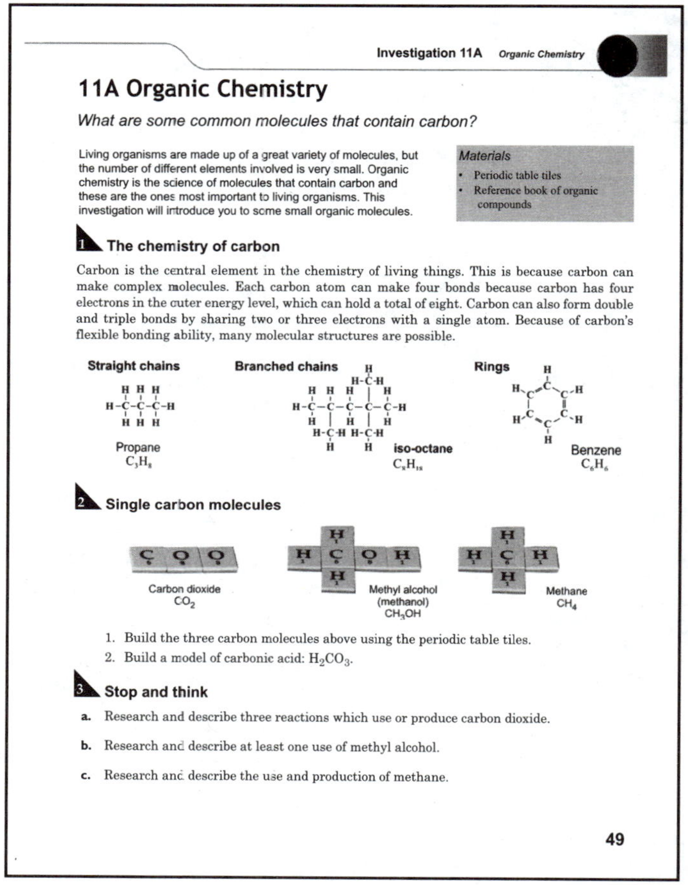
Investigation 11A Organic Chemistry

11A Organic Chemistry

What are some common molecules that contain carbon?

Living organisms are made up of a great variety of molecules, but the number of different elements involved is very small. Organic chemistry is the science of molecules that contain carbon and these are the ones most important to living organisms. This investigation will introduce you to some small organic molecules.

Materials
- Periodic table tiles
- Reference book of organic compounds

1 The chemistry of carbon

Carbon is the central element in the chemistry of living things. This is because carbon can make complex molecules. Each carbon atom can make four bonds because carbon has four electrons in the outer energy level, which can hold a total of eight. Carbon can also form double and triple bonds by sharing two or three electrons with a single atom. Because of carbon's flexible bonding ability, many molecular structures are possible.

2 Single carbon molecules

1. Build the three carbon molecules above using the periodic table tiles.
2. Build a model of carbonic acid: H_2CO_3.

3 Stop and think

a. Research and describe three reactions which use or produce carbon dioxide.

b. Research and describe at least one use of methyl alcohol.

c. Research and describe the use and production of methane.

49

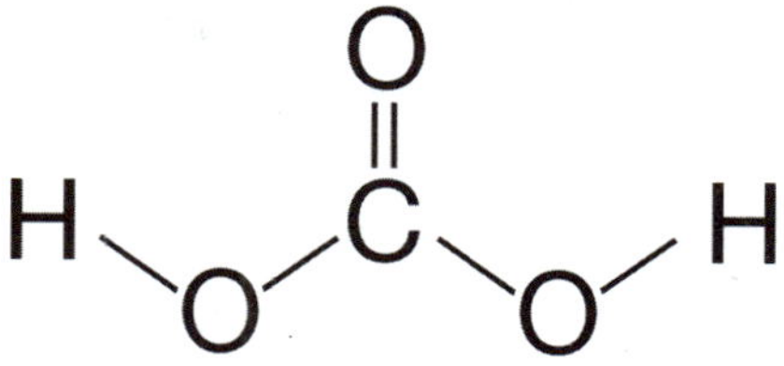

Figure 11.1: Carbonic acid structure.

3 Stop and think

Students conduct research to answer questions a-c. Discuss answers after students have completed their inquiry.

4 Molecules with two carbon atoms

Many molecules contain more than one carbon atom. As the number of carbon atoms increases, the molecules become more complex. Many molecules contain numerous carbon atoms bonded together. Build models of each molecule described in part four. Be sure that each carbon is connected in four places. Remember bonds may be double or even triple.

Students use tiles to build molecules. See Figure 11.2 for structures of acetic acid, ethane, and acetylene.

5 Stop and think

Students investigate to find answers to these questions. Discuss correct responses before proceeding with the investigation.

6 Biological molecules

Proteins are important molecules for living organisms. They are very large molecules and contain many carbon atoms. Proteins are a type of macromolecule. The prefix macro- means large. The large protein molecules are made up of smaller units, called amino acids. Some examples of amino acids are shown on your handout.

Students view amino acid structures.

7 Stop and think

Look at each of the three structures shown. What do they have in common? How are they different?

Students should observe that these amino acids are made up of common atoms: nitrogen, hydrogen, carbon, and oxygen. They may also notice that each amino acid differs only in one place (the R or side group). See more about amino acids and proteins at right.

Proteins are built from 20 different amino acids. All living organisms are constructed from these same twenty amino acids. Each of these amino acids has structural similarities with the exception of a side group. Compare the structures of glycine, alanine, and serine. Circle the part of each structure that is different.

Students locate the side groups of each amino acid.

Now examine the structure shown in part seven of your handout. Could it be an amino acid? How can you tell?

Students evaluate the structure and look for the amino acid backbone. Have students explain why they classified (or did not classify) the structure as an amino acid.

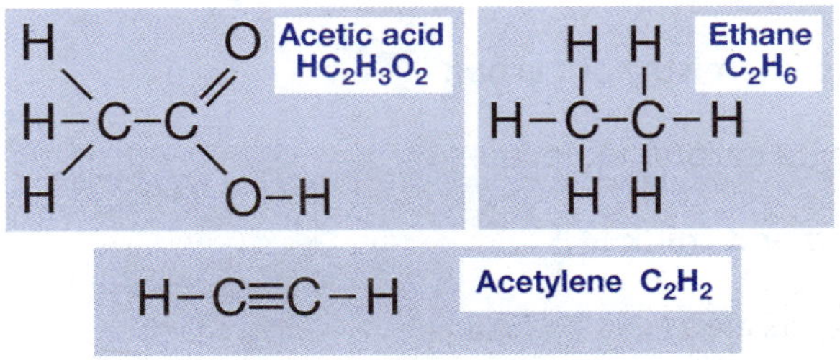

Figure 11.2: Structure of organic compounds.

Amino Acids and Proteins

Proteins are long chains of amino acid sequences held together by a specific type of bond called a *peptide bond*. Hemoglobin, an important protein, contains 584 amino acids! Amazingly these chains are made from only 20 different amino acids. What makes each amino acid unique is its *R*-group or side chain. Each side chain has specific chemical properties.

Amino acids have distinctive parts: an amino group, a hydrogen atom, a carboxyl group, and a specific side chain. Each of these parts is bonded to a central carbon atom.

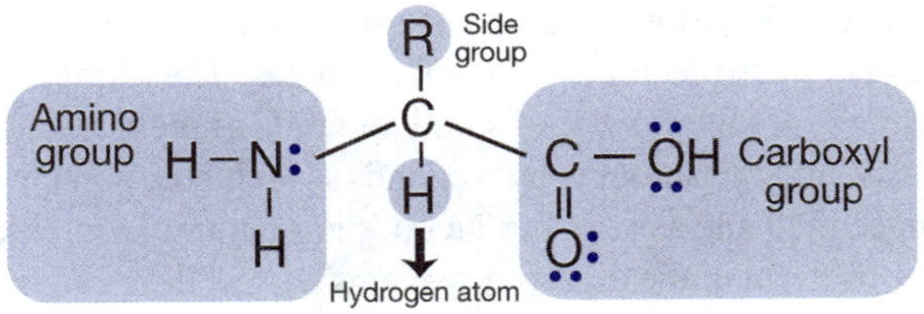

The repeated pattern of the amino group, hydrogen atom, carboxyl group and central carbon atom form a polypeptide chain, sometimes referred to as the "backbone" of the protein.

Investigation 11A: Data and Answers

1 The chemistry of carbon

2 Single carbon molecules

3 Stop and think

a. Answers are:

Dissolving effervescent tablets in water produces CO_2.
$H_2O + NaHCO_3 \rightarrow NaOH + CO_2 + H_2O$

You can refer back to investigation 10B, which used this reaction.

Burning butane produces CO_2
$2C_4H_{10} + 13O_2 \rightarrow 8CO_2 + 10H_2O$

Mixing baking soda ($NaHCO_3$) and vinegar ($HC_2H_3O_2$) produces CO_2.
$NaHCO_3 + HC_2H_3O_2 \rightarrow NaC_2H_3O_2 + H_2O + CO_2$

This was reaction #3 in investigation 10A.

b. Methyl alcohol is also known as methanol or wood alcohol. It is used as fuel for racecars, dragsters, radio controlled model airplanes, and some model trains. In smaller amounts it can be added to gasoline to reduce smog production, and is an ingredient of the fuel additive MTBE. It is also used by industry as an antifreeze for pipelines and a solvent, which makes it useful in the manufacture of other chemicals like formaldahyde, plastics, paints, and explosives.

c. Methane is the primary component of natural gas, and is often found in deposits of petroleum, coal, and natural gas. The refinement of most fossil fuels involves the release of small amounts of methane. Methane is also released in large amounts by wetlands, which host populations of bacteria that produce methane as a by-product of their natural living biological process. Other bacteria in the digestive systems of buffalo, camels, sheep, cattle, and goats produce methane. Wastewater treatment, rice cultivation, and livestock manure management also are large sources of methane emissions. Methane hydrate deposits under polar ocean regions also contribute to methane emissions by releasing methane gas. As a fuel, it is used globally in the form of natural gas for home heating, cooking, vehicle fuel, along with large and small scale electricity production.

4 Molecules with two carbon atoms

5 Stop and think

a. Acetic acid is also called ethanoic acid, and is found in vinegar.

b. Ethyl alcohol is also called ethanol, and is mainly used as a fuel and fuel additive for gasoline. It is also the alcohol present in alcoholic beverages.

c. Acetylene and oxygen produce a flame used for welding and cutting metal that measures over 3300 degrees C. This is a two stage process where actylene burns with oxygen. The first combustion reaction takes place with the oxygen that it is mixed with in a supply tank in a one to one ratio. The second combustion reaction takes place with the oxygen in the air and involves the by-products from the first step and the acetylene from the tank;

First step: $C_2H_2 + O_2$ -----> $2CO + H_2$

Second step: $2CO + O_2$ -----> $2CO_2$

also: $2H_2 + O_2$ -----> $2H_2O$

and: $2C_2H_2 + 5O_2$ -----> $4CO_2 + 2H_2O$

6 Biological molecules

7 Stop and think

a. All three amino acids share a similar structure in their general make up;

b. Yes, this chemical shares the same common structure as the other three that were all amino acids.

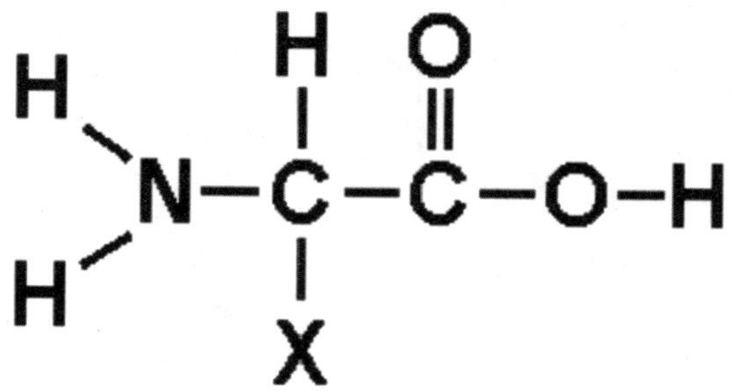

Common stucture

Notes:

11.2 Proteins, Fats, and Nucleic Acids

About the Lesson

In this lesson students continue to learn about organic molecules, with an emphasis on proteins, fats, nucleic acids.

What's in your Combo Meal?

Ask students to describe their favorite fast food meal. Is it a fully dressed hamburger, french fries, and a soft drink? Maybe it is a slice of pepperoni pizza. Ask students to classify each part of their favorite meal as a source of carbohydrates, fats, or proteins. Students may be unsure of the primary macromolecule which comprise each food item. Encourage students to visit the web pages of their favorite restaurants to access the menu and the nutritional value of foods.

Students are introduced to biologically important molecules in grade seven. Use this activity to assess prior knowledge. For example, you may ask students to tell which elements make up each molecule or to explain how each molecule functions to maintain living organisms.

Students complete Investigation 11B before reading section 11.2. In this investigation students learn about the structure and function of DNA. Students use gumdrops to create strands of DNA and apply Chargaff's rules to build complementary DNA strands.

Investigation 11B: The Structure of DNA

Setup	Materials	Investigation Vocabulary
1. Allow one class period to complete the investigation. 2. Students work in pairs. 3. Gather a collection of biochemistry references for student use (part four). Students may also use the Internet or visit the school library to learn more about DNA.	• licorice sticks (about 6" long) • round toothpicks • colored gumdrops	DNA nitrogen base

The Pathway to a Protein

Cell division is an important and necessary process for living organisms to survive. Before a cell divides, it must copy its DNA. This process, known as DNA replication, is driven by enzymes.

Unzipping the double helix

DNA consists of two strands of nitrogen bases held together by hydrogen bonds. Enzymes "unzip" or separate the strands by breaking these hydrogen bonds. Each unzipped strand is used as a template for creating a new strand of complementary bases. In the end two identical DNA molecules are created. One side of each DNA molecule contains the original strand, while the other side is made up of the complementary strand.

Passing along the code

DNA is stored within the nucleus of a cell. Proteins are constructed in the cytoplasm. In order for the genetic information to reach the ribosomes in the cytoplasm it must be transferred to messenger RNA (mRNA). This process is called *transcription*. During transcription, RNA polymerase (an enzyme) separates the strands of DNA and copies a single complementary mRNA strand.

Translation

Proteins are synthesized by ribosomes in the cytoplasm, where transfer RNA picks up the amino acids and transports them to mRNA. Transfer RNA contains anticodons (a series of three nitrogen bases) which bind to complementary codons in mRNA. As tRNA continues to carry amino acids to mRNA, chains of amino acids link together, forming proteins. The order in which these amino acids connect depends on the sequence of codons, which is like a genetic message. The message is *translated* as each codon is read.

Guided Review: Biologically Important Molecules

Use the table below to review the properties of each macromolecule with students. Focus on the presence of carbon in each molecule. Explain how the ability of carbon atoms to form diverse and complex molecules impacts its structure and function.

Facts About Macromolecules

	Structure	Function	Examples
CARBOHYDRATES	• contain carbon, hydrogen, and oxygen atoms • made up of smaller sugar molecules, like glucose	energy storage, provide support and shape	sugars and starches, such as fructose in fruit, galactose in milk, sucrose, rice, potatoes, corn, breads, pasta, glycogen
PROTEINS	• amino acid sequences linked by peptide bonds • made up of carbon, hydrogen, oxygen, nitrogen, and sulfur atoms	movement, structure, transport, nutrition, defense, regulation	enzymes, hemoglobin, collagen, silk, human hair, meats, fish, dried beans and peas, and tofu
NUCLEIC ACIDS	• nucleotides made up of carbon, hydrogen, oxygen, nitrogen, and phosphorus atoms	contains genetic information in cells, make proteins	DNA and RNA
LIPIDS	• carbon, hydrogen, and oxygen atoms arranged in chains of fatty acids and glycerol molecules	long-term energy storage, insulation, repelling water, protection, control functions in organisms	fats, oils, steroids, waxes

1. Have students keep a food diary for a week. Provide each student with a copy of the food pyramid. Instruct students to calculate the percentage of their diet which falls into each category.
2. Direct students to learn more about careers in biochemistry.
3. Teach students to analyze a food label.

Teaching Tip . . .Enzymes

Have students choose one enzyme to learn more about. They might remember that catalase was the enzyme from the potato slice in reaction #1 of investigation 10A that decomposed the hydrogen peroxide. Remind students that enzymes are not limited to chemical reactions within the human body. For example many industrial processes, household cleaning applications, as well as plants use enzymes.

Students complete section 11.2 review questions.

Investigation 11B: The Structure of DNA

1 About the DNA molecule

Have you ever seen a model of a DNA molecule? What does it look like?

Students share experiences. Display a model of a DNA molecule if one is available.

A DNA molecule is put together like a twisted ladder, or double helix. Each side of each rung of the ladder is made of a sugar that contains five carbon atoms, a phosphate group, and a nitrogen base. Let's take a moment and look at each of these components of the DNA molecule more closely.

Point out each major component of the molecule to students as you describe it.

The five carbon sugar is called deoxyribose. This is where the D in DNA comes from. DNA is an acronym for deoxyribonucleic acid.

Write the word on the board and underline the *deoxyribo-* portion.

Phosphate groups are made up of a central phosphorus atom with four oxygen atoms bonded to it. The final component is the nitrogenous base. Nucleic acids contain sequences of four nitrogen bases. The bases are adenine (A), thymine (T), guanine (G), and cytosine (C).

Write the capital letter symbols for each base on the board.

Each of these bases has a partner to which it specifically bonds. In a DNA molecule, *A* bonds only with *T* and *G* bonds exclusively with *C*. These are called base pairs. Two nitrogen bases are paired in the center of the ladder so each rung is composed of two similar groups.

Model for students.

The genetic code in a DNA molecule is stored in the sequence of nitrogen bases. Look at the DNA molecule shown in part one. What is the sequence of nitrogen bases found on the left side of the molecule?

The sequence is AATGCA.

Very good. What is the sequence of the strand on the right side?

The sequence is TTACGT.

You can see that each nitrogen base is paired with its proper complement. For this reason, scientists refer to these strands as complementary strands.

Explain to students that *complement* means to "belong with." Be sure students are able to distinguish this from "compliment" which means to express admiration.

Now you try this with your partner. What is the complementary sequence for GTCATG?

The complementary sequence is CAGTAC.

2 Building a DNA molecule out of gumdrops

Let's build a model of a DNA molecule. You will use gumdrops to complete your model. First, each nitrogen base needs a unique color. Which gumdrop color will you use to represent each base?

11B The Structure of DNA

How does a DNA molecule carry information?

DNA is found in all living creatures we know. This complex molecule contains the information on how to build every protein used in your body. DNA is like a blueprint for building a living creature. The DNA molecule itself is a fascinating structure, as you will see in this activity.

Materials

- Licorice sticks (about 6" long)
- Round toothpicks
- Colored gum drops

1 About the DNA molecule

A DNA molecule is put together like a twisted ladder, or double helix. Each side of each rung of the ladder is made of a 5-carbon sugar called deoxyribose, a phosphate group, and a nitrogen base. Two nitrogen bases are paired in the center of the ladder so each rung is composed of two similar groups.

The genetic code in a DNA molecule is stored in the sequence of the nitrogen bases. For example the sequence AATGCA is coded in the DNA molecule in the diagram.

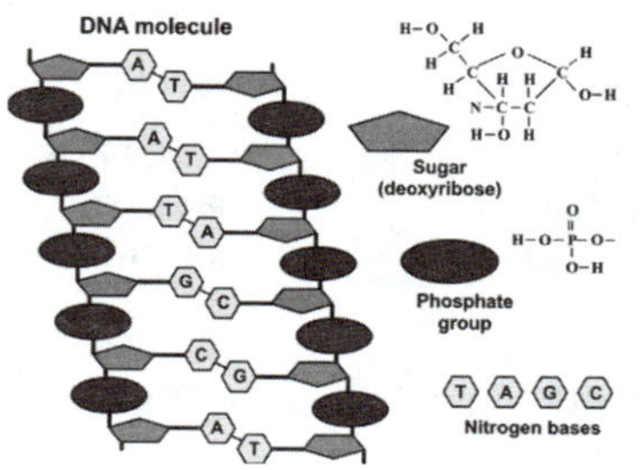

2 Building a DNA molecule out of gum drops

There are four different nitrogen base groups in a DNA molecule. Each should be represented with a different color of gum drop. Choose the color you want to use to represent each nitrogen base (A, G, C, T).

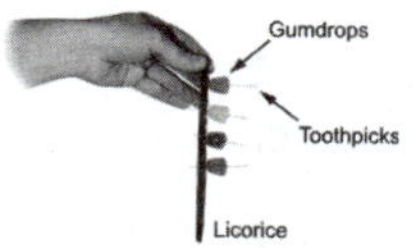

Table I: Nitrogen base colors

Adenine (A)	Guanine (G)	Cytosine (C)	Thymine (T)

1. The licorice will be the backbone of sugar and phosphate. Put 8 toothpicks through the licorice (evenly spaced) and push a gum drop past the middle of each toothpick.
2. The color sequence of your gum drops will be your DNA code!

Students choose colors. If you desire to have consistency with all DNA models throughout the class then choose a common color bases among the groups.

Write the color selected to represent each base in Table 1.
Students record colors in the table.

Now it is time to build your model. Use the licorice as the backbone of sugar and phosphate. Put eight evenly spaced toothpicks through the licorice. Then push a gumdrop past the middle of each toothpick. The color sequence of your gumdrops is your DNA sequence.
Students build the model.

3 Pairing up the nitrogen bases

Each team should have a sequence of eight nitrogen bases. Write the color of each gumdrop in your sequence in row one of Table 2. Start at the top of the sequence. Double check that you have recorded the correct sequence.
Students record the sequence. You may also share with students that enzymes do this job of "checking the sequence" as they unzip the molecule.

The next step is to record the letters of the bases that go with each color gumdrop. Write these in row two.
Students record the bases in the sequence.

Which bases form the complementary DNA strand? Write the letter of each base which pairs with those from your original strand in row three. Then use gumdrops to build the complement. Push each gumdrop up against its partner. Complete the DNA molecule by attaching the opposite side of the ladder. You can give the ladder a gentle twist so it look like a double helix.
Students complete the model. If time permits allow students to build a second model with different bases. Or you may have students switch models with different groups and have them identify the base sequences in both the original and complementary DNA strands.

4 How DNA is reproduced

DNA replication is an important part of the cell cycle in living organisms. Read about the stages of the cell cycle before the investigation to refresh your memory about DNA replication. It would be particularly helpful to review the subphases of interphase (part of the cell cycle) with special emphasis on the synthesis (S) subphase. This is the point at which DNA is replication occurs. To learn more about the role of nucleic acids in protein synthesis read teacher's guide page 180, *The Pathway to a Protein*.

When a living organism reproduces, its genetic information is passed on to its offspring. Use any available resources to conduct research to help you answer the questions in part four.
Students conduct research. Discuss students' answers and offer clarification of concepts that may be confusing.

Chargaff's Rules for Base Pairing

Students may question why adenine and thymine (uracil in RNA) or guanine and cytosine are always paired together in DNA sequences. The answer lies in research conducted by Dr. Erwin Chargaff, a former Columbia University professor.

Through his studies, Chargaff observed that the concentration of A and T molecules in any DNA sample within an organism were relatively equal. He observed the same trend between G and C. This discovery was a key component of Watson and Crick's understanding of DNA structure and supported the idea of base-pairing. Therefore, if an original DNA strand is known, the complementary DNA strand can be inferred.

Base pairs are held together by hydrogen bonds. These bonds are important for structural stability and explain why A and T are complementary instead of A and C (or G and T). Two hydrogen bonds exist between adenine and thymine; and three bonds join guanine and cytosine.

Investigation 11B: Data and Answers

1 About the DNA code

2 Building a DNA molecule out of gum drops

Adenine (A)	Guanine (G)	Cytosine (C)	Thyamine (T)
red	yellow	green	black

3 Pairing up the nitrogen bases

1 Colors	red	green	black	yellow	green	black	yellow
2 Letters (A,G,C,T)	A	C	T	G	C	T	G
3 Letters (A,G,C,T)	T	G	A	C	G	A	C
4 Colors	black	yellow	red	green	yellow	red	green

4 How DNA is reproduced

a. Once the DNA molecule splits, RNA primase inserts starter RNA nucleotides to match with exposed bases. DNA polymerase begins to bring in matching bases to connect to exposed bases. These connections are built off of the RNA that has attached to the mother strands and continue to grow along the exposed bases until they run into another RNA primase. Once all the bases are matched, the RNA primase detaches from the mother strands and DNA polymerase fills in the gaps made where the RNA used to be with the appropriate matching bases.

b. ATCCGGT.

c. More than 3 billion base pairs!

Notes:

Chapter 11 Answers

11.1 Section Review

1. Carbon is commonly found because carbon is the lightest element that can make up to four bonds at the same time, including bonds with itself. It easily bonds with other elements.
2.

Biological Molecule	Composed of what atoms?	Example	Importance
carbohydrate	C, H, O	sugar	store and transfer energy
fat	C, H, O	oils, cholesterol	long term storage of energy
protein	C, H, N, O and trace elements	found in skin and muscle tissue	building block of cells
nucleic acid	C, H, O, N, P	makes up DNA	protein synthesis and heredity

3. Photosynthesis is the foundation of the food chain on Earth. In photosynthesis, energy from the sun is converted into the stored energy of sugars in plants.
4. Photosynthesis is a chemical change.
5. Answers are:
   a. $6H_2O + 6CO_2 + \text{light energy} \rightarrow C_6H_{12}O_6 + 6O_2$
   b. $C_6H_{12}O_6 + 6O_2 \rightarrow 6H_2O + 6CO_2 + \text{energy}$
6. Most of the chemical reactions that sustain life only work in solutions of water.
7. (1) Water is a good solvent. (2) Water exists as a liquid over a large range of temperatures. (3) Water has a very high specific heat, which helps living organisms maintain a stable body temperature even though outside temperatures may fluctuate a great deal.

11.2 Section Review

1. A saturated fat has two hydrogens for every carbon. An unsaturated fat is partially hydrogenated, meaning some of the carbon molecules have double bonds with each other instead of with hydrogen. Chemical processing of food removes some hydrogen, making the fat molecule have a longer shelf life before spoiling. Saturated and partially hydrogenated fats are considered unhealthy, because they can clog your arteries.
2. amino acids
3. They all contain NH_2
4. Enzymes are proteins that act as catalysts for chemical reactions. Enzymes allow the chemical reactions in living things to have lower activation energies and to proceed in many small steps, with each step using only the thermal energy provided by ordinary body temperature.
5. In its nitrogen bases, DNA stores the code on how to assemble proteins from amino acids.
6. DNA is able to make exact replicas of itself. When a cell reproduces, enzymes split the DNA molecule down the center. Each half of the DNA molecule contains the complete genetic information for how to make proteins. At the end of the replication process, there are two identical DNA molecules.
7.

Vitamin	Examples of why it is important	Foods to eat that supply this vitamin
C	synthesis of chemicals in brain and nervous system	citrus fruit
D	strong bones	fish, butter, milk
B	bolsters metabolism, helps maintain healthy skin, enhances immune and nervous system function, promotes cell growth and division	whole grain cereals, nuts, milk, eggs, meat, fish, fruits, leafy green vegetables
folate	needed to make RNA, DNA, and red blood cells	yeast, liver, green vegetables, whole grain cereals

8. The molecule on the left is correct, AT are pairs and GC are pairs.
9. B
10. a

Connection

1. LDL transports cholesterol from the liver to the rest of the body. If there is too much LDL cholesterol in the blood, it can be deposited on artery walls and is thus called "bad cholesterol." HDL transports cholesterol from the blood back to the liver for elimination from the body. Because HDL makes it less likely for extra cholesterol to be deposited in the arteries, it is referred to as "good cholesterol."
2. Unsaturated fats (or "good fats") lower LDL and raise HDL cholesterol levels, leading to a lowered risk for heart disease. Saturated (or "bad fats") are easily converted by the liver to LDL cholesterol, which increases our risk for heart disease. Trans fats (also "bad fats") raise LDL cholesterol levels and also lower HDL cholesterol levels.
3. The FDA is requiring that food labels list the amount of unhealthy trans fats. The FDA hopes that a more detailed label will give consumers a complete picture of the foods they eat.

NOTE: Activity answers are given at the end of this chapter.

Chapter 11 Assessment

Vocabulary

Section 11.1

1. organic chemistry
2. photosynthesis
3. cellular respiration
4. carbohydrates
5. nucleic acid

Section 11.2

6. fat
7. proteins
8. partially hydrogenated
9. unsaturated fat
10. amino acids
11. enzymes
12. mutations
13. catalyst
14. protein synthesis
15. nitrogen bases

Concepts

Section 11.1

1. carbon
2. Carbon is the lightest element with four electrons in its outer shell. This allows it to bond with up to four other atoms, including itself.
3. Answers are:
   a. Carbohydrates are made of carbon, hydrogen, and oxygen in a ratio of about 1:2:1. Examples include sucrose and glucose.
   b. Proteins are large molecules composed of carbon, hydrogen, oxygen, nitrogen, and trace elements. Skin and muscle tissue are composed primarily of protein.
   c. Fats are medium to large, nonpolar molecules that do not dissolve in water. Structurally, fats are long chains of carbon and hydrogen with different elements added every so often. Examples include cooking oils and cholesterol.
   d. Nucleic acids store the genetic code that allows organisms to reproduce. Examples include DNA and RNA.
4. Carbohydrates are made of carbon, hydrogen, and oxygen in a ratio of about 1:2:1.
5. Sugars are broken down quickly because they are the smallest types of carbohydrate molecules. The more complex types of carbohydrate molecules, starches, take longer to break down and digest.
6. Cellulose and starch are both made from chains of thousands of glucose molecules. In starches, all the glucose molecules are the same orientation. In cellulose, alternate glucose molecules are inverted.
7. Water's high specific heat means it takes a lot of energy to raise the temperature a small amount. This property of water helps living organisms maintain a stable body temperature even though outside temperatures may fluctuate a great deal.
8. Photosynthesis takes energy from sunlight and stores it as chemical bonds in molecules of glucose and other simple sugars. Cellular respiration breaks down glucose into water and carbon dioxide, extracting energy in the process. Respiration is almost the reverse of photosynthesis, releasing energy that originally came from the sun.
9. a

10. a
11. cellular respiration and the ATP-ADP cycle
12. Answers are:
    a. amino acids
    b. carbon dioxide and water
    c. fatty acids and glycerin

Section 11.2

13. Your body uses enzymes, which are special proteins that are catalysts for chemical reactions. The enzymes allow reactions to have lower activation energies and to proceed in many small steps, with each step using only the thermal energy provided by ordinary body temperature.
14. Fats provide long term energy storage for the body.
15. A catalyst is a chemical that allows a reaction to have a much lower activation energy than it normally would. Catalysts allow a reaction to proceed in many small steps instead of all at once. Enzymes act as catalysts in the body, allowing chemical reactions to occur at lower temperatures.
16. c
17. Changes in DNA lead to new proteins which make changes in living organisms that are passed in in successive generations. This is the chemical basis for evolution.
18. Certain enzymes compare the old and new DNA strands during reproduction, checking for errors and correcting them by replacing nitrogen bases where necessary.
19. b
20. a

Problems

Section 11.1

1. Answers are:
    a. cellulose
    b. starch
    c. sugar
    d. cellulose
    e. sugar
2. a
3. The plant gained weight because it used photosynthesis to convert carbon dioxide into $C_6H_{12}O_6$ which was then converted to wood.
4. The energy released by cellular respiration is used by cells for life processes.
5. Answers are:
    a. organic
    b. organic
    c. not organic
    d. not organic
    e. not organic
    f. organic

Section 11.2

6. Answers are:
    a. carbohydrate
    b. nucleic acid
    c. fat
    d. carbohydrate
    e. fat
7. The molecule could not be a nucleic acid because nucleic acids contain phosphorus atoms. The molecule could not be a fat because fats do not contain nitrogen.
8. Nucleic acids, DNA and RNA, synthesize and reproduce proteins.
9. c
10. c
11. c
12. Your body cannot synthesize vitamins, they must be obtained through food.
13. A

Activity

Food	Calories	Fat (g)	Protein (g)	Vit A % dr	Vit C % dr	Vit D % dr	Calcium % dr	Iron % dr
mac and cheese	250	12	5	4	2	0	20	4
fruit yogurt	240	3	9	2	4		35	0
puff cereal	369	6	11	10	10	10	10	45
sports drink	50	0	0	0	0	0	0	0
popcorn	140	8	2	0	0	0	0	4
carrots	52	0	1.2	308	13		4	2
spaghetti	200	1	7	0	0	0	0	10
grapes	62	0.3	0.6	2	6		1	1
orange	85	0.2	1.7	8	160		7	1
milk	102	2.4	2.2	10	0	25	26	5
frozen yogurt	37	0	0	0	0	0	0	0
meatballs	259	16.3	26	1			1	17
sauce	70	2	2	15	8	0	6	4
chocolate covered animal crackers	137	6.6	1.6	0	0	0	2	6
apple juice	117	0.3	0.1	0	172	0	2	5
bread	176	2	5	0	0	10	4	8
butter	140	16	0	20	0	50	0	0
tuna fish	190	9	16	4	8	100	3	11

Applying your knowledge: sample answers

a. frozen yogurt, chocolate chip cookies, candy

b. meatballs and macaroni & cheese

c. meatballs and macaroni & cheese

d. I choose carrots because they are low in fat and calories and have lots of vitamin A.
I choose an orange because it is high in vitamin C.
I choose tuna salad because it is high in vitamin D and iron.

Chapter 12: Distance, Time, and Speed

Learning Goals	California Content Standards	Activities and Resources	Assessment
12.1: Distance, Direction, and Position The student will: • Explain the meaning of motion. • Describe an object's position relative to a reference point. • Describe and compare motion of common objects in terms of distance, direction, and position. • Contrast one-dimensional motion with motion in two dimensions.	PS 1a	**Laboratory Investigation 12A:** Distance, Time, and Speed **Teaching Illustrations:** X and Y Coordinates **Skill and Practice Sheets:** Position on the Coordinate Plane	**Exam View Test Bank** Chapter Pre-Test **Section 12.1 Review**
12.2: Speed The student will: • Define speed as it relates to distance and time. • Calculate an object's speed and velocity using proper units. • Differentiate between speed and velocity.	PS 1b PS 1c PS 1d PS 1e PS 2e PS 9f	**Laboratory Investigation 12B:** Motion with Changing Speeds **Teaching Illustrations:** Forms of Speed Equation **Graphic Organizer:** Steps to Solve a Physics Problem **Skill and Practice Sheets:** Velocity	**Section 12.2 Review**
12.3: Graphs of Motion The student will: • Construct and analyze graphs of position versus time and speed versus time. • Interpret graphs of distance, direction, and position. • Recognize and explain how the slope of a line describes the motion of an object.	PS 1f PS 9d PS 9e PS 9f PS 9g	**Teaching Illustrations:** Graphs: Speed vs. Time and Position vs. Time, Drawing a Speed vs. Time Graph **Skill and Practice Sheets:** Analyzing Graphs of Motion with Numbers, Analyzing Graphs of Motion with and without Numbers, Calculating Slope From a Graph **Connection:** Roller Coasters: The Physics Behind the Thrills **Chapter Activity:** How Fast are You?	**Section 12.3 Review** **Chapter Assessment** **Exam View Test Bank**

Estimated Instruction Time	Language Tools	Differentiated Strategies	Materials
Three 45-minute class periods	**Literary Picks** **Vocabulary:** position, origin **Word Origins:** origin	**ELL Strategies:** Listed in ELL Ancillary **Learning Strategies:** • Active Learning - Motivate and Extend • Cooperative Learning - Explore: Treasure Hunting • Guided Practice - Explain: Human Coordinates **Teaching Tip:** Desktop Coordinates	**Motivate:** globe or map
Four 45-minute class periods	**Vocabulary:** speed, constant speed, velocity, average speed **Word Origins:** velocity	**Learning Strategies:** • Cooperative Learning - Investigation 12A • Discussion - Motivate • Guided Practice - Speed Calculations • Active Learning - Extend **Teaching Tip:** Speed Limits **Addressing Misconceptions:** Speed and Velocity	**Investigation 12A:** energy car and track, timer and photogates, rubber bands, clay, graph paper **Motivate:** tennis ball **Extend:** marble, meter stick, stopwatch, other items to propel and measure
Four 45-minute class periods	**Vocabulary:** slope **Closure Cards**	**Learning Strategies:** • Cooperative Learning - Investigation 12B • Active Learning - Motivate: Carlotta's Travel • Guided Practice - Explain: Analyzing Graphs of Motion • Active Learning and Research - Extend	**Investigation 12B:** energy car and track, timer and photogates, rubber bands, clay, graph paper **Chapter Activity:** tape measure, meterstick or ruler, stopwatch or watch with a second hand, tape

12.1 Distance, Direction, and Position

About the Lesson

In this lesson, students learn and apply the meanings of the terms, *distance, direction,* and *position*. Students also contrast one-dimensional motion (and position) with motion in two dimensions.

Ask a student volunteer to locate the state of California on a globe or a map of the US. Ask students questions like, "where is the White House relative to California?" "Which direction must you travel to go from California to Texas?" You may also want to use a map of the state and ask questions like, "which is farther north, Los Angeles or Oakland?" If local maps (of the city or town) are available, tell students to find their street and describe the location of their home relative to the school.

Visit http://www.ed.gov/pubs/parents/Geography/free.html for a listing of places where you can obtain maps inexpensively or free.

Treasure Hunting

Divide the class into groups of three to five students. Instruct each group to "bury" an item on the school campus. Be sure to select the location for this activity in advance. Direct each group to create a map with instructions for finding their hidden treasure. Have students exchange maps and directions with another group. Encourage students to provide specific directions for finding their buried treasure. For instance, "walk five steps north" is more appropriate and useful than simply saying "walk five steps."

A second option is to have students use a map to locate items or various places within the school or neighborhood.

Extension: Create your own treasure map. Then have your student groups use it to locate the items you have hidden. As students locate each item, tell them to explain how the object is used to describe distance, direction, or position. Some suggestions for objects to hide are a tape measure, a small globe, a pedometer, a compass, a ruler, or a small weathervane.

Literary Picks...

Reading Level	Title
3-5	**What's Faster than a Speeding Cheetah?** *by Robert E. Wells* The author compares the speed of different animals, including humans and cheetahs. A comparison of faster objects is also included, such as rockets, meteoroids, and the speed of light.
6-8	**Objects in Motion: Principles of Classical Mechanics** *by Paul Fleisher* This book uses real-life examples to explain the universal laws of science in a reader-friendly language. Topics include, Newton's law of motion, universal gravitation, and falling objects.
9-12	**Galileo's Daughter** *by Dava Sobel* A historical memoir of Gailleo that includes a glimpse of his personal life and his close relationship with his daughter, Marie Celeste, a Franciscan nun. Many of Galileo's scientific achievements are covered, such as his discoveries about motion and gravity.

Guided Practice: "Human Coordinates"

A fun way for students to practice plotting coordinates is to play a game, like *Human Coordinates*. Take your students to the gymnasium or to the playground (if a large cement slab is available). Use tape (or chalk, if on cement) to create a large two-dimensional plot. Label the axes. Include positive and negative values along each axis. Divide the class into groups and send members from each group to specific positions. For example, you might say to students, "I need John to go to the coordinate point (-6, 2) and Susie to go to the point (4, -1)." Another option is to place items at random locations on the plot. Then have students identify the items by their coordinates.

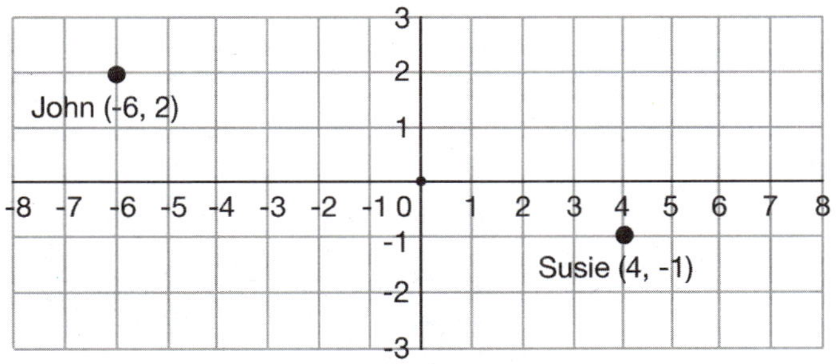

> **Teaching Tip . . .Desktop Coordinates**
>
> Setup: Assign each student a partner. Have students use pieces of colored tape to make a grid on their desks. Then instruct students to label the *x*- and *y*-axes on the tape.
>
> Play the game: One student places a common object at one position on the desk. The partner then writes down the coordinates at which the object was placed. Each student takes a turn.

> **Word Origins...origin *(from Latin stem origin- or oriri, meaning "to arise")***
>
> The term, *origin*, refers to the beginning, starting point, or place from which something arises.
>
> Engage students in a discussion about their (or ancestors) country or state of origin. Students are most comfortable if you (teacher) begin by sharing an interesting story about yourself and from where your family "arose", or originated. (This is a great discussion for promoting pride within your students. It also helps students to learn about one another's history and culture.)
>
> Ask students what it means to be original? Have students describe a time when they were (or did something) original. (to be the first, creative, or authentic)

1. Show students how to use online mapping resources, like Mapquest, to obtain directions from their home to different places.
2. Plan a field trip to the nearest airport so that students can learn how airline pilots rely on navigational tools. Have students note the position of their destination relative to the school.
3. Read *Treasure Island* by Robert Louis Stevenson.
4. Play a battleship game, where students must detect their opponents position by coordinates.

Students complete Section 12.1 review questions.

12.2 Speed

About the Lesson

In this lesson students learn to define speed by relating distance and time. Students use proper units to describe speed and apply a formula to solve speed problems. Students also differentiate speed and velocity; as well as constant speed and average speed.

Place a tennis ball on a table or desk in your classroom. Be sure all that students are able to see the ball. Ask students to describe its motion. Students should see that the tennis ball is not moving, or is at rest. Slowly roll the tennis ball on the floor or across the table. Roll the ball again, but quickly this time. Then bounce the tennis ball on the table. Ask students to describe the motion of the ball in each circumstance. Help students to infer that motion is often described in terms of how fast or slow an object moves.

Students complete Investigation 12A before reading section 12.2. This investigation focuses on how scientists describe motion, as well as the relationship among distance, time, and speed. Students use the car and track to observe the motion of an object at constant speed, and to measure position versus time. Students also create a position versus time graph and analyze it to determine the average speed of the car.

Investigation 12A: Distance, Time, and Speed

Setup	Materials	Investigation Vocabulary
1. Allow one class period to complete the investigation. 2. Students work in groups of three to five. 3. Refer to the *Energy Car and Track* equipment setup.	• energy car and track • timer and photogates • rubber band • clay • graph paper	speed constant speed average speed

Teaching Tip . . .Speed Limits

Ask students what the term, *speed limit*, means. Have students identify places they have seen speed limit signs. Talk about the information provided on speed limit signs, like 55 miles per hour and what it means. You may also discuss other units which are used to describe speed.

Word Origins...velocity *(from Latin stem veloc- meaning "quick")*

Velocity is a measure of the rate of change of an object's position with respect to time. Ensure that students understand that direction is a key component when expressing velocity and that it is not the same as speed. *See Misconceptions on the next page.*

Have students use the word velocity in a sentence. Then ask students to describe how the literal meaning of velocity relates to its definition.

Addressing Misconceptions: Speed and Velocity

Students are often confused when asked to differentiate speed and velocity. A common question is, Aren't they the same? Of course, the answer is no. The key point for students to remember is that speed makes *no* reference to direction.

Suppose a baseball pitcher is asked how fast he threw a ball. He could respond by saying that he threw it 95 miles per hour. While this is an acceptable response for gauging speed, it does not describe velocity. In order to describe velocity, the pitcher would have to designate direction along with the speed, for example, 95 miles per hour, north.

Speed is a scalar. A *scalar* is a quantity that has magnitude, but no direction. Both distance and time are scalar quantities. Velocity is a *vector* because it has both magnitude and direction.

Use the examples below to help students distinguish speed and velocity.

1. The pilot flew the airplane 500 mph, southeast. (velocity)
2. Tammy rode her bicycle 15 m/s. (speed)
3. A car traveled 45 km/hr, west. (velocity)
4. The van went past the pedestrian at 70 mph. (speed)

1. Invite a police officer to your classroom to discuss how radar guns are used in law enforcement.
2. Take your students outdoors or to the gymnasium. Set up four or five *Speed Stations*. Divide the class into teams. Create a speed-related task for students to complete at each station. For example, you could have students propel a marble; then measure how far the marble traveled over time to determine its speed. Have team members record one another's data. Once each team member has completed all stations, direct students to find the average speed at each station by group. Distribute extra credit points to all participants and bonus points to the teams with the fastest speed at each station.

Students answer section 12.2 review questions.

Guided Practice: Speed Calculations

Try these speed problems to help your students understand how to use all three forms of the speed equation.

1. Lourdes swam 50 meters in 10 seconds. What was her speed? (Answer: 5 m/s)
2. An object traveling at a speed of 45 m/s went a distance equal to 75 meters. How much time was needed? (Answer: 1.7 s)
3. JIm's dog chased a neighbor's cat a distance of 30 feet for 12 seconds. What was the dog's speed? (Answer: 2.5 ft/s)
4. How far does a car travel if moves at a speed of 65 mph for 42 minutes? (Answer: 45.5 miles) Hint: Remind students to convert minutes to hours. Forty-two minutes equals 0.7 hour.
5. A man paddles a canoe 568 kilometers at a speed of 36 km/hr. How long does it take for the man to travel this distance? (Answer: 15.7 hours)

Investigation 12A: Distance, Time, and Speed

1 The relationship between distance, time, and speed

Complete the motivational activity for this section before introducing the investigation. Once students have used terms, such as slow or fast to describe motion, explain to them that those terms are not enough to describe speed. Write the formula for speed on the board. Point out each variable in the equation. Ask students to provide definitions of distance and time. Have students give examples of units of both. Then tell students that speed is the ratio of distance traveled divided by the time taken.

Let's try a few calculations using the speed equation.

Use questions a-c as guided practice.

2 Setting up for constant speed

What does the term, *constant*, mean?

Review that constant means not changing. An object moving at constant speed has the same speed all the time. Its speed does not change; it is like being on cruise control. Almost nothing moves at constant speed for very long, however, it is useful as a teaching concept.

Follow the instructions provided in part two to set up the track.

Circulate and observe students as they assemble the track. This should not take longer than three to five minutes. Guide students to proper positioning of the photogate.

Put a single twist in the rubber band on the launching end of the track. To launch the car, rest your palm on the wooden part of the stop block and use your index finger on the tab (on the top of the car) to pull the car back against the rubber band. Use the adjustable screw to set how much the rubber band is deflected. Gently snap your fingertip back and up to launch the car. This is a tricky technique. Practice it until you get the car to pass through a photogate three times with a time that changes by no more than 0.0010 second.

Help students to master the technique of launching the car. Many will be able to launch the car with variations of less than 0.0005 second from one run to the next. This is a fun (and useful) activity for the students. Allow at least five minutes for students to practice and get comfortable with this technique.

3 Making the car move at constant speed

Now for the tricky part. If there are two photogates on the track, and the car is moving at constant speed, what should the times be from each individual photogate?

Since the time (A or B) is the time needed for the flag to pass through the light beam, both photogates should measure the same time if the car is traveling at the same speed.

See how close you can get the times from photogate A and B. You will have to adjust the slope of the track by raising or lowering the screw feet on either end. Once you get the times to be identical, you know the car is traveling at constant speed. It will be almost impossible to get the times exactly identical. Try to get them to within 0.0010 second of each other.

Investigation 12A *Distance, Time, and Speed*

12A Distance, Time, and Speed

How do scientists describe motion?

The average speed is the ratio of the distance traveled divided by the time taken. This is an idea you already use. For example, if your car is moving at a speed of 60 miles per hour that means the ratio of the distance the car moves divided by the time it takes is always equal to 60 miles/hour. If you go 120 miles in 2 hours, your average speed is 60 miles per hour because the ratio 120 miles/2 hours equals 60 miles/hour.

Materials
- Energy car and track
- Timer and photogates
- Rubber band

1 The relationship between distance, time, and speed

Speed

$$v = \frac{d}{t}$$

Speed (m/sec); Distance traveled (meters); Time taken (seconds)

In science, you will mostly be working with speeds in units of centimeters per second (cm/s) or meters per second (m/s).

a. A duck swims 10 meters in 5 seconds. What is the average speed of the duck?

b. A car on cruise control is moving at a constant speed of 55 mph. How far does the car travel in three hours?

c. A jet aircraft is moving at a constant speed of 500 kilometers per hour. How long does it take the jet to travel 2,000 kilometers?

2 Setting up for constant speed

Setting up the straight track

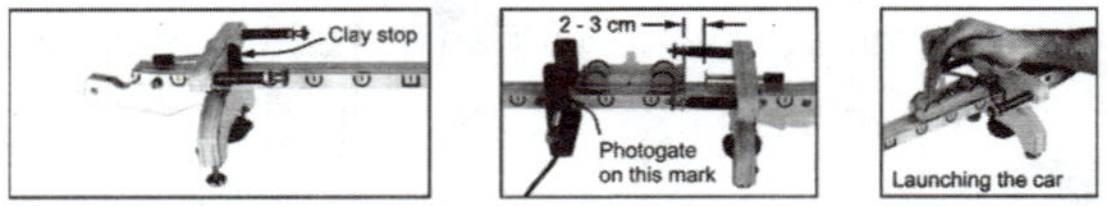

1. Put the track together as shown in the diagram. Use one rubber band on the launching end and a ball of clay on the catching end to stop the car.

53

4 Stop and think

How do the measured time intervals from the photogates prove that the car has constant speed, or nearly constant speed? How would you calculate the speed of the car in meters per second?

Lead the class in a discussion. The width of the flag on the car is one centimeter, or 0.01 meter. The easiest way to calculate speed is to divide 0.01 meter by the time it takes to pass through a single photogate. You can also calculate that speed using the distance between the photogates and the time interval between the photogates (from A to B). If the students placed the photogates on the marks on the track, as shown in the diagram, the distance between the light beams is sixty centimeters, or 0.6 meter. The speed can be calculated by dividing 0.6 meters by the time from A to B. Both techniques should give about the same value.

5 Measuring position versus time

Now you will measure position versus time. Set up the track as shown in part five of the investigation. Put photogate A just ahead of the car on the launching end. Don't move photogate A once you get it set. You will use photogate B to measure the speed and position of the car at five centimeter intervals along the track. You will not be able to measure some positions around where the two sections of the track are joined together. Don't worry about this. To do the experiment, you will need to record all three times (A, B, and A to B) for each run of the car. Remember, you don't have to run the car three times for each position of photogate B. You can get all three times by turning the A and B lights on and off without pressing *Reset*.

Circulate and assist students as they collect data.

6 Making the position versus time graph

The graph you want to draw shows position on the *y*-axis (vertical) and time on the *x*-axis (horizontal). The time is the time from A to B. The position is the distance from A to B.

Students construct the graph of position versus time. They should get a straight line.

7 Stop and think

What does a straight line on a position versus time graph tell you about the speed of the car?

Explain that a straight line means the car always travels an equal distance in an equal time interval anywhere on the track. This is what is meant by constant speed. Constant speed means an object moves the same distance in the same time interval anywhere along its motion.

Use your graph to calculate the speed of the car.

Speed is distance traveled divided by time taken. The distance traveled is the amount that the graph changes in vertical direction. The time is the amount the graph changes in the horizontal direction. This is equivalent to calculating the slope of the line on the graph.

How would you use the information in this experiment to predict the time needed for a car to travel two meters if it kept the same speed?

Review the three forms of the speed, distance, and time relationship. Show students that time is distance divided by speed.

> **Teaching Tip . . .Time from the speed equation**
>
> As students answer question 7c, a common example is often useful. For instance, ask you students how long it would take to travel 120 miles at a speed of 60 mph. Most students recognize the answer is two hours. Then ask them how they arrived at the answer (2) from the numbers 120 and 60. This exercise reinforces the truth (and use) of the equation, that the time it takes to complete a trip is really the distance divided by the speed.

Investigation 12A: Data and Answers

1 The relationship between distance, time, and speed

a. 2 m/sec

b. 165 miles

c. 4 hours

2 Setting up for constant speed

3 Making the car move at constant speed

4 Stop and think

a. Since the times measured by both photogates are the same or extremely close, the speed of the car must be constant, or nearly constant.

b. The speed of the car is 1.47 m/sec.

5 Measuring position versus time

Position (cm)	Time through photogate A (sec.)	Time through photogate B (sec.)	Time from photogate A to B (sec.)
10	0.0068	0.0069	0.0675
20	0.0067	0.0069	0.1380
40	0.0067	0.0070	0.2754
50	0.0067	0.0070	0.3427
60	0.0066	0.0069	0.4147
70	0.0067	0.0071	0.4850

6 Making the position versus time graph

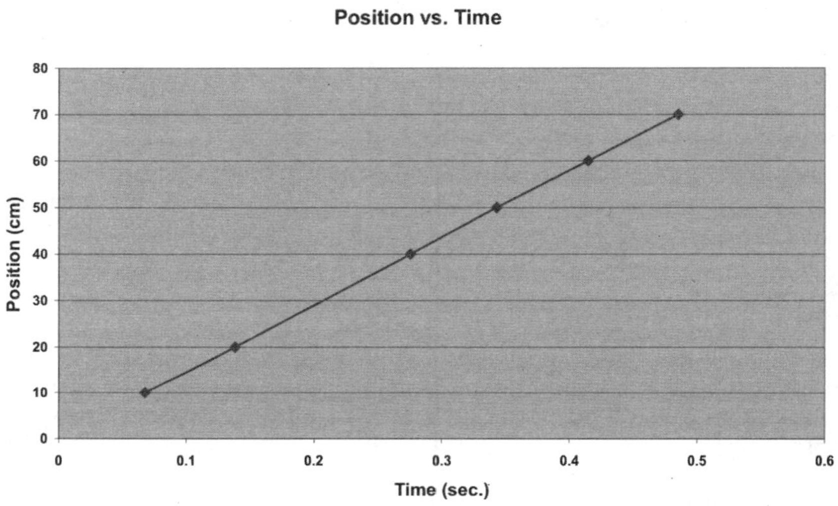

7 Stop and think

a. The shape of the graphed data is a straight line.

b. The average speed of the car is 144 cm/sec.

c. It would take the car 1.39 seconds.

d. The speed would increase as it moved down the track.

e. The graph would be a curve that got steeper as the car car moved down the track.

f. The graph would still be a straight line, but it would be steeper than the one we made.

g. A car with a constant speed will have a straight line graph. A car that is speeding up or slowing down will have a curve. The steeper the line, the faster the car is moving.

Notes:

12.3 Graphs of Motion

About the Lesson

In this lesson students learn to create and analyze graphs of motion. Students also explain how the slope of a line can be used to describe the motion of an object.

Distribute graph paper to your students. Have each student draw *x*- and *y*-axes; and label each axis ten units from the origin in each direction. Tell students to mark directions (N, S, E, and W) on the graph paper. Instruct each student to listen very carefully as you read a short story. Tell students to place their pencil at the origin. Then move the correct number of spaces in the proper direction each time you make a reference to distance and direction (position). Once you have finished reading the story, ask students to tell you the coordinates of their final position. A sample story, followed by the plot of Carlotta's trip, is shown (Figure 12.1).

This activity will help you to determine if any of your students need remediation as far as understanding distance, direction, position, or how to plot points.

Students complete Investigation 12B after reading section 12.3. In this investigation students create a system that is used to analyze the motion of a car with varying speed. Students collect data and make graphs of position versus time and speed versus time. By the end of the investigation, students are able to distinguish between graphs of constant speed and changing speed.

Investigation 12B: Motion with Changing Speed

Setup	Materials	Investigation Vocabulary
1. Allow one class period to complete the investigation. 2. Students work in groups of three to five. 3. Refer to the *Energy Car and Track* equipment setup.	• energy car and track • timer and photogates • rubber band • clay • graph paper	system variable average speed constant speed slope

Carlotta's Travels

Carlotta was so excited that the weekend had finally come. She could not wait to visit with her friend, Sam. Since it was such a beautiful day, Carlotta decided to walk to Sam's house. Before she left home, her mother reviewed the directions. When you walk out the door, go five paces north, then turn right and walk three more paces east. You will notice a blue house with white shutters. Go south ten paces. Turn right. The playground will be on your left. Walk six paces and you will arrive at Sam's house.

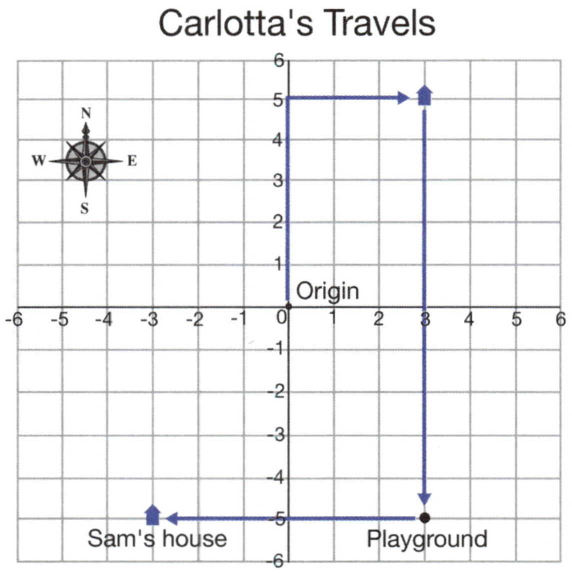

Figure 12.1: Carlotta's trip to visit her friend, Sam.

Guided Practice: Analyzing Graphs of Motion

Draw a position versus time graph like the one shown below on the board. Lead a discussion to help students answer each of the questions shown at right.

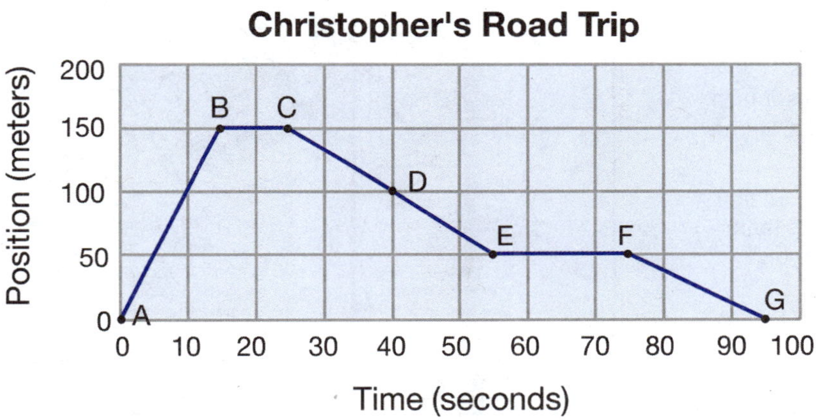

1. Direct students to create graphs of motion and write questions to accompany their graph. Then have each student exchange graphs and answer the questions that go along with their partner's graph.
2. Have students make a presentation about biomechanics. Discuss how graphs of motion are important in this field of study.
3. Use the graph of *Christopher's Road Trip* to create a speed versus time graph.

Students complete section 12.3 review questions.

Guided Practice: Questions and Answers

Christopher borrowed his mother's car to run a quick errand. Use the graph to answer questions about his trip.

1. How far did Christopher's car travel between points A and B? (150 meters)
2. How much time did it take for Christopher to travel from point A to point B? (15 seconds)
3. Describe the motion of Christopher's car between points B and C. (Christopher's car is at rest between points B and C. The car did not move any distance from its point of origin, so the speed of the car is at zero.)
4. What is the speed of the car between points A and B? (The car traveled 150 meters in a 15-second period. The speed of the car is 10 m/s.)
5. How would you describe the slope of the graph between points D and E? (The slope is negative.)
6. What does a negative slope tell you about the direction the car is traveling? (A negative slope means the car is traveling opposite its original direction.)
7. Is Christopher traveling faster between points A and B or points F and G? How can you tell? (Points A and B; because the slope of the line is steeper.)
8. What happens at point G? (Christopher is back at the position, or place, he started from.)

Investigation 12B: Motion with Changing Speed

1 Making a system

Almost nothing in the universe stays constant for very long. Things are always changing. This change may involve moving from here to there, which is a change is your position. Slowing down or moving faster are changes in speed. Warming up or cooling down are changes in temperature. Each of these changes are described by principles of physical science. In fact, many of the natural laws relate to exactly how much change is possible in a given circumstance. Here is an example. If I drop this marble to the floor, do you think it will be going one hundred miles per hour when it hits the floor?

Do a demonstration like this. The goal is to get the students discussing how change is described. Also, the demonstration introduces the idea that the amount of change possible in a given circumstance is limited by something. The limit is the height from which you drop the marble.

If you drop the marble from a height of one meter then the maximum speed it can have when it hits the floor is 4.4 m/sec, or 9.9 miles per hour. Why? What is it about the universe that limits the speed of the marble? In order to answer these questions we have to think about two big ideas. The first is energy. What is energy?

Engage students in a discussion about energy. Students at this level may respond that energy is the ability to do work. Accept all reasonable responses.

Energy measures the ability of a system to change itself or to change other systems. When I raise the marble off the floor I give it some energy. As it falls down, the energy of height is converted of energy of speed. Both the height and the speed of the marble change as it falls. Both changes involve energy, as do all changes. If there is no energy then no change is possible. The amount of speed the marble can have is limited by the amount of energy it can release by falling. The farther the marble is allowed to fall, the more energy gets converted from height to speed, and the faster the marble goes.

Drop the marble from various heights to make this point.

The second big idea that we have to focus on is the idea of a system. The universe is huge and it is impossible to understand everything all at once. A system is a group of objects and influences that are related, and which we wish to understand. We choose what to include in our system. We include things that we want to understand and the factors that affect what we want to understand. We exclude everything else. To understand the motion of the falling marble, we only need to concern ourselves with the factors that are relevant to the system. Choosing the system allows us to focus on what is important. Of course, we don's always choose correctly at first. We often choose a simple system to include as few things as possible. We can include more things as our understanding gets better. In today's investigation you will make a system using the energy car and track. Use the diagram shown in part one to help you.

Students follow the instructions shown in part one of the investigation to assemble the car and track and record data in Table 1. Assign each group of students a different hole to attach to the track. This will give each group a different angle for their track.

12B Motion with Changing Speed

How do we describe motion for which the speed is not constant?

Almost nothing moves at constant speed for very long. Think about your own speed as you drive somewhere, speeds are always changing. What kind of graphs can be made that show changing speeds? How do you tell that speed is changing from a graph? These are questions for this investigation.

Materials
- Energy car and track
- Timer and photogates

1 Making a system

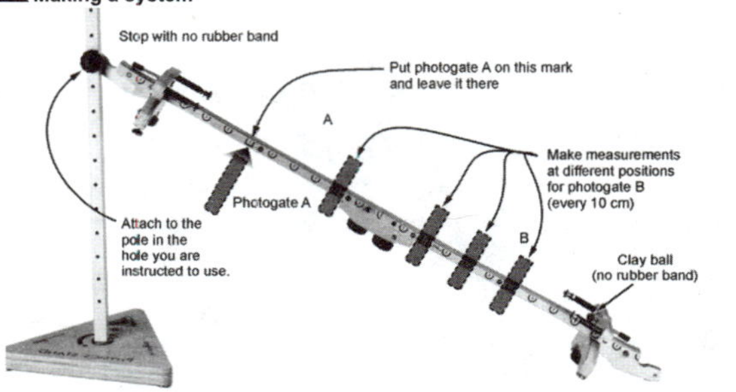

1. Set up the track as a long straight hill. Your teacher will tell which hole in the stand to attach the track.
2. Move photogate B to different positions 10 cm apart along the track (measure position).
3. For every position of photogate B, record the time through the beam at photogates A and B and also the time from A to B. Leave the "Speed at B" column blank for now.
4. Take at least 6 data points along the track being careful to start the car the same way every time. Use photogate A to test whether you should keep the data from a trial or do it over

2 Making the position versus time graph

1. Start the graph by putting the position of the car on the vertical (y) axis. Position is where you placed photogate B.
2. Put the time from photogate A to B on the horizontal (x) axis.
3. Finish the graph by plotting the points from Table 1 and giving the graph a title.

56

Put two photogates on the track. Put photogate A in the position shown on your diagram. Move photogate B to different positions 10 centimeters apart along the track. For every position of photogate B, record the time through the at photogates A and B, and also from the time from A to B. Leave the "speed at B' column blank for now. Take at least eight data points along the track, Develop a method and stick to it. It is important that you start the same way each time. Use photogate A to test whether you should keep the data from a trial or do it over.

Students have attached the track to the pole at different heights, and therefore will have variation in their data. Use this time to have students make predictions about how this may affect the speed of the car.

Which car do you expect to be the fastest? Which car should have the shortest time between photogates? Which car should have the longest time?

Students should recognize that the steepest track should have the fastest car and the shorter time.

2 Making the position versus time graph

Make a graph that shows the position of the car as a function of time. Start by putting the position of the car on the *y*-axis. Position is where you placed photogate B. Put the time needed for the car to travel from photogate A to photogate B on the *x*-axis. Use the data you recorded in Table 1 to finish the graph. Be sure to give your graph a title.

Students complete the procedure in part 2 of the investigation.

3 Stop and think

Think about the position versus time graph you made in the last investigation. How is this graph similar to the graph you made today? How is it different?

Facilitate a class discussion to compare and contrast the graphs students made it this investigation to the position versus time graph in investigation 12A.

4 Calculating the speeds

Calculate the speed of the car at photogates A and B. During the time the light beam is broken, the car travels a distance of one centimeter, the width of the wing on top of the car. You can calculate the speed by dividing one centimeter by the time it takes to pass through a single photogate.

Students calculate speeds for each position and record their answers in Table 1.

5 Making the speed versus time graph

Students create the speed versus time graph using the data from Table 1. Observe that students have correctly labeled each axis and given the graph an appropriate title.

6 Stop and think

Students complete answers to all questions in part six of the investigation. Discuss how the graphs of constant speed differ from graphs of motion with changing speed.

Investigation 12B: Data and Answers

1 Making a system

2 Making the position versus time graph

All data and graphs derived from track placed at 10th hole.

Position of Photogate B (cm)	Time through Photogate A (sec.)	Time through Photogate B (sec.)	Time from A to B (sec.)	Speed at Photogate A (cm/sec.)	Speed at Photogate B (cm/sec.)
10	0.0116	0.0083	0.0989	86.2	120.5
20	0.0116	0.0068	0.1761	86.2	147.1
40	0.0116	0.0053	0.2997	86.2	188.7
50	0.0116	0.0049	0.3507	86.2	205.0
60	0.0116	0.0045	0.3985	86.2	222.2
70	0.0116	0.0042	0.4440	86.2	238.1

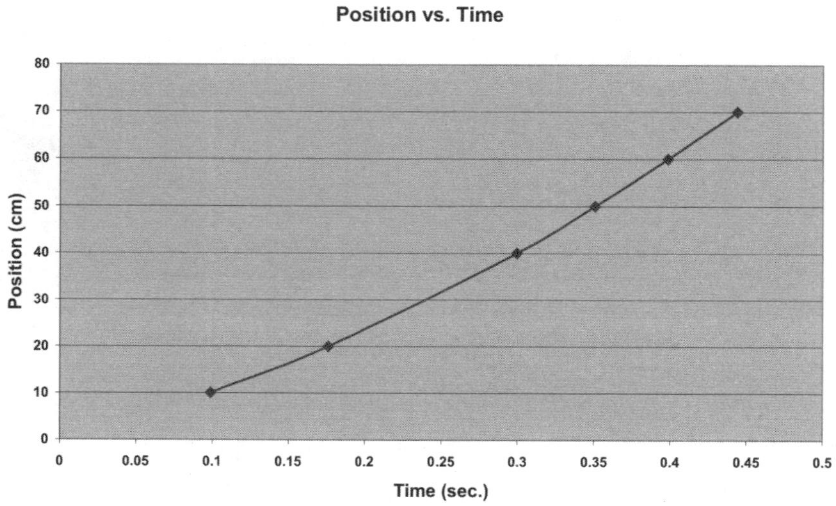

3 Stop and think

a. The graphs are similar because the both slope upwards. They are different because this graph is a curve while the other one is a straight line.

b. The speed is increasing, and this can be seen by each segment of the graph getting progressively steeper.

4 Calculating the speeds

5 Making the speed versus time graph

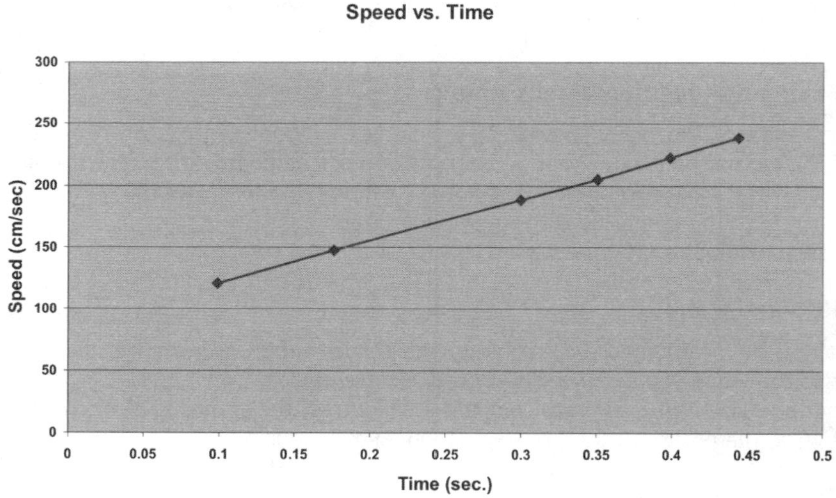

6 Stop and think

a. The graph is a straight line.

b. This is the speed of the car when the elapsed time from A to B is zero. We measured this speed at the A photogate. The intercept on our graph looks to be very close to the speed of 86.2 that we measured at A.

c.

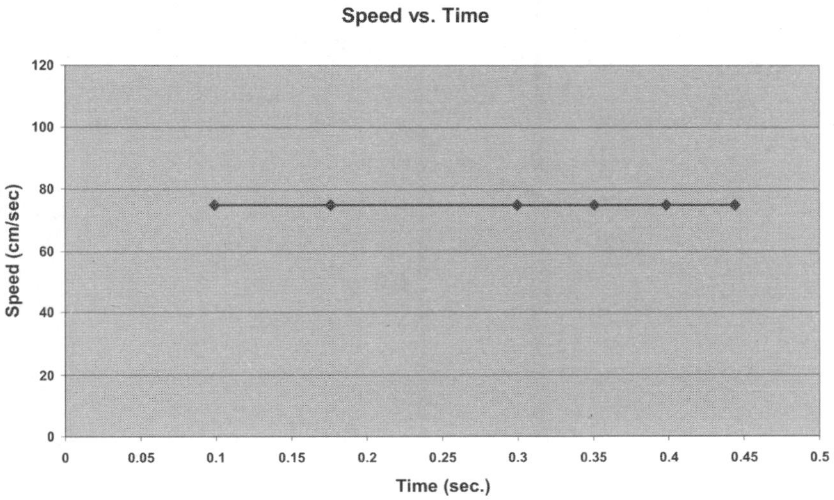

d. A graph of constant speed will be a straight horizontal line, and a graph of changing speed will be a straight line angled up or down.

e. According to our Speed versus Time graph it would reach 200 cm/sec in 0.335 seconds.

f. According to our Position versus Time graph this would happen at 43 cm.

g. Yes. Our prediction was within 5 cm/sec.

Notes:

Chapter 12 Answers

12.1 Section Review

1. distance and direction
2. Distance is the amount of space between two points; distance is a quantity. Position is the location of an object compared to a reference point; position is a distance with a direction.
3. 2 meters east
4. Sample answers:
   one dimension: the position of a car moving in a straight line.
   two dimensions: the position of a car moving in a circle.
   three dimensions: the position of a satellite relative to your house.
5. Answers are:
   a. (+4 km, -2 km)
   b. (-1 km, +1 km)
6. The origin is usually fixed at zero by convention, but could be set anywhere. When the origin is zero, relative positions to the object are usually more easily calculated. Graphs are often drawn with a non-zero origin when one of the coordinates is much larger than the other, i.e. the point (2, 5004) might be easier to represent from an origin of (0, 5000) rather than trying to scale the 5000 to fit on the graph.

12.2 Section Review

1. It moves at a constant distance each second.
2. Speed tells you how fast something is moving, and velocity tells you how fast and in what direction it is moving.
3. The ball rolls in a straight line for the same distance and in the same direction each second.
4. average speed
5. $v = d \div t = (280\text{ km}) \div (4\text{ h}) = 70\text{ km/h}$
6. $d = v \times t = (15\text{ km/h})(2\text{ h}) = 30\text{ km}$
7. $t = d \div v = (100\text{ m}) \div (5\text{ m/s}) = 20\text{ s}$
8. two days = 48 hours
   $d = v \times t = (20\text{ km/h})(48\text{ h}) = 960\text{ km}$

12.3 Section Review

1. a bigger slope (steeper line) means a faster speed
2. The x-values represent time and the y-values represent position.
3. ***Graph of Kayaker's Position vs. Time***

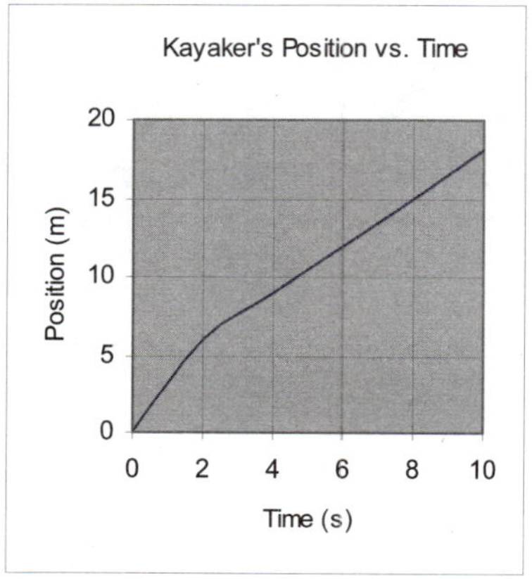

4. Robin has the faster speed, because the slope of the position vs. time graph is steeper.
5. Robin's speed = rise/run = 100/12 = 8.3 m/s
   Joel's speed = rise/run = 80/12 = 6.7 m/s
6. Answers are:
   a. Graph of Maria's position vs. time

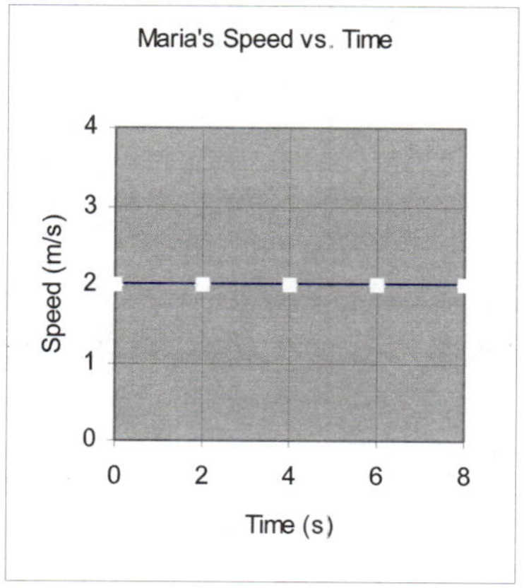

   b. Maria's distance = $v \times t = (2\text{ m/s})(8\text{ s}) = 16\text{ m}$;
   Alternatively, the area under the line is equal to $(2\text{ m/s})(8\text{ s}) = 16\text{ m/s}$.

Connection

1. Velocity is the rate at which an object's position changes.
2. Inertia causes the feeling of being thrown around. When the roller coaster goes around a curve, the your body want to keep going forward and you feel like you're being thrown to the side. You feel like you're falling when the roller coaster is in free fall. During free fall, the roller coaster seat is exerting no force on you, and you feel a sensation of weightlessness
3. Gravity applies a constant downward force on the cars. When the track slopes down, gravity pulls the front of the car toward the ground resulting in acceleration.
4. The first hill has to be highest, because the roller coaster is powered by gravity.

Chapter Activity

Applying your knowledge: sample answers

a. The graph with the steeper slope will be the faster average speed.

b. The partner who finished first will have the faster average speed.

c. If the lines on the graph are straight, then I moved at constant speed. Neither of us moved at constant speed.

d. Jane = 0.89 m/s = 88.9 cm/s
Will = 1.01 m/s = 101 cm/s

Position (m)	Will's somersault Time (s)	Jane's somersault Time (s)
0	0	0
2	1.0	2.2
4	3.7	5.6
6	6.3	7.1
8	7.9	9.0

Graph of Sample Data:

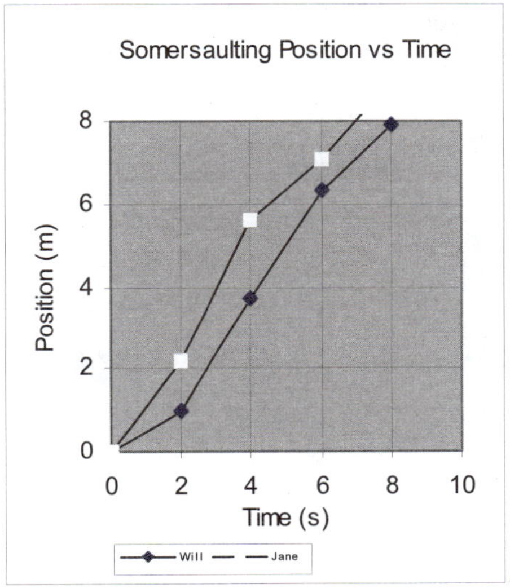

Chapter Assessment

Vocabulary

Section 12.1

1. origin
2. position

Section 12.2

3. constant speed
4. average speed
5. speed

Section 12.3

6. slope

Concepts

Section 12.1

1. all of them
2. (-4 miles, 2 miles)

Section 12.2

3. A speedometer can tell you that the car is moving at a constant *speed*. However, the speedometer cannot tell you what *direction* you are moving. If the car's direction is changing then you are not going a constant *velocity*. The speedometer cannot differentiate between speed and velocity. You'd need a compass in addition to the speedometer to know your velocity.
4. Sample answer: mph, kph, m/s, ft/s
5. $v = d \div t$
   $d = v \times t$
   $t = d \div v$

Section 12.3

6. The steeper line on a position vs. time graph indicates the faster speed.
7. The middle graph.
8. The graph on the left.

Problems

Section 12.1

1. -5 meters; to the left of the origin
2. +25 cm
3. (130 m, -20 m)
4.

distance (m)	speed (m/s)	time (s)
60	10	6
45	5	9
100	50	2

5. $t = d \div v = (1{,}500 \text{ km}) \div (300 \text{ km/h}) = 5$ hours
6. $d = v \times t = (33.63 \text{ km/h})(4 \text{ h}) = 134.52$ km
7. Answers are:
   a. $v = d \div t = (300 \text{ cm}) \div (1 \text{ h}) = 300$ cm/h
   b. 1 hour = 60 min; $v = d \div t = (300 \text{ cm}) \div (60 \text{ min}) = 5$ cm/min
   c. 1 m = 100 cm; 300 cm = 3 m; $v = d \div t = (3 \text{ m}) \div (1 \text{ h}) = 3$ m/h

Section 12.2

8.

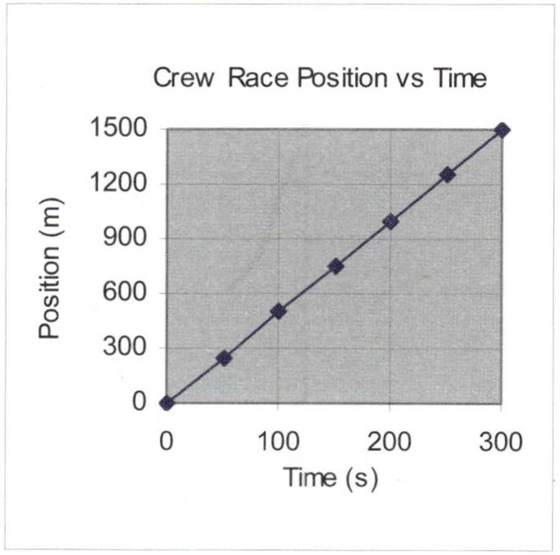

Section 12.3

9.

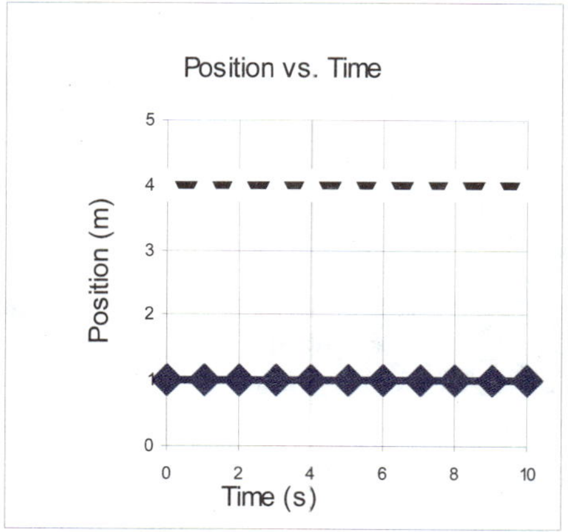

10. Answers are:
    a. speed = rise/run = (2 m)/(1 s) = 2 m/s
    b. speed = rise/run = (2 m)/(2 s) = 1 m/s

11. Answers are:

a.

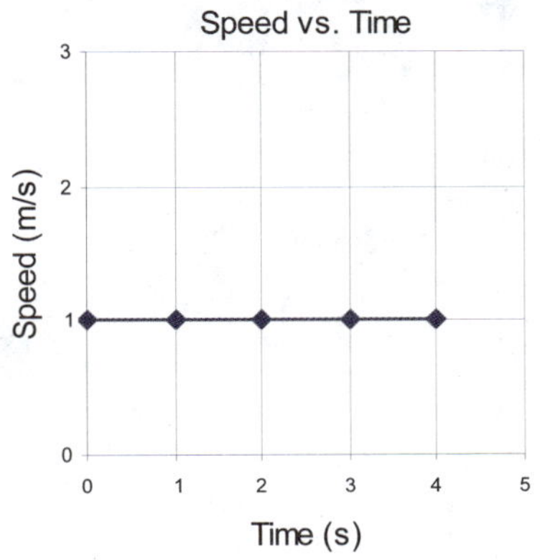

b.

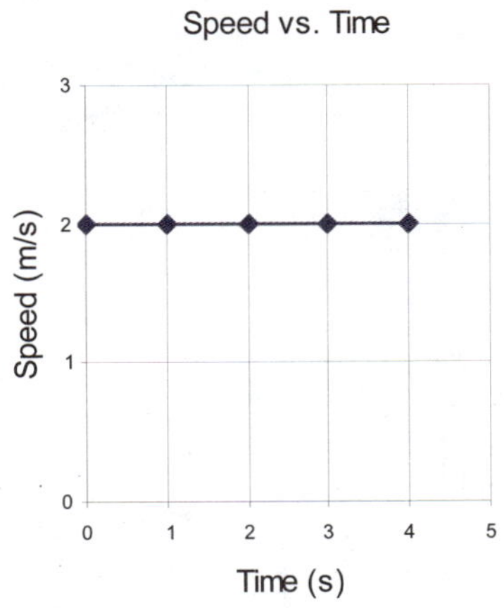

12. distance = $v \times t$ = (4 m/s)(10 s) = 40 m;
Alternatively, the area under the line is equal to (4 m/s)(10 s) = 40 m/s.

13. Answers are;

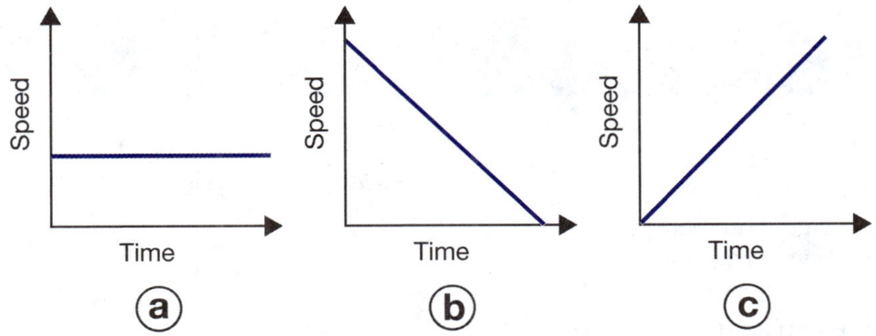

Chapter 13: Forces

Learning Goals	California Content Standards	Activities and Resources	Assessment
13.1: Forces The student will: • Define force. • Explain the relationship between gravity and weight. • Apply a rule to calculate weight. • Differentiate frictional forces. • Discuss applications of friction.	PS 2b PS 2d PS 9f	**Laboratory Investigation 13A:** Forces **Teaching Illustrations:** Four Elementary Forces, Calculating Weight, Friction, Airplane Forces **Graphic Organizer:** Forces **Biographies:** Chien-Shiung Wu **Skill and Practice Sheets:** Mass vs. Weight	**Exam View Test Bank** Chapter Pre-Test **Section 13.1 Review**
13.2: Addition of Forces and Equilibrium The student will: • Distinguish scalars and vectors. • Define equilibrium. • Draw free-body diagrams. • Describe spring and tension forces. • Calculate the net force on an object.	PS 2a PS 2b PS 2c PS 2d	**Laboratory Investigation 13B:** Equilibrium **Teaching Illustrations:** Free-Body Diagram **Skill and Practice Sheets:** Equilibrium **Connection:** Defy Gravity? It Can Be Done **Chapter Activity:** Kilograms to Newtons: Make Your Own Measurements!	**Section 13.2 Review** **Chapter Assessment** **Exam View Test Bank**

Estimated Instruction Time	Language Tools	Differentiated Strategies	Materials
Five 45-minute class periods	**Literary Picks** **Vocabulary:** friction, static friction, sliding friction **Word Origins:** friction	**ELL Strategies:** Listed in ELL Ancillary **Learning Strategies:** • Cooperative Learning - Investigation 13A • Active Learning - Motivate • Guided Practice - Explain: Weight Calculations • Active Learning - Extend **Teaching Tip:** Art and Friction	**Investigation 13A:** force scales, washers, loops of string, friction block, mass balance
Six 45-minute class periods	**Vocabulary:** vector, magnitude, net force, balanced forces, equilibrium, normal force, free-body diagram, tension, compression **Word Origins:** Equilibrium **Closure Cards**	**Learning Strategies:** • Cooperative Learning - Investigation 13B • Active Learning - Motivate: Scalars and Vectors • Demonstration - Explain: Confusing Words • Active Research - Extend **Teaching Tip:** Positive and Negative Numbers, Drawing free-body diagrams and determining net force	**Investigation 13B:** force scales, key ring, loops of string, rubber band **Chapter Activity:** masses of 50, 100, 200, and 500-grams, 30 1/2 inch steel washers, spring scale calibrated in newtons

13.1 Forces

About the Lesson

In this lesson, students learn about forces, such as gravity and friction. Students distinguish mass and weight and apply appropriate units of measure to both quantities. Other types of forces, such as the electromagnetic force and strong nuclear force are also discussed.

Have a student volunteer sit on top of a tall stool so that his or her feet will not touch the floor. Say to the student, “Without touching the stool with your feet or using your thighs to push off of the stool, I want you to try and stand up.” Give the student a few attempts at standing. The student will not be able to stand if he follows your instructions.

Ask the student, “Why weren't you able to stand up?” A common response is that the she had nothing to push off of. Then have all students stand. Ask students to describe what their legs and feet do. Students should observe that their feet pushed the ground and they were able to stand. They should also infer that the applied force (push) is what enabled them to stand.

Students complete Investigation 13A before reading section 13.1 of the student text.The investigation consists of two experiments: the first allows students to differentiate mass and weight while the second experiment is a lesson about friction. Try the investigation ahead of time so you know what to expect.

Investigation 13A: Forces

Setup	Materials	Investigation Vocabulary
1. Allow one class period to complete the investigation. 2. Students work in groups of three to five. 3. Check the condition of the scales before the investigation. 4. Buy the washers at a home improvement store. It is cheaper to buy them by the box. Have extras on hand as washers get lost easily.	• force scales (four per group • washers • loops of string • friction block • mass balance	force weight mass calibrate friction coefficient of friction

Literary Picks...

Reading Level	Title
3-5	***Science Experiments: Gravity* by John Farndon** Explores the study of gravity as a force and includes a historical perspective of Isaac Newton and Albert Einstein. Includes real-world applications of gravity and related concepts.
6-8	***Baloney, Henry P.* by Jon Scieszka** A tale spun by Henry, an alien schoolboy who is pressed into telling a whopper about being late or risk a lifelong detention from his teacher. The full of baloney story includes an explanation of the law of gravity that is humorous and fun.
9-12	***The Physics of Baseball (3rd Edition)* by Robert Adair** This book is a mix of scientific fact coupled with sports trivia. Topics include effects of ball stitching on air resistance, fundamentals of force, velocity, pitch formulas, and bat vibration. Adair is Sterling Professor Emeritus of Physics at Yale University.

Guided Practice: Weight Calculations

Write the equation for determining the weight of an object on the board. Model how to apply the equation *W*= *mg* to solve weight problems. Refer students to the three forms of the weight equation shown on page 275 of the student text. Then use the problems shown below as guided practice.

1. The weight of an object is 600 N. What is the mass of the object in kilograms? (Answer: 61.2 kg)
2. A weight lifter can lift 300 N. What mass in kilograms can she lift? (Answer: 30.6 kg)
3. Calculate the weight of an object with a mass equal to 150.0 kg. (Answer: 1,470 N)
4. Anna weighs 98 N. If her mass is 10 kg then how much is gravity pulling on Anna? (Answer: 9.8 N/kg)
5. A wooden toy has a mass equal to 0.43 grams. Determine its weight in newtons? (Answer: 4.2 N)

1. Have students work in groups to design and conduct experiments to demonstrate examples of rolling, sliding, and static friction.
2. How does friction impact sports? Have students discuss how friction is involved in sliding into home plate, winning a relay race, or other sports like curling.
3. Turn your classroom into a bowling alley. Set up a variety of lanes, each made up of a different type of material. Some examples include wood, carpet, plastic, metal, and sandpaper. Have students roll a hard ball along each surface. Which ball travels farther? Ask students to explain what they observe.

Students complete section 13.1 review questions.

Word Origins...friction (from Latin *fricar* meaning "to rub")

When students are asked to describe or demonstrate friction, a common response is to rub their hands together or to say that it is the result of objects rubbing against one another. Ask students to list three examples of friction.

Teaching Tip . . .Art and Friction

Ask your students what they know about Leonardo DaVinci. Most students will associate DaVinci with the *Mona Lisa*, and will be surprised to learn of his pioneering studies of friction. Have students research to learn more about how DaVinci studied and arrived at conclusions about friction.

Investigation 13A: Forces

Complete the "Motivate" activity before starting the investigation.

Think of times when you had to use a force. When you apply a force, what are you really doing?

I was pushing or pulling.

Good answer. A force is a push or a pull. Have you ever thought about how to measure a force?

Students often respond by saying that forces cannot be measured since they are simply a push or pull. This is a common misconception among students.

Think about how you determine your weight. When you step onto a bathroom scale your body presses down on it because gravity is pulling you down. The scale measures that pull of gravity and tells you how much you weigh.

Stand on a scale and demonstrate this concept as you lead the discussion.

In the Unites States pounds are used to measure forces. In the metric system newtons (N) are used to express weight.

1 Measuring forces

Forces have two important properties: strength and direction. In the laboratory, you can measure force with a spring scale. Before you can use the scale it must be calibrated t. You can do this by turning the nut at the top of the scale until the pointer reads zero.

Hold up a spring scale for the students to see to so they can identify it in their equipment. Demonstrate this procedure then help the students who may be having difficulty.

Now, I want you to experience a force. Pull on the spring scale until you have applied a force of 2 N's. Is one newton larger or smaller than one pound?

The newton is smaller.

2 Weight: the force of gravity

Now we are going to look at weight more closely. Earlier we defined weight as the pull of gravity on an object. What does an object have to have in order for gravity to exert a pull?

The object must have mass.

Who can recall the meaning of mass?

Mass is the amount of matter an object has.

Good answer. Let's find out if a relationship exists between mass and weight.Take a piece of string and attach five washers to it. Now attach the string with the washers to the spring balance you have just calibrated. Measure the weight of the washers and write down your answer in the data table

Students record data.

Be sure your mass balance is also calibrated. Once your balance is calibrated, place the string and five washers on the pan. Record the mass in kilograms in Table 1. Use the calibrated force scale to measure the weight of the washers. Record this measurement in the data table.

Repeat the procedure to find the weight and the mass for 10, 15, 20, 25 and 30 washers.

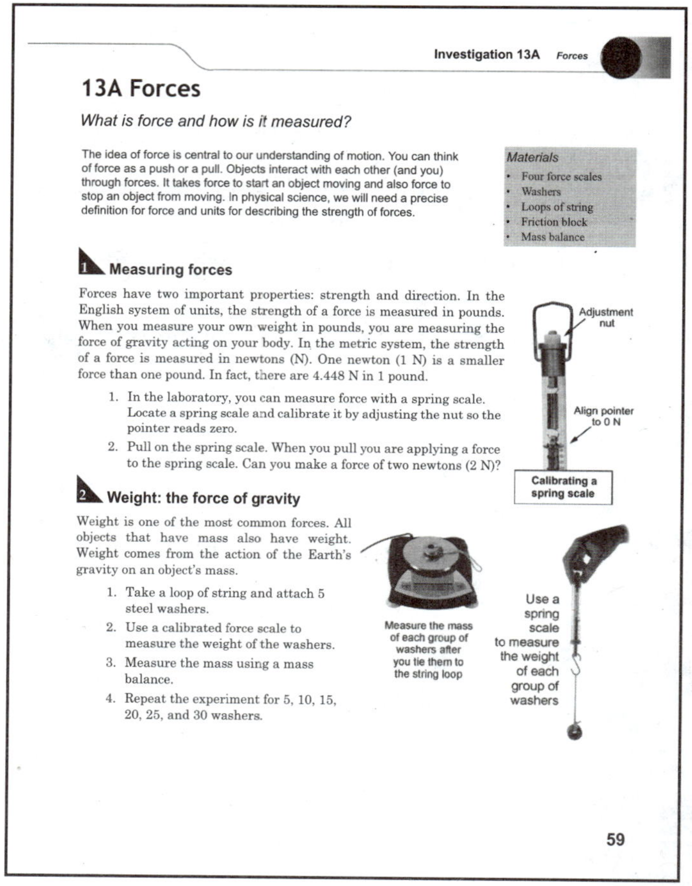

Investigation 13A Forces

13A Forces

What is force and how is it measured?

The idea of force is central to our understanding of motion. You can think of force as a push or a pull. Objects interact with each other (and you) through forces. It takes force to start an object moving and also force to stop an object from moving. In physical science, we will need a precise definition for force and units for describing the strength of forces.

Materials
- Four force scales
- Washers
- Loops of string
- Friction block
- Mass balance

1 Measuring forces

Forces have two important properties: strength and direction. In the English system of units, the strength of a force is measured in pounds. When you measure your own weight in pounds, you are measuring the force of gravity acting on your body. In the metric system, the strength of a force is measured in newtons (N). One newton (1 N) is a smaller force than one pound. In fact, there are 4.448 N in 1 pound.

1. In the laboratory, you can measure force with a spring scale. Locate a spring scale and calibrate it by adjusting the nut so the pointer reads zero.
2. Pull on the spring scale. When you pull you are applying a force to the spring scale. Can you make a force of two newtons (2 N)?

Calibrating a spring scale

2 Weight: the force of gravity

Weight is one of the most common forces. All objects that have mass also have weight. Weight comes from the action of the Earth's gravity on an object's mass.

1. Take a loop of string and attach 5 steel washers.
2. Use a calibrated force scale to measure the weight of the washers.
3. Measure the mass using a mass balance.
4. Repeat the experiment for 5, 10, 15, 20, 25, and 30 washers.

Measure the mass of each group of washers after you tie them to the string loop

Use a spring scale to measure the weight of each group of washers

59

3 Stop and think

Make a graph to compare the data you collected for mass and weight. Which variable belongs on the *x*- and *y*- axes?

The *y*- axis is labeled with weight in Newtons and the *x*-axis with mass in grams. Students use data from the table to create the graph and answer questions in part three.

4 Friction

Let's look at a different type of force called friction. What comes to mind when I say "friction?"

Students share in the discussion. A common response is that it is necessary to overcome friction in order to move with greater ease.

Unfortunately that is usually the case. Have you ever walked across an icy road or a wet floor? What was it like?

Students share experiences. Some may share that it was slippery and hard to move.

The reason is that there was very little friction. The amount of friction between objects depends on several factors which include the roughness of a surface and the force or weight pressing the objects together. Scientists call this the coefficient of friction.

Students complete the procedure in part four. Assist as needed

5 Applying what you have learned

What happened to the friction force as you added weight to the friction block?

The amount of force needed to move the block increased.

Why do you think this happened?

As weight is added to the block, the amount of force pushing down on the block increased. A greater force must be applied to the block in order to compensate for the added weight.

You may have observed the ratio you calculated changed each time you changed the weight of the friction block. How did it change?

As more weight was added the ratio became larger.

This ratio is called the coefficient of friction. It allows us to compare the weight of a pulled object with the amount of force it takes to pull the object. This means that the larger coefficient of friction, the more force is needed to move the object. Suppose we were to make our friction measurements on sand paper instead of our lab area. What do you think would happen to the coefficient of friction?

It increases because more force is needed to pull the friction block across the sandpaper.

Why would it take more force to pull the friction block over the sandpaper?

The sandpaper is rougher and will grab onto the friction block more. In order to compensate for this, we will need to use a larger force.

Let's find out if your hypothesis is correct. Flip your friction block over and find the coefficient of friction for the rough side. What happened?

The coefficient of friction increased.

Teaching Tip . . .Calibrating the balance

Calibration techniques depend on the type of balance you are using. However, most have a screw under the sample pan that you can use to tare with. Electronic balances have a tare button.

Investigation 13A: Data and Answers

1 Measuring forces

2 Weight: the force of gravity

Number of Washers	Mass (kg)	Weight (N)
5	0.139	1.4
10	0.272	2.8
15	0.404	4.1
20	0.538	5.4
25	0.666	6.6
30	0.804	8.0

3 Stop and think

a.

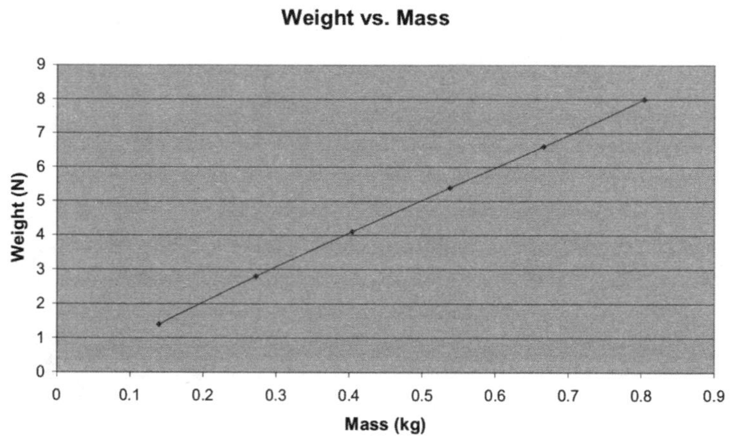

b. The graph is a straight line that is angled up to the right.

c. It means that 1 kilogram feels a force of 9.8 newtons pulling it toward the center of Earth.

d. It would weigh 98 newtons.

4 Friction

Mass (kg)	Weight (N)	Friction Force (N)	Ratio
0.106	1.04	0.20	0.192
0.288	2.82	0.55	0.195
0.481	4.71	0.90	0.191
0.677	6.63	1.25	0.189

5 Applying what you have learned

a. The friction force increased. The greater the weight of an object the harder it pushes down on a surface.

b. The ratio basically stayed the same, no matter what the weight of the block.

c. The coefficient of friction increases on rougher surfaces like sandpaper.

Notes:

13.2 Addition of Forces and Equilibrium

About the Lesson

In this lesson students continue their studies of forces, with emphasis on force as a vector. Topics discussed include equilibrium, normal forces, and spring forces. Students also learn to draw free body diagrams.

Scalars and Vectors

Students have already been introduced scalars and vectors in chapter 12. Provide them with a list of measurable quantities such as length, volume, force, mass, speed, temperature, and velocity. Have students classify each as a scalar or a vector. Then have students explain their responses.

Students complete Investigation 13B before reading section 13.2. In this investigation students use force scales to learn about equilibrium and balanced forces. They also learn to describe the direction of forces by drawing free body diagrams and how to determine the unknown force.

Investigation 13B: Equilibrium

Setup	Materials	Investigation Vocabulary
1. Allow one class period to complete the investigation. 2. Students work in groups of three. 3. Check the condition of the force scales before the investigation.	• force scales (four per group) • key ring • loops of string • rubber bands	equilibrium free body diagram balanced force

Teaching Tip . . .Positive and Negative Numbers

Try these quick problems to find out if your students have mastered how to solve mathematical operations with negative numbers. Answers are provided in blue.

1. 5 +(-3) = **+2**
2. -24-12 = **-36**
3. 16-18 = **-2**
4. 2-(-5) = **+7**
5. -8+12 = **+4**

If students are still unclear about positive and negative numbers, then draw a number line with values from -10 to +10. Label the origin. Model addition and subtraction with positive and negative numbers along the line.

Then create an *x*-*y* grid, showing four quadrants on the board. Place a dot anywhere in each quadrant and ask students, “Is the value of *x* positive or negative in the first quadrant?” and “What about the value of *y*?” Ask these questions for each quadrant. Record students’ responses on the board using plus and minus signs for *x* and *y*.

Confusing Words

Students may already be familiar with the terms *force*, *spring*, *tension*, and *net*, but not in terms of physics. Explain to students that words often have multiple meanings and that it is important to understand the context in which a word is used. Use each word in a sentence to show how it may have different meanings.

1. He pushed the ball with great ***force***.
2. Kim decided to ***force*** Mai into the swimming pool.
3. ***Spring*** is Alberto's favorite season.
4. The hikers drank water from the ***spring***.
5. The mattress ***spring*** provides cushioning as you sleep on the bed.
6. The basketball went through the ***net***.
7. The ***net*** charge on the atom is zero.
8. Mr. Hall's students measured ***tension*** in the rope.
9. The conflict created ***tension*** between the two friends.

1. Use play money to teach students about positive and negative balances in an account. At the beginning of the week, distribute an equal sum of money to each student. Assign a cost to tasks like sharpening a pencil or speaking without recognition. Then assign a value for rewards when students turn in their homework on time or do other things that you designate. Instruct students to keep a register of all their losses and gains to determine their balance at the end of the week.
2. Have students research Hooke's law and explain its meaning. Encourage students to work in teams to demonstrate examples of the law.

Students complete section 13.2 review questions.

Word Origins...equilibrium (from Latin *aequilibrium* meaning "equal balance")

As students learn about equilibrium with respect to forces remind them of their initial introduction to equilibrium in this course (studying chemical reactions). Ask students to compare the meanings of equilibrium in both situations.

Many other words that students may be familiar with share the same origin as equilibrium. For example, the abbreviation for pound (*lb*), and the sign of the zodiac, *Libra* which is also the name of a constellation. Show students a picture of the symbol for Libra and ask them to explain what it means. The pair of scales implies balance.

Another interesting connection exists to the word *deliberate*. Have students look up the meaning of this term. To deliberate means to carefully weigh both sides of an issue. As a person deliberates, he or she seeks to reach a point of balance.

Teaching Tip . . .Drawing free-body diagrams and determining net force

Provide ample opportunities for practicing these two important skills. Students understand these concepts better as they continue to practice. Guide students as they draw free-body diagrams and solve net force problems.

Investigation 13B: Equilibrium

What wants to play tug of war?

Choose two students who appear to have approximately equal weight and strength. Provide the students a rope with a flag tied in the middle. Instruct them to play "tug of war" with one student winning, then the other. Be sure they understand the need for periods where it appears that students are at a stalemate.

Who won the battle?

Students observe that each player won a turn.

Was there ever a time when neither was winning?

Yes, when the flag in the middle did not move.

What can you say about the forces when the flag was not moving?

The forces were equal; but in opposing direction.

You are describing balanced forces. When forces are balanced they add up to a net force of zero. This is called *equilibrium* and it is the subject of today's investigation.

1 Making equilibrium

Calibrate your spring scale. Hold the spring scale vertically and turn the nut at the top until the scale reads zero. Now attach the three scales to the loops of string with the key ring in the middle. Make sure you note which numbered scale you are using. Each person needs to pull on their scale while keeping the ring motionless. Be sure to keep your scales even distances from the ground.

Number the scales 1, 2 and 3. This will help with data recording.

Now, record the forces when two people are pulling against one. Be sure to record your forces in the data table.

Students record data.

Go ahead and experiment. Try several different positions and forces for each balance. Remember the goal is to keep the key ring still. Write your data in Table 1.

Students record data. Walk around the classroom and monitor students as they carry out the experiments.

2 Stop and think

Look at your data table. Do you notice anything special about the forces for each trial?

Yes, if scale one subtracts the forces for scales two and three the answer is zero. Bear in mind that spring scales get stretched and are not always 100% accurate.

Very good. We can represent this by using a free body diagram. A free body diagram is a picture showing the forces acting on an object. The special thing about a free body diagram is that the length of each arrow is proportional to the strength of each force.

Write the term *free body diagram* on the board.

13B Equilibrium

Are there still forces even when things are not moving?

When nothing is moving it does not mean there are no forces acting. It means things are in equilibrium. In equilibrium the total (net) force is zero. You can have as many forces as you wish, they just have to be arranged so the total is zero. This Investigation explores the concept of equilibrium using force scales.

Materials
- Four force scales
- Key ring
- Loops of string
- Rubber bands

1 Making equilibrium

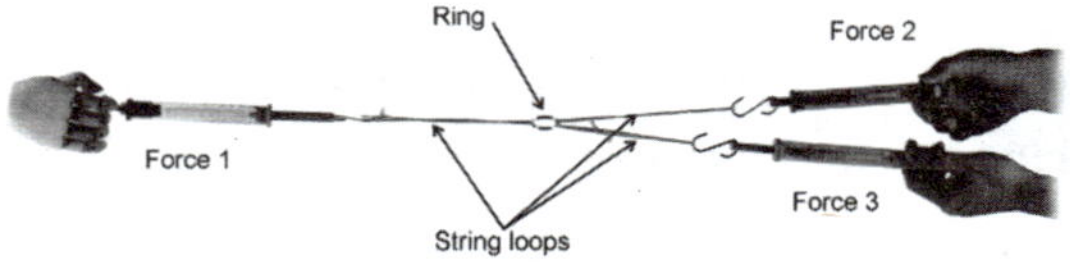

The ring is in equilibrium when it is not moving

Try to keep your spring scales as parallel as possible, not at an angle as in this illustration

1. Hold each spring scale vertically and check that it is properly calibrated. If it needs to be adjusted, turn the nut at the top of the scale until it reads zero.
2. Attach three scales to the loops of string with the key ring in the middle.
3. Have three people each pull a scale, keeping the ring motionless with all the scales in a line.
4. Record the forces when 2 people are pulling against 1.
5. Try a few different combinations of scales and forces.

Table 1: Force data

Scale #1 (N)	Scale #2 (N)	Scale #3 (N)

62

Let me show you an example. I will draw a free body diagram for the ring and forces. I am going to use 1 cm for every newton of force. I have 15 N of force going left, so I am going to draw a 15 cm arrow to the left of the ring. I also have a 10 N force going to the right, so I am going to draw a 10 cm arrow to the right. Then I also have a 5 N force to the right, so I will draw another arrow to the right which is 5 cm long.

Draw the diagram for question *b* on the board. Use a clear ruler in order to model how to determine the length process.

Now, I want each of you to draw free body diagrams for the data you collected in Part 1.

Students draw the diagrams. Assist as needed.

If an object is not moving, what do you now know about the forces acting on the object?

The forces are balanced and in equilibrium. The sum of all forces is zero.

Very good. Using that information, if you replaced one of the spring scales with a rubber band and repeated the experiment, would you be able to identify the unknown force?

Yes. We would just have to use subtraction to find the unknown force.

3 Determining the unknown force

Excellent. Let's give it a try. Set up the experiment again. Except this time, I want you to replace spring scale three with a rubber band. Make sure that spring scale one is on one side of the key ring and spring scale two and the rubber band are on the other side. Pull on both spring scales and the rubber band until the key ring no longer moves. Record the two forces when the system is in equilibrium.

Students experiment and refer to the diagram in part three.

Well done. Now, how can you find the amount of force that was being pulled on the rubber band?

Take the force from scale one and subtract the force from scale two. Write the following on the board: *scale 1 force - scale 2 force = rubber band force*

4 Applying what you have learned

Now that we know all three forces acting on the key ring, draw a free body diagram of the three forces acting on the key ring.

Students draw the free body diagrams. Students may need your assistance with the diagrams. Choose one diagram to model the drawing for students.

Very good. Think about the forces acting on a bridge. Can you identify them?

Sample answers include wind and the weight of cars and trucks.

What about the weight of the bridge itself. What holds that up?

The strength of the materials that the bridge is made of.

What types of materials are needed to build a bridge?

Things like steel that are strong.

Draw a free body diagram of a bridge. Be sure to label all forces acting on the bridge.

Review students' diagrams.

Investigation 13B: Data and Answers

1 Making equilibrium

Scale #1 (N)	Scale #2 (N)	Scale #3 (N)
12	3	9
8	2	6
10	5	5
14	8	6

2 Stop and think

a. The two forces on one side together are equal to but in the opposite direction of the one force on the other side.

b. The diagram will look like the picture in part 1. Force amounts will vary, but should be equal and in opposite directions.

c. The forces acting on the object must be in equilibrium.

3 Determining the unknown force

4 Applying what you have learned

a.

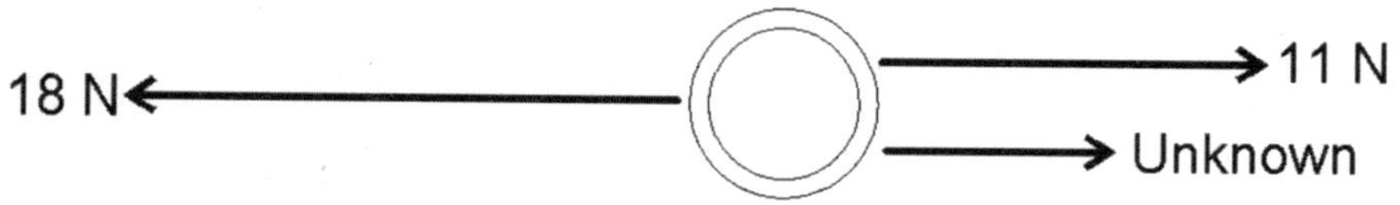

b. The force from the rubber band is 7 N. This is correct, because in order to be in equilibrium, the forces on both sides of the keyring must be equal and in opposite directions. 18 N - 11 N = 7 N.

c.

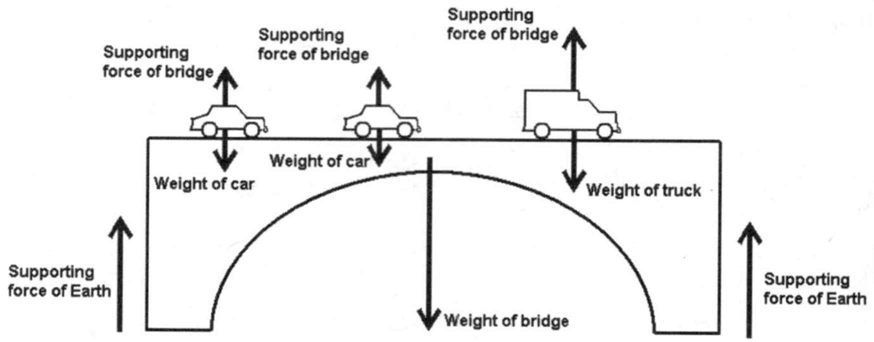

Notes:

Chapter 13 Answers

13.1 Section Review

1. Sample answers:
   Gravitational forces are created between the sun and Earth..
   Forces that move a soccer ball are created by your muscles when you kick the ball.
   Forces that pull at thrown football toward the ground are created by gravity.
2. b
3. 5 pounds is a greater force.
4. Mass never changes by moving to another planet because it is a fundamental property of matter. Weight will change depending on the force of gravity.
5. $W = mg = (3 \text{ kg})(9.8 \text{ N/kg}) = 29.4 \text{ n}$
6. D
7. Sample answers: ball bearings in a bicycle; oil lubricant in a car; wheels (substituting the lower rolling friction for sliding friction)
8. Sample answers: the braking system on an elevator; using a "sticky" material like rubber; gymnasts use chalk on their hands to help them from slipping on the apparatus.
9. b
10. $g = (9.8 \text{ N/kg}) \div 6 = 1.63 \text{ N/kg}$
    $W = mg = (100 \text{ kg})(1.63 \text{ N/kg}) = 163 \text{ N}$

13.2 Section Review

1. c
2. Decibels are not vectors because they only specify a quantity and not a direction.
3. net force = sum of all forces = 9 N + 6 N - 6 N = 9 N
4. a
5. Both forces are in the same direction: 150,000 N + 25,000 N = 175,000 N. The train's engine needs to produce 175,000 N in the opposite direction (uphill) for the train to be in equilibrium.
6.

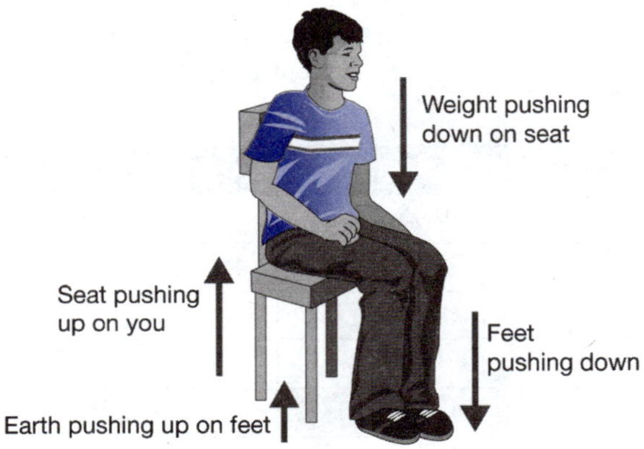

7. b
8. Each rope supports 1/2 the child's weight, so one rope has a tension of 75 N

Connection

1. Weight is caused by the gravitational pull of Earth. It pulls the airplane downward, perpendicular to the body of the airplane.
2. Lift goes in the opposite direction of weight, perpendicular to the body of the airplane. Lift is caused by the wing, and allows the airplane to fly when it overcomes weight.
3. Thrust is a force that provides the forward movement of airplane. It pushes horizontally towards the front of the airplane, opposite the direction of drag.
4. Drag is a force mainly caused by air resistance when an airplane is in flight. It pushes horizontally towards the back of the airplane, opposite the direction of thrust.

Chapter Activity

Applying your knowledge: sample answers

1. Multiply the mass in kilograms by 10 (or 9.8 to be more exact).
2. Divide the weight by 10 (or 9.8 to be more exact).

Spring scales are difficult to read with great accuracy, so it is easiest to tell students to round the values to even readings. When discussing the relationship between mass and weight, point out that the true force of gravity on Earth is found by multiplying the mass in kilograms by 9.8, but 10 is a close enough approximation.

Object	Mass (kilograms)	Force of gravity (newtons)
50-gram mass	0.05	0.5
100-gram mass	0.1	1.0
200-gram mass	0.2	2.0
500-gram mass	0.5	5.0
10 washers	0.0091	0.09
20 washers	0.0181	0.18
30 washers	0.0272	0.27

Chapter 13 Assessment

Vocabulary

Section 13.1

1. friction
2. sliding friction
3. static friction

Section 13.2

4. net force
5. vector
6. weight
7. newtons
8. mass
9. tension
10. normal force
11. free-body diagram
12. equilibrium

Concepts

Section 13.1

1. A 10-N object on the moon has more mass than a 10-N object on Earth.
2. Ice generally has very little friction when it is in contact with other materials for two reasons. First, it is relatively smooth, and has few bumps to cause friction. Second, there is usually a layer of low-friction water between the ice and the object sliding across it.
3. On a converyer belt, friction usually pushed forward on the objects. The conveyer belt moves with a forward force and the static friction pushes in the same direction in order to counteract the inertia that the objects already have.
4. Sliding friction occurs when an object is in motion across another object, and static friction occurs when one oject is being pushed but does not move across the other object.
5. Friction keeps a nail in place because the wood around the nail pushes in on it, and when you try to pull the nail out, friction acts against the nail's motion.

Section 13.2

6. a, d, and f are vectors
7.

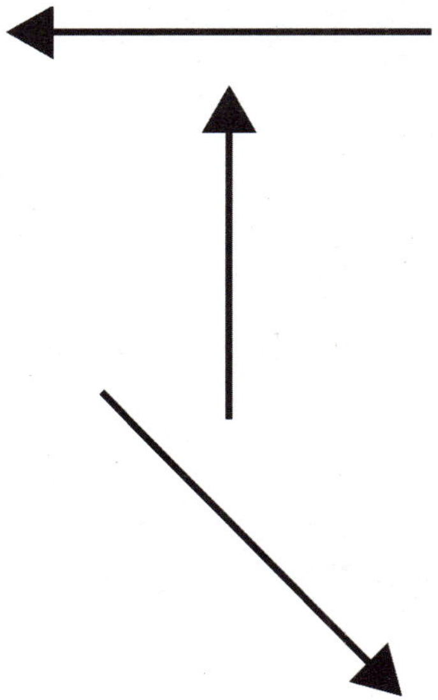

8. The forces acting on a plane are weight, lift, air friction, and engine force. To accelerate forward, the engine force must be bigger than the air friction force, and to stay in the air, the lift must equal the weight.
9. The race car is slowing down because the force backwards is greater than the force forward.
10. d
11.

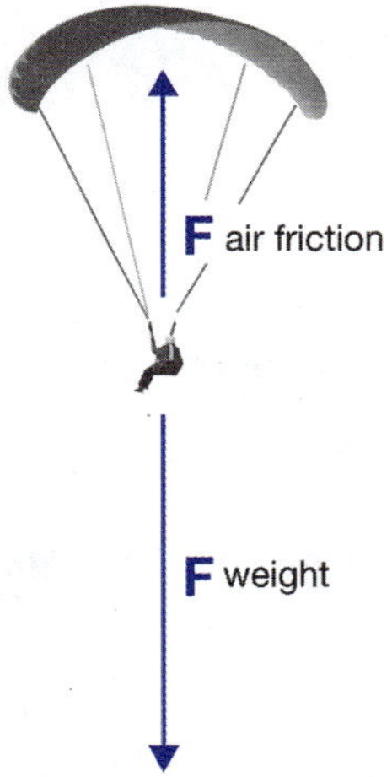

Problems

Section 13.1

1. $W = mg$ = (40 kg)(9.8 N/kg) = 392 N
2. $m = W \div g$ = (50,000 N) ÷ (9.8 N/kg) = 5,102 kg
3. When the 100-kg refrigerator is filled with 25-kg of food, there will be 1-1/4 times as much friction as when it is empty. The frictional force is 250 N.

Section 13.2

4. 8 N east
5. Weight of the book = $W = mg$ = (4 kg)(9.8 N/m) = 39.2 N down. The normal force is 39.2 N pointing upward on the book.
6. Answers are:
   a. Each friend pulls with a force of 200 N, and the tension in the rope is 200 N.
   b. The single friend needs to pull with 200 N to get the same workout. The tension in this case is still 200 N, because the tree exerts a 200 N force making the net force equal to zero.
   c. In both cases above, the net force is zero.

Chapter 14: Force and Motion

Learning Goals	California Content Standards	Activities and Resources	Assessment
14.1: Newton's First and Third Laws The student will: • Define inertia. • Explain action and reaction forces. • Describe the effects of collisions.	PS 2a PS 2b PS 2e PS 2f	**Laboratory Investigation 14A:** Newton's Third Law **Teaching Illustrations:** Newton's First Law, Newton's Third Law **Graphic Organizer:** Newton's Laws of Motion **Biographies:** Isaac Newton	**Exam View Test Bank** Chapter Pre-Test **Section 14.1 Review**
14.2: Newton's Second Law The student will: • Define acceleration. • Apply a formula to calculate acceleration. • Describe the relationship among force, mass, and acceleration. • Relate Newton's second law and gravity. • Apply principles of gravitational force and acceleration to explain orbital motion.	PS 1e PS 2a PS 2c PS 2d PS 2e PS 2g PS 9f	**Laboratory Investigation 14B:** Newton's Second Law **Teaching Illustrations:** Example of a Newton, Units of Acceleration, Acceleration Formula, Newton's Second Law, Equation for Newton's Second Law, Object in Free Fall **Skill and Practice Sheets:** Applying Newton's Laws, Acceleration, and Newtons' Second Law **Connection:** Skateboarding with SIr Isaac Newton **Chapter Activity:** Making a Spool Car	**Section 14.2 Review** **Chapter Assessment** **Exam View Test Bank**

Estimated Instruction Time	Language Tools	Differentiated Strategies	Materials
Six 45-minute class periods	**Literary Picks** **Vocabulary:** force, Newton's first law, inertia, unbalanced forces, Newton's third law, **Word Origins:** inertia	**ELL Strategies:** Listed in ELL Ancillary **Learning Strategies:** • Cooperative Learning - Investigation 14A • Active Learning - Motivate: Does Everything Move? and Extend • Discussion - Explain: Inertia **Teaching Tip:** Spinning Egg	**Investigation 14A:** energy car and track, CPO timer, rubber bands, weights for car **Teaching Tip:** raw egg, boiled egg, stopwatch
Six 45-minute class periods	**Vocabulary:** acceleration, Newton's second law, free fall, acceleration due to gravity, Newton's law of universal gravitation, orbit **Word Origins:** acceleration **Closure Cards**	**Learning Strategies:** • Cooperative Learning - Investigation 14B • Demonstration - Motivate: Newton's Second Law • Paired Learning and Active Learning - Extend **Teaching Tip:** Newton's Second Law, Newton's Laws	**Investigation 14B:** energy car and track, CPO timer, photogate, rubber bands, mass balance **Chapter Activity:** thread spool, assorted rubber bands, metal washer, masking tape, pencil, paper clip, rug or carpeted floor

14.1 Newton's First and Third Laws

About the Lesson

In this lesson, students learn about how the forces acting on an object are related to the object's mass and motion. Students are introduced to inertia, action and reaction forces, and the result of collisions.

Does Everything Move?

Ask students to make a list of everything they can think of that does not move. Students usually generate interesting lists. Then ask students to explain how applying a force to the objects on their list will affect its motion. For example, students commonly list a rock as an item that does not move. Say to students, "Will this rock move if I apply a force to it?" Use an eraser to push the rock along a table or desk. As students, "Did the rock move?"

Forces are constantly acting on objects, even when they do not appear to be in motion. Explain that all matter moves when forces are applied but effects are not always easy to see. An example of this is what happens when students push on a wall during the introduction to Investigation 14B.

Students complete Investigation 14A before reading section 14.1 of the student text.The investigation helps students to infer the meaning of action and reaction forces and that these forces act in opposing directions. Students use the energy car and the CPO timer in this investigation. Review the proper technique for setting up and using the equipment before starting the activity.

Investigation 14B: Newton's Third Law

Setup	Materials	Investigation Vocabulary
1. Allow one class period to complete the investigation. 2. Students work in groups of three to five.	• energy car and track • CPO Timer • four rubber bands • weights for the car	collision action-reaction forces

Literary Picks...

Reading Level	Title
3-5	***Forces and Motion: From Push to Shove* by Christopher Cooper** *Science Answers* is the title of this new science series by Heinemann Library that provides answers and connections to everyday life such as why submarines float. Includes experiments that lead to discovery learning and address student curiosity.
6-8	***Forces & Motion* by Peter Riley** Part of the *Science Topics* series which is specifically designed to supplement middle school curriculums. Key topic concepts are explored and examples of real life applications are explained. Includes colored photos and glossary.
9-12	***The Story of Science: Newton at the Center* by Joy Hakim** Newton is the focus in this second book series by Hakim as she describes the discoveries of motion, gravity, and light. Includes sidebar information and full color illustrations and graphics.

Inertia

Ask students to imagine pushing an empty grocery cart across a parking lot. Is a lot of force needed to get the cart moving? Once the cart is in motion, when does it stop moving? Not a great deal of force is required to start the cart moving. Students commonly respond that the cart stops moving when it collides with a car or some other object in the parking lot.

Say to students, "Now envision the shopping cart filled with groceries." How does the mass of the groceries affect its motion? Is it easier to initiate movement of the empty cart or the filled cart? Why? Once both carts are moving, which one stays in motion longer without interference from any outside force? Use this example to illustrate the relationship between mass and inertia.

1. Have students design an experiment to test the effects of mass on the motion of objects. For example, students may think about the braking distance for various vehicles traveling at different speeds. They can investigate how mass impacts stopping distance. Students may test toy cars with different weights added to them. Then have students obtain a copy of a driving manual from the local Department of Public Safety to find out about recommended braking distances.
2. Bring two skateboards to class. Have students stand on the skateboards and push off of one another (gently) to demonstrate action-reaction forces.
3. Use Newton's cradle to demonstrate action-reaction forces or have students stand against a wall while wearing skates or roller blades. Push against the wall and describe what happens.
4. Blow up a balloon and let it go. Have students identify the action and reaction forces involved as air is released from the balloon.

Students complete Section 14.1 review questions.

Word Origins...inertia *(from Latin* *iners* meaning "inactivity" or "lacking skill"*)*

Inertia is defined as the ability of an object to resist changes in motion. Ironically, inertia shares its origin with *inert*, a term often used to describe noble gases on the periodic table which are for the most part "inactive."

When studying about inertia, students learn that "a body in motion stays in motion" or that "a body at rest remains at rest" unless acted upon by an outside force. One way of expressing this is to say that a body "lacks the skill or ability" to change its pattern of motion on its own. Only force can effect change in motion.

Teaching Tip . . .Spinning Eggs

Use one raw egg and one boiled egg to demonstrate the effects of forces. Spin each egg (on its side). Have students use a stopwatch to determine how long each egg spins. Which egg spins longer? Ask students to describe how friction affects the amount of time each egg spins.

Spin the eggs once more. This time gently tap the egg with your finger. Do both eggs stop moving? Students should observe that the raw egg resumes motion. The liquid portion of the egg also has inertia. Therefore it does not stop moving until a great enough force is applied to overcome the effects of its inertia.

Investigation 14A: Newton's Third Law

I need everyone to stand up and find a spot along the wall. Put both hands on the wall and lean into the wall. Now, push off of the wall and stand up straight again.

Give students time to comply.

What happened?

I pushed on the wall to stand upright again.

Do it again and this time pay special attention to what is happening between the wall and your hand.

Students repeat the activity. Target response: I pushed the wall and the wall pushed back. Students may not "see" the wall push back but Newton's third law justifies that it does.

Good observation. What you just experienced is Newton's third law. It is also known as the action-reaction law. It states that when one object exerts a force on a second object, the second object exerts an equal and opposite force. That is exactly what you have experienced. You pushed on the wall with a certain force, and the wall pushed back with the same amount of force.

Sketch the example of equal but opposite forces on the board.

1 Making a collision

Let's explore this law in a little more detail. In this part of the investigation, you will make a collision. What is a collision?

Students share opinions. Collisions occur when objects bump into each other or make contact.

Good answer. Set up the long track with the rubber band launcher on one end and a clay ball on the other end. Make sure the track is level by using the bubble level. Set the adjustment screw to get 3 cm of deflection with the rubber band. Now, wrap one car with 2 rubber bands. This is will be the moving car that we launch. When you place it on the track, the nose of the car should be facing the launcher rubber band and the "V" should be facing the center of the track.

Model the setup for students.

Set the other car in the center of the track with the "V" end facing the launcher. This will be the target car when the cars collide. Now, launch the moving car and observe what happens. Make sure you record your observations.

Assist students with the setup as needed.

There are four questions after step six in part one. Discuss these questions with the members of you group.

Allow time for students to discuss these questions with group members.

2 Thinking about what you observed

What happened? What did the moving car do?

It hit the target car and stopped.

14A Newton's Third Law

Why do things bounce back when they collide?

When you apply a force to throw a ball you also feel the force of the ball against your hand. That is because all forces come in pairs called action and reaction. There can never be a single force (action) without its opposite (reaction) partner. Action and reaction forces always act in opposite directions.

Materials
- Energy Car and Track
- CPO Timer
- 4 rubber bands
- Mass balance

1 Making a collision

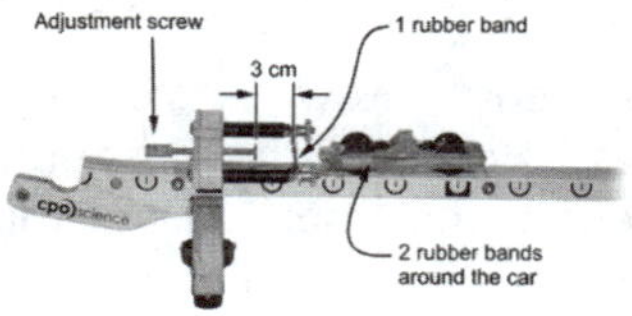

Launching the car

Rest your palm on the wood and pull the car against the screw with your finger on the tab nearest the far end of the car.

1. Set up the long straight track with a rubber band launcher on one end and a clay ball on the other end. Use the bubble level to set the track level.
2. Set the adjustment screw to get 3 cm of deflection from the launcher rubber band.
3. Place one steel marble in each car.
4. Wrap one car with two rubber bands. This is the moving car. The nose of the car should be facing the launcher rubber band and the "V" should be facing the center of the track.
5. Set the other car in the center of the track with the nose end facing the launcher. This is the target car. The cars will collide with each other at the center of the track.
6. Launch the moving car and observe the motion of both cars before and just after they collide. Discuss with your group what you think the answers are to the following questions as you observe the collision.
   Does the moving car bounce back after the collision?
   Does the moving car keep going forward after the collision?
   Does the moving car stop at the collision?
   How does the target car behave?

64

Good. Now, what happened to the target car?

When it was hit, it moved forward on the track.

How strong is the force the target car exerts on the moving car to stop it?

The target car exerts the same amount of force as the moving car did when the moving car hit the target car.

How do you know this?

I know this because the moving car stopped and the target car went forward with the same speed as the moving car had.

So, how can we explain this using Newton's third law?

When the moving car hit the target car toward the end of the track, the target car responded with an equal force, but in the opposite direction, toward the beginning of the track.

3 Changing the masses

Well said. Now look at this experiment from a different perspective. Let's see what happens when we change the masses of one or both of the cars. There are four different combinations shown on your handout for part three. Use the marbles on your table and test those 4 different combinations.

Students experiment with various weights added to the car and measure the effect of these changes. Have students record what they observe as the target car collides with the moving car.

4 Applying what you have learned

Good job! Now let's look at what you observed. What happened to the motion of the cars when the target car had more mass than the moving car?

The moving car stopped and the target car moved forward slightly and more slowly.

That is correct! What happened when the target car had less mass than the rolling car?

When the target car was hit, it still moved forward, but the moving car did not stop. It also moved forward.

How do these observations explain the idea that there are action and reaction forces?

Every time the cars collided, forces were exchanged between them causing the motion of each car to change.

If the action-reaction forces are equal but opposite, why does one car move at a different speed after the collision of cars with unequal masses?

This has to do with Newton's second law which is the topic of the next lesson. Students may infer that when two objects of different mass have the same force applied to them the object with the less mass will have a larger acceleration. You may need to steer students toward this response.

Excellent answer. Now, draw a free-body diagram of our collision with cars of equal masses. Make sure you include the action-reaction forces on each car.

Students draw the free-body diagram of the collision. Review the diagram with students.

Investigation 14A: Data and Answers

1 Making a collision

2 Thinking about what you observed

a. The moving car leaves the rubber band and moves down the track. It collides with the target car in the center of the track. The moving car stops completely after the collision, and the target car moves down the track with about the same speed that the moving car had. The target car stops when it hits the clay at the end of the track.

b. The force is equal, and the evidence is the speed of the target car after the collision. The speed is the same of the moving car just before the collision.

c. The stopping force the target car exerts on the moving car to stop it is equal to the force the moving car exerts on the target car to get it rolling.

3 Changing the masses

4 Applying what you have learned

a. When the target car had more mass than the moving car, the moving car bounces backward toward the rubber band at a slow speed and the target car rolls down the track in the other direction, also at a slow speed.

b. When the target car has less mass than the moving car they both keep moving in the original direction of the moving car after the collision. The target car moves with a speed greater than the moving car's original speed, and the moving car slows down quite a bit as they both roll down the track.

c. The action of the moving car hitting the target car is paired with the reaction of the target car pushing against the moving car and stopping it. Because the forces are equal, the target car rolls down the track with just about the same speed of the moving car just before the collision.

d. A larger or smaller mass will be accelerated at a different rate when the same force is applied. Since the two cars are different masses, the same force propels them at different speeds.

e.

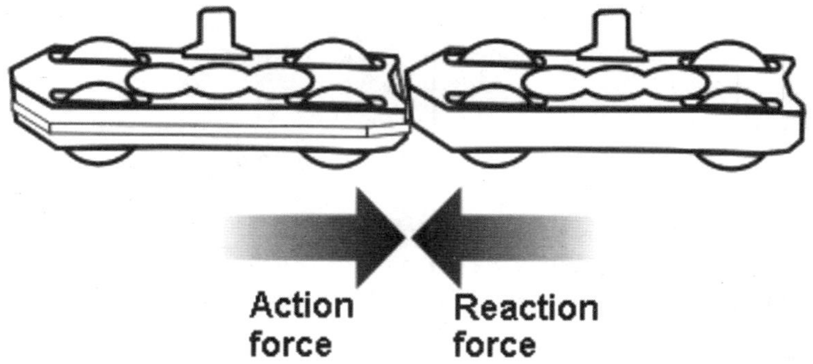

Notes:

14.2 Newton's Second Law

About the Lesson

Students have already learned that applied force causes changes in motion. In this lesson students learn about the result of changes in motion: acceleration. Students learn to apply Newton's second law of motion.

Teacher Demonstration: Newton's Second Law

Hold up notebook or pad of paper and a student textbook. Ask students, "Which will fall faster, the notepad or the textbook?" Students often mistakenly assume that the objects fall at different rates. Then ask, "Which of these will have the greatest impact when dropped?"

Drop both the notepad and the book. Explain to students that both objects actually fell at the same rate based on the acceleration due to gravity (neglecting air resistance). However, the impact of each item falling was different. Then ask students to explain why this is so. The answer lies in the force of each object's impact due to mass. Since the mass of the textbook is greater, its force of impact as it hits the ground is greater. See the Teaching Tip (right) for a second demonstration.

Students complete Investigation 14B before reading section 14.2. In this investigation students use the energy car and track to infer the relationship between force and motion and how the forces acting on an object affect its speed and mass.

Investigation 14A: Newton's Second Law

Setup	Materials	Investigation Vocabulary
1. Allow one class period to complete the investigation. 2. Students work in groups of three.	• energy car and track • CPO timer • photogate • three rubber bands • mass balance	acceleration

Teaching Tip . . .Newton's Second Law

In the motivational example, the phrase *force of impact* is used to describe the falling objects. Here is another way to demonstrate this concept. Ball up a piece of paper. Drop the piece of paper (from a specified height) into a container of sand. What was the impact? Was there a depression created in the sand?

Try the experiment again, this time with a heavy rock or piece of concrete. Drop it from the same height. What do students observe? The greater force of impact is evidenced by the bigger depression in the sand.

Word Origins...acceleration *(from Latin* ***acceleratus*** *meaning "to quicken")*

Acceleration is the rate at which the velocity of matter changes. Ask students to share what comes to mind when thinking about acceleration. A common response is "going fast" or "moving quickly."

Record a television advertisement for an automobile which claims that the car can go from zero to 60 mph in a given amount of time. Use the amount of time stated to determine the acceleration of the car.

Acceleration

Help students make the connection between unbalanced forces and acceleration. Explain to students that velocity changes as a result of unbalanced forces acting on an object. These changes in velocity are reflected by a change in speed (increasing or decreasing) or direction. The rate at which this change in velocity occurs is the *acceleration*.

With more advanced students, you may choose to focus on applying the second law quantitatively to solve acceleration problems. On the other hand, eighth graders must be able to differentiate acceleration and deceleration and to identify the direction and relative magnitude of the force(s) causing the acceleration. It is important for students at this level to make qualitative predictions about a system based on their understanding of how variables like mass, velocity, acceleration, and forces are related.

Also, ensure that students understand the factors that influence the rate of change of velocity. One factor is the mass of the object. Velocity changes slower with more massive objects. Another factor is the size of the force. The greater the unbalanced force, the more rapidly the velocity changes.

Teaching Tip . . .Newton's Laws

Have students apply Think-Pair-Share to solve these conceptual problems.

1. A landscaper decided to create a stone walkway. He drove to a home improvement store to pick up the stones he ordered. His speed was 45 miles per hour. As he returned home, the landscaper observed that it took a longer time to reach the same speed. What caused the difference in the time needed to reach the same speed on the way home from the store? What could the landscaper do to reach the 45 mph speed in the same amount of time?
2. Which airplane requires more force to take off: a small airplane or a large 300 passenger jet? Why?
3. Two vehicles take off from rest at a stop light. One vehicle is a tiny sports car; the other a large SUV. Describe factors that affect each vehicle's acceleration. Which would you expect to have the greater force of impact in a collision? Explain your answer.

1. Have students create a chart describing the different types of forces they have learned about.
2. Divide students into teams. Have each team demonstrate an example of Newton's first, second, or third law. Allow classmates to guess which law is being demonstrated. Then have students explain what is happening with respect to the law.
3. Write a paragraph explaining the role of gravity in the "holding together" the solar system.

Students answer section 14.2 review questions.

Investigation 14B: Newton's Second Law

1 Changing force with constant mass

Today, we are going to investigate the relationship between force, mass and acceleration. First thing we need to do is set up our equipment.

Students set up the track. Assist as needed.

Set up the long straight track with a rubber band on one end and a clay ball on the other. Adjust the screw with the rubber band so that the rubber band can be pulled back about 2 cm. Now, attach a photogate about 20 cm away from the rubber band. Put one marble in the center of the car and place the car on the track. Using the screw as a stop guide, pull the car back into the rubber band and then launch the car.

Demonstrate the launch technique.

Practice launching the car until you get three times that are within 0.0015 seconds of each other. Record those three closest results on the data table.

Students practice the launch and record their results.

Now that you are all experts in launching the car with one rubber band, I want you to repeat this process with 2 and then 3 rubber bands.

Walk around the classroom and monitor students as they carry out the experiments.

Let's examine your data so that we can compare what happened. We will need to find the average speed of the car for each trial of rubber band cord data. Who remembers the equation for speed?

Speed is the distance divided by time. Write the equation on the board.

Very good. We know the distance is 20 cm, but we have 3 different times. Is there a way in which we can represent all 3 times?

Yes, find the average.

Very good. Who remembers how to find an average?

Students calculate the average. You may need to review the process of averaging with students.

Now use the speed formula find the speeds of all 3 trials. Be sure to use the average time in your calculations. Remember to write these values in your data table.

Students complete the data.

2 Stop and think

Well done. Now let's take a look at what we did and what we learned. During which part of the car's motion did the rubber band affect its speed?

At the beginning, when the car was just being released.

Is there a way in which we can compare the different data for the 3 trials we just ran?

Yes, we can draw a graph.

14B Newton's Second Law

What is the relationship between force and motion?

The relationships between force and motion are known as Newton's laws. These are among the most widely used relationships in all of physics. This investigation is about the second law which relates the force acting on an object to the object's change in speed and its mass.

Materials
- Energy Car and Track
- CPO Timer
- 3 rubber bands
- Mass balance

1 Changing force with constant mass

For the first part of the investigation you will use different amounts of force to launch the car. What do you think will happen? As the force gets larger, what happens to the speed of the car?

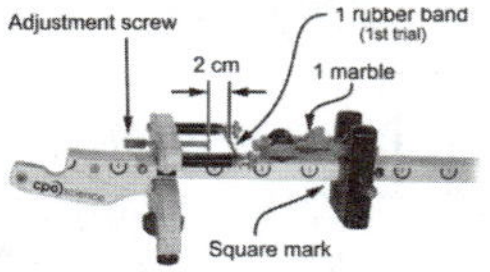

Rest your palm on the wood and pull the car against the screw with your finger on the tab nearest the far end of the car.

1. Set up the long straight track with a rubber band on one end and a clay ball on the other end. Adjust the screw so the rubber band deflects about two centimeters.
2. Put one photogate about 20 cm away from the rubber band.
3. Put one marble in the center of the car. Use the screw to launch the car using the same deflection of the rubber band each time. This means the same force is applied to each launch. Try to get three launches with times that are within 0.0015 seconds of each other. Record your three closest results.
4. Repeat the experiment with two and three rubber bands.

Table 1: Changing force data

Number of rubber bands	Time through photogate (sec)	Speed (cm/sec)

2 Stop and think

a. During which portion of the car's motion is the rubber band affecting its speed?

b. Make a graph showing the speed of the car on the y-axis and the number of rubber bands on the x-axis. As the force was increased, what happened to the speed of the car?

c. Why was the same mass used for all trials (with different force)?

66

Good idea. Let's all draw a graph. Should we put the number of rubber bands or the speed on the *x*-axis?

We should put the number of rubber bands on the *x*-axis because it is the most consistent variable. Students graph the data.

Use your graph to explain what happened to the speed of the car?

As the number of rubber bands increased, so did the speed of the car.

Excellent. Now, think for a moment. Why did we use the same mass each time as we changed the force?

If we would have changed the mass each time, we could not have been sure that it was the force that was causing a change in the speed or acceleration.

3 Changing mass with constant force

Well said. Now let's take a look at the effect mass has on our experiment. Make sure you leave the photogate at the 20 cm mark. Remove two of the rubber bands from the last experiment so there is only one rubber band at the launching end of the track.

Allow time for students to alter the setup.

Now, place the empty car on the track. Launch the car and record the time it took for the car to pass through the photogate. Next, place a marble in the car. Then repeat the process and find out how much time it takes for the car with one marble to pass through the photogate. When you have the time for the car and one marble, add another marble and find its time. Follow the same procedure for 3 marbles. Be sure to record your times on your data table.

Students record data.

Next, we need to find the mass of the car for each trial. Calibrate your balance so that it is reset to zero. Place the empty car on the weighing pan and find its mass. Record this information in the data table.

Students record data.

Now add a steel marble to the car. Find the mass of the car with one steel marble. Record this information in the data table. Find the mass of the car with two marbles and then with three marbles. Be sure to record this information in the data table.

Students experiment and note observations.

Next, we need to find the speed of the car as it passed through the photogate each time. Use the equation for speed and make your calculations.

4 Applying what you have learned

Create a graph of your data from table two. Then use the information from your graph and your observations to answer the questions in part four.

Student create the graph and Newton's second law. Discuss the answers to the questions with students. See Teaching Tip at right.

Teaching Tip . . .Applying what you have learned (part four)

Have students analyze their graphs. Ask, "What does the graph tell you is happening to the speed as the mass increases?" Students should observe that the speed decreases. Ask, "Is this a direct relationship or an inverse relationship?" It is an inverse relationship because as the mass increases, the speed decreases. They are opposite of each other.

Emphasize the fact that force causes a change in speed. Revisiting the law of inertia, tell students that an object will travel at a constant speed until an outside net force acts on it.

Then revisit the motivational activity. Remind students that they discussed the change in speed of the ball, which is called acceleration. Use students observations about increases in force and mass to infer the second law. Students observed that acceleration increases as force increases and that acceleration decreases as mass increases. Using proportionality statements, students can see acceleration is equal to the force divided by mass.

Investigation 14B: Data and Answers

1 Changing force with constant mass

Number of rubber bands	Time through photogate (sec)	Speed (cm/sec)
1	0.0121	82.6
2	0.0092	108.7
3	0.0066	151.5

2 Stop and think

a. Only the very first 2 cm, because that is the only time the car is contact with the rubber band.

b. The speed of the car increased as the force was increased.

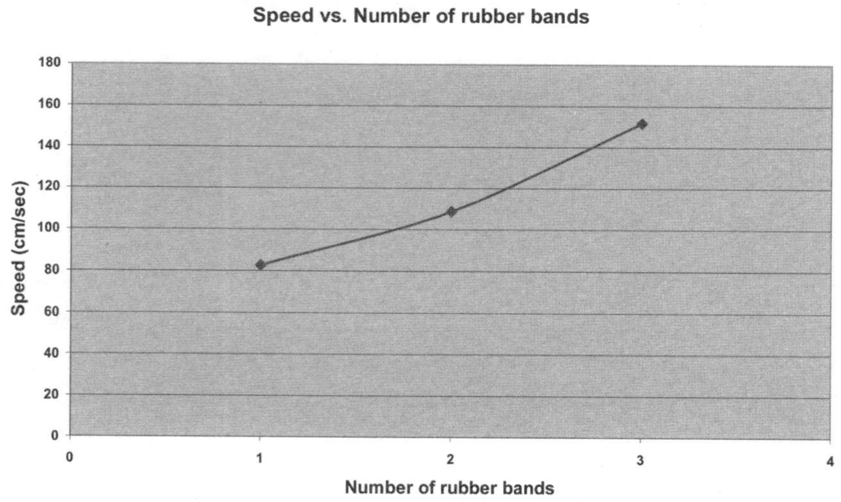

c. The mass of the car was one of the control variables. The experimental variable in this part was the number of rubber bands.

3 Changing mass with constant force

Mass of car (kg)	Time through photogate (sec)	Speed (cm/sec)
0.0568	0.0101	99.0
0.0851	0.0123	81.3
0.1133	0.0142	70.4
0.1415	0.0156	64.1

4 Applying what you have learned

a. The graph shows an indirect relationship. The greater the mass of the car, the lower the speed.

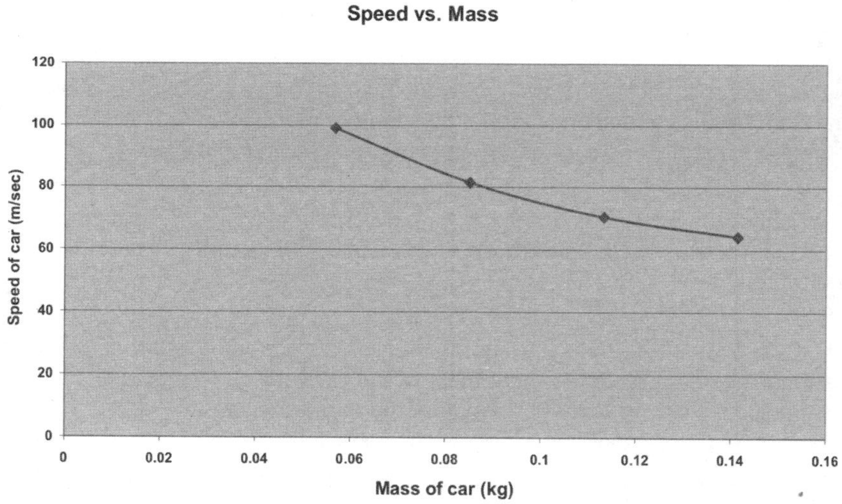

b. The speed changed because more mass requires more force to achieve the same speed. We observed that as we increased mass with the same amount of force, the speed decreased.

c. The force causes a change in speed. Once the car left contact with the rubber band, it did not seem to go any faster. The force applied by the clay stopped the car, a clear change in speed.

d. Force = mass × acceleration

Notes:

Chapter 14 Answers

14.1 Section Review

1. Answers are: a. wind b. the foot kicking the ball c. gravity d. gravitational pull of the moon
2. a shopping cart full of groceries
3. Answers are: a. gravity and friction b. gravity c. friction
4. The action force of Emilio pushing against the canoe creates a reaction force that pushes the canoe away from Emilio. The action and reaction forces are equal in strength and opposite in direction. Emilio clearly misjudged the effect of the action/reaction forces.
5. Each swing chain supports half of Molly's weight, or 250 N.
6. The team with the greater force pair of feet pushing against the ground/ground pushing against feet will win the contest. The force pairs involving the rope are identical between both teams because tension in the rope is the same everywhere in the rope, no matter how hard either team pulls on the rope. The winning team will push hardest against the ground.

14.2 Section Review

1. Yes, an object is accerating if its velocity is changing. Since velocity is a vector with both magnitude and direction, a change in direction while going around a curve is a change in velocity.
2. One unit of time is part of the speed unit, and the other is the time over which the speed changed. If the change in speed is in meters per second and the time is in seconds, then the unit for acceleration is m/s/s or meters per second per second. This is just another way of saying that acceleration is the speed change per second.
3. acceleration = change in speed ÷ time = 6 m/s ÷ 3 s = 2 m/s^2
4. Answers are:
   a. change in speed = 5 m/s - 3 m/s = 2 m/s
   b. acceleration = change in speed ÷ time = 2 m/s ÷ 4 s = 0.5 m/s^2
5. The acceleration of an object is directly proportional to the applied force, and inversely proportional to mass. So increasing the force will increase the acceleration, but increasing the mass will decrease the acceleration. If you push an object harder, its speed will increase. If you increase the mass of an object, inertia slows down its acceleration.
6. $a = F/m$ = 4,000 N ÷ 1500 kg = 2.6 m/s^2
7. The potato with the least mass will have the acceleration - the 100-gram potato.
8. gravity
9. The orbiting moon falls around Earth. But as it falls, Earth curves away beneath it, so it doesn't get pulled into Earth. As Earth pulls on it, the moon's direction changes but it does not come any closer to Earth.
10. The acceleration of the motorcyle is 2.5 m/s^2. The mass is 250 kg.
    $F = ma$ = 250 kg × 2.5 m/s^2 = 625 N

Connection

1. No, skateboarders do not defy the laws of physics. While it may appear as though they do, skateboarders actually use the laws of physics to perform their tricks.
2. When a skater pumps, he or she stands upright on the skateboard while it is moving. This suddenly shifts his or her center of mass upward. The sudden shift in the center of mass in the direction of the motion causes an energy gain. The skateboarder accelerates.
3. Torque is a twisting force that starts rotation. Skateboarders use torque in order to spin around in mid-air.

Activity

a. Turning the pencil gives the car energy. Energy is stored in the rubber band.

b. Winding the rubber band a greater number of times did not always make the car go farther. For each rubber band, turning the pencil more than a certain number of turns made the car go a shorter distance.

c. Friction from the floor moved the car forward.

d. The car just spun in place and then turned in a circle on the smooth floor. There was not enough friction for the wheels to get a grip on the floor and move the car.

e. First law: The car stayed at rest until a force (the friction of the spinning wheel) acted on it. Once it was moving, it kept moving until the friction of the spinning wheel stopped pushing it and the dragging pencil's friction slowed it down. Second law: The greater the force of friction on the floor, the greater the acceleration. The car could not get a large force when it was on the smooth floor, so there was not much acceleration.
Third law: The spinning wheel pushed backward on the floor, and the floor (friction) pushed the car forward.

Chapter 14 Assessment

Vocabulary

1. force
2. Newton's first law
3. Newton's third law
4. inertia
5. acceleration
6. Newton's second law
7. free fall
8. law of universal gravitation
9. orbit
10. acceleration due to gravity

Concepts

Section 14.1

1. c
2. mass or inertia
3. inertia is decreased
4. either moving at constant velocity or not moving at all
5. The net force is the sum of all the force vectors acting on the object added together as if they were one single force acting on the object. Acceleration is the motion due to the net force.
6. b
7. b
8. the force on the bat from the baseball
9. The table is pushing up on the brick.

Section 14.2

10. a
11. You could increase its speed, decrease its speed, or change its direction.
12. Answers are: A; X; X; A; A; A
13. If you are applying the brakes on your bicycle, you are accelerating because you are changing your speed.
14. d
15. When one quantity is "directly proportional" to another quantity, if one quantity increases, the other quantity increases, and if one quantity decreases, the other also decreases.
16. c
17. c
18. A satellite doesn't fall into Earth even though it is being pulled toward Earth by gravity because its velocity is great enough that it "falls around" Earth. As Earth pulls on it, the satellite's direction changes but it does not come any closer to Earth.
19. Examples include:
    a. kilogram, gram
    b. newton, pound
    c. newton, pound
20. Mass is a measure of the amount of matter an object has, while weight is a measure of the force of gravity on the object. They are two different measurements and use two different units.
21. Their masses and the distance between them.
22. gravity and air resistance

Problems

Section 14.1

1. 150 N to the right
2. 20 m/s; since the net force is zero, the object continues moving at a constant velocity.
3. 10-kg ball
4. 1000 N (the same force)
5. (1) Water pushes on the hose - hose pushes on the water
   (2) Firefighter pulls on hose - hose pulls on firefighter
   (3) Firefighter pushes against Earth - Earth pushes against firefighter
6. Car: $F = ma = (3{,}000\text{ kg})(2\text{ m/s}^2) = 6{,}000\text{ N}$
   The truck and car collide with the same force = 6,000 N
   acceleration of truck $= F/m = 6{,}000\text{ N} \div 5000\text{ kg} = 1.2\text{ m/s}^2$
7. 100 N

Section 14.2

8. After 1 second: 3 m/s
   After 2 seconds: 6 m/s
   After 15 seconds: 45 m/s
9. Answers are:
   a. $a = (v_1 - v_2) \div t = (29\text{ m/s} - 5\text{ m/s}) \div 4\text{ s} = 6\text{ m/s}^2$
   b. It's acceleration would be -6 m/s^2; the car is slowing down or decelerating rather than speeding up.
10. $a = F/m = 10{,}000\text{ N}/2{,}000\text{ kg} = 5\text{ m/s}^2$
11. $m = F/a = 20\text{ N} \div 140\text{ m/s}^2 = 0.143\text{ kg}$
12. Assuming east is positive: The net force is 50 N - 100 N = -50 N
    $a = F/m = (-50\text{ N})/(10\text{ kg}) = -5\text{ m/s}^2$ (to the west)
13. $F = ma = (40\text{ kg})(7\text{ m/s}^2) = 280\text{ N}$
14. $W = mg = (75\text{ kg})(9.8\text{ m/s}^2) = 735\text{ N}$
15. 9.8 m/s^2

Chapter 15: The Solar System

Learning Goals	California Content Standards	Activities and Resources	Assessment
15.1: The Solar System The student will: • Explain how the solar system is organized. • Identify the planets in the solar system. • Describe the effects of orbital motion. • Compare properties of the planets. • Distinguish terrestrial and gas planets. • Calculate astronomical distances.	PS 2g PS 4b PS 4c Ps 4e	**Laboratory Investigation 15A:** The Size of the Solar System **Teaching Illustrations:** The Solar System, Planet Size Relative to the Sun, Planet Overview **Graphic Organizer:** The Solar System **Math Skills:** Ratios/Proportions **Biographies:** Copernicus, Galileo Galilei, Johannes Kepler, Benjamin Banneker **Skill and Practice Sheets:** Touring the Solar System, Gravity Problems	**Exam View Test Bank** Chapter Pre-Test **Section 15.1 Review**
15.2: The Planets The student will: • Learn interesting facts about each planet in the solar system.	PS 4c PS 4d PS 4e	**Laboratory Investigation 15B:** The Size of the Planets **Teaching Illustrations:** Mercury, Venus, Earth, Mars, Jupiter, Saturn, Uranus and Neptune, Pluto **Connection:** Jupiter's Volcanic Moon **Chapter Activity:** Alien Design	**Section 15.2 Review** **Chapter Assessment** **Exam View Test Bank**

Estimated Instruction Time	Language Tools	Differentiated Strategies	Materials
Six 45-minute class periods	**Literary Picks** **Vocabulary:** planet, solar system, planetary satellite, Newton's law of universal gravitation, terrestrial planets, gas planets, asteroid, comet, meteor, meteorite **Read Aloud:** Motivation: *How Much is a Million* by David Schwartz	**ELL Strategies:** Listed in ELL Ancillary **Learning Strategies:** • Cooperative Learning - Investigation 15A • Demonstrate - Explain: Johannes Kepler and the Ellipse • Paired Learning - Extend: Orbital Size of Planets • Mnemonics - Planetary Mnemonic • **Teaching Tip:** Planetary Mnemonic	**Investigation 15A:** **Explain:** two styrofoam cones, two pieces of tag board - one with a cut out circle, and the other with a cut out ellipse
Six 45-minute class periods	**Closure Cards**	**Learning Strategies:** • Cooperative Learning - Investigation 15B • Active Learning - Motivate: Planetary Limerick • Active Research - Extend: Is There a 10th Planet? **Teaching Tip:** NASA Educational Resource Centers in California **Multicultural Experience:** Women in Space **Addressing Misconceptions:** Reflected Sunlight, Earth's Seasons	**Investigation 15B:** metric tape, measuring sewing tape, lacrosse ball, balls of various sizes, calculator **Chapter Activity:** planet fact sheet

15.1 The Solar System

About the Lesson

In this lesson, students learn about the organization of the solar system and that gravity is the force that keeps the planets in orbit around the Sun. They will compare size and distance in the solar system, orbital motion, asteroids, and comets.

How Much is One Million?

It is often difficult for students to conceptualize the size and distance of objects in space. To help them make comparisons within the solar system it is important to understand relative size. Engage students by showing them a piece of chalk and asking them to formulate a hypothesis about how long one million pieces of chalk would be. Then read aloud the story *How Much Is a Million?* by David Schwartz. The story will help students visualize astronomical numbers.

Students complete Investigation 15A before reading section 15.1 of the student text.The investigation answers the question, "How big is the solar system?" Students use proportions to determine scale distances among the planets and set up a scale model.

Investigation 15A: The Size of the Solar System

Setup	Materials	Investigation Vocabulary
1. Allow one class period to complete the investigation. 2. Students work in groups two to three. 3. Remind students to bring a calculator to class on the day before the investigation. You may also choose to invest in a class set of inexpensive calculators. 4. Meet with your school's facilities administrator and reserve the football or track field for the day so you will not interfere with any of the gym classes.	• metric track and field tape measure (25 m or more) • simple calculator	scale model proportion

Literary Picks...

Reading Level	Title
3-5	***Magic School Bus: Lost in the Solar System* by Joanna Cole** Ms. Frizzle and her class take off for a real tour of the solar system when the local planetarium closes for repairs. Bus stops include the moon, the sun, Mercury, Venus, and Mars. Includes full color illustrations and plenty of facts.
6-8	***Exploring Our Solar System* by Sally Ride and Tam O'Shaughnessy** Another tour of the solar system, but this time it is navigated by astronaut Sally Ride and educator Tam O'Shaughnessy. This is a well-written book with straightforward language that is filled with facts and full colored photographs.
9-12	***NightWatch: A Practical Guide to Viewing the Universe* by Terence Dickson** NightWatch, the third edition is easily considered one of the best field guides for beginning astronomers. Includes discussions on light pollution, buying a telescope and binoculars, pronunciation of stars and constellations and photographs. This edition provides planet visibility through 2010.

Johannes Kepler and the Ellipse

Prior to Kepler, astronomers believed the planets had a circular orbit. Kepler was successful in showing the orbit of Mars by using an oval. He later fit the orbit to a curve called an ellipse. An ellipse, a closed curve that is shaped like an egg, is in the conic family. When one celestial body moves about another body under the influence of gravity, its orbit is conical. You can observe this with an upside down cone and a paper with a diameter cut. If the paper is perpendicular to the cone, the shape is a circle. If the paper is at an angle, the shape is an ellipse.

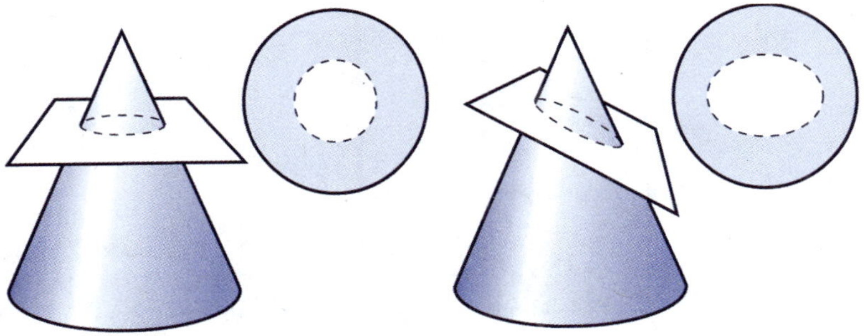

Orbital Size of Planets

Students need to understand that not all orbits are the same size. When the axis or diameter of the planet increases in length, the orbit increases in size. Pair students together and provide time for them use the media center and/or the Internet to do research. Have students compare and contrast the planets to explain which ones have the largest and smallest orbits. Then ask students to explain the reasons for their answers.

Students complete Section 15.1 review questions.

Teaching Tip . . .Planetary Mnemonics

"My Very Educated Mother Just Served Us Nectarines"

The letters in blue represent the eight planets in the solar system with respect to relative distance from the sun. Mercury, Venus, Earth, Mars, Jupiter, Saturn, Uranus, and Neptune.

Far-out orbits

Scientists know that Neptune and Pluto orbit the Sun at tremendous distances. It takes Pluto, a dwarf planet, about 249 years to make one full revolution around the Sun. Since Pluto was discovered in 1930, we have only watched it make a fraction of its orbit around the Sun.

It takes Neptune about 165 years to make one full revolution around the Sun. So how could Neptune ever be consider the farthest planet if it takes Pluto longer to go around the Sun? It's all about the *eccentricity*!

One can think of eccentricity as a measure of how much a planet's orbital shape deviates from a circle. The smaller the eccentricity the more circular the orbit. Pluto's eccentricity is about 25, the most of all the planets in the solar system. The eccentricity of Neptune is about 01. Due to the eccentricity, Pluto's orbit results in an overlap of the orbiting planets. We know this is a fact because Pluto was the 8th planet from the Sun between 1979 and 1999.

Footnote: On August 24, 2006, the International Astronomical Union (IAU) passed a new definition of a planet. According to the new definition, Pluto is classified as a "dwarf planet."

Investigation 15A: The Size of the Solar System

Let's try something a little different today. Using the art materials I have provided, I want you to draw a picture of your bedroom. Since I have never seen your bedroom I want you to make your drawing as accurate and neat as you can.

Each students draws his or her bedroom.

Who would like to share their drawing with the class?

Allow a students to share their pictures.

As I look around I see some really great drawings. Do any of you really live in a room the size that you have drawn?

No, the drawings are too small.

You are correct. The drawings are much too small for someone to actually live in a room that size. What would you call your drawing so people know that it is just a representation of your actual room?

It is a model.

A model is used to represent an object or a relationship among variables. Can you name any other models you may have seen or used?

Students answers will vary and may include cars, graphs, maps, planes, trains, or buildings.

There are many types of models. For instance, a map is a model that contains a scale. A scale model allows you to see the details of a larger object; but it is much smaller than the actual object. In today's investigation you will build a scale model of the solar system.

1 Using proportions to determine scale distances

Before you make the scale model you need to set up a scale. A *proportion* will help you to do this. Does anyone know what a proportion is?

Answers will vary but responses should be along the lines of a ratio or relationship used to make comparisons between quantities.

We use proportions to help us understand quantities that are too large to handle. It allows us to make a smaller but accurate model of the idea we are tying to represent. For this investigation, you will use the proportion shown in part one since Pluto is 5,900,000,000 kilometers from the sun. Using this example, every meter we mark will represent 59 million kilometers.

Write the relationships shown in part one on the board.

Knowing that our ratio is set up in the form of x *(the measurement we plan to use) ÷ the distance from the sun to the planet*, you can set this equal to the original ratio. The result is an equation that you can use to solve for the distances of all of the planets for our model.

Write the relationships on the board. Guide students through these proportions slowly.

Now that you have an equation, let's make it easier to use. How can you simplify this equation?

Isolate x to one side of the equation.

If you isolate x before you start to use the equation, you will only have to do it once. You can then use the new equation for all of your work. How can you isolate x?

15A The Size of the Solar System

How big is the Solar System?

It is difficult to comprehend great distances. For example, how great a distance is 150,000,000 kilometers, the distance from Earth to the sun? One way to get a sense of these distances is to create a scale model. A globe is a scale model of Earth and road maps are scale models of geographic regions. Scale models help us visualize the true sizes of objects and the distances between them. In this Investigation, you will make a scale model representing distances in the Solar System. The results may surprise you.

Materials

- Metric track and field tape measure (25 m or more)

1 Using proportions to determine scale distances

Pluto is an average distance of 5.9 billion kilometers from the sun. We can use a *proportion* to determine a scale distance for our model. Assume the largest distance you can measure is 100 meters. The length of a soccer field is usually between 90 and 120 meters long. For this Investigation, we will use 100 meters as the scale distance between the sun and Pluto.

$$100 \text{ m} = 5,900,000,000 \text{ km}$$

If the distance from the sun to Pluto equals 100 meters, where would you find the other planets? You can answer this question by setting up the following proportion where x is the distance from the sun to any planet, in meters:

$$\frac{x}{\text{Distance from the sun to planet}} = \frac{100 \text{ m}}{5,900,000,000 \text{ km}}$$

Mercury is 58,000,000 kilometers from the sun. Using our proportion, we can find the scale distance:

$$\frac{x}{58,000,000 \text{ km}} = \frac{100 \text{ m}}{5,900,000,000 \text{ km}}$$

Cross-multiply and rearrange the variables to solve for x:

$$x = \frac{100 \text{ m}}{5,900,000,000 \text{ km}} \times 58,000,000 \text{ km} = 0.98 \text{ m}$$

Mercury is 0.98 meters from the sun using this scale.

2 Determining scale distances for the other planets

1. Based on the example in Part 1, you would place Mercury 0.98 meters or 98 centimeters from the sun in your 100-meter scale model.
2. Calculate the scaled distances from the sun to the other planets. Write the distance in meters for each planet in Table 1

68

Multiply both sides of the equation by the distance from the sun to the planet, so it will cancel out on the side with *x*.

Let's try the new equation and see if we can find the scale distance from the sun for Mercury. Knowing that Mercury is 58 million kilometers from the sun, we will substitute 58,000,000 km for distance from planet to sun in our equation. Then we will use the calculator and do the math. What is the answer?

The answer is 0.98 meters.

2 Determining scale distances for the other planets

Now that we all know how to find the distance we will actually be using, I want you to complete Table 1 for part two using the using the following equation.

Write the equation $x = \frac{100 \text{ meters}}{5{,}900{,}000{,}000 \text{ km}} \times \text{actual distance from planet to sun}$ on the board.

Then work one more problem as an example for the students on the board. Be sure to circulate the room and assist the students as they figure the distances between the planets and the sun.

3 Setting up the scale model

Ok, class, let's get into our groups so that we can go outside and build our model. Make sure everyone has a meter stick. When we get outside, I will mark the spot for the sun. Group 1 will measure 10 meters from our sun and we will spray paint that mark. Then group 2 will measure 10 meters from the 10 meter mark, and we spray paint that mark. We will continue this until all 10 groups have measured 10 meters so that we have an area that is 100 meters long. Once you have finished measuring your share of the field, each group will then need to see me for a sign that represents a planet. You will then go and find the spot to place your planet. Once you have your planet in place, turn around and look to see where your planet is relative to the sun and the other planets.

See Teaching Tip at right.

4 Stop and think

What do you think about the size of our solar system?

Answers will vary, but should students should infer that it is very large.

Who would like to explain some of the advantages of the scale model we just created?

It allowed us to get a feel for how far apart the planets are and how far away they are from the sun. It also helped us to better understand the size of our solar system.

Very good. It does help us to visualize the size of the solar system. Now, what are some of the disadvantages?

Our model was too big to really work with. Being part of the model also made it more difficult to see the model as a whole. Also, our model would only benefit the people that were with us when it was built.

Teaching Tip . . .Setting up the scale model

Draw a picture of what the field will look like as the students start to measure off their 10 meter marks. Make sure you have signs for each of the planets and the sun. Also, write down the distance from the sun to that particular planet on each card for easy reference once you are outside. If possible, laminate them so they will hold up for future use.

Divide the class into 10 groups: one for each planet and one for the sun. Go to one end of the field and have the sun group stand at the zero mark. Spray paint a mark to indicate the zero mark.

Then have the first group measure 10 meters. Make a mark with the spray paint. Do this with all of the remaining groups until you have 11 spray painted marks on the field that are 10 meters from each other. Number them sequentially, 0, 10, 20, and so forth. This will help the students find the location for their planet and it will be easier for you to monitor their accuracy. Also, by using the spray paint, you will be set up for the rest of your classes.

Once the field is set up, instruct the students to find the scale location of their planet. After the model has been created instruct the students to look around and notice the distances between them and the sun and other planets.

Investigation 15A: Data and Answers

1 Using proportions to determine scale distances

The planets compared to the solar system

2 Calculating the scales sizes of the planets

Planet	Actual diameter (km)	Scale diameter (m)	Scale diameter (mm)
Sun	1,391,980	0.0236	23.6
Mercury	4,880	0.000083	0.083
Venus	12,100	0.000205	0.205
Earth	12,800	0.002017	0.217
Moon	3,475	0.000059	0.059
Mars	6,800	0.000153	0.153
Jupiter	142,000	0.00241	2.41
Saturn	120,000	0.00203	2.03
Uranus	51,800	0.000878	0.878
Neptune	49,500	0.000839	0.839
Pluto	2,300	0.000039	0.039

3 Stop and think

a. 2.36 cm

b. The Sun's diameter is about 109 times bigger than Earth's. Earth's diameter is about 3.7 times larger than the moon's.

c. All but Mercury and Pluto. Earth may be seen but the moon can not be seen.

d. The size of the planets is amazingly small compared to the size of the solar system.

4 Stop and think

Planet	Actual diameter (km)	Scale diameter (cm)
Sun	1,391,980	696
Mercury	4,880	2.4
Venus	12,100	6.05
Earth	12,800	6.4
Moon	3,475	1.74
Mars	6,800	3.40
Jupiter	142,000	71.0
Saturn	120,000	60.0
Uranus	51,800	25.9
Neptune	49,500	24.8
Pluto	2,300	1.15

Notes:

15.2 The Planets

About the Lesson

In this lesson, students learn that every planet is unique and has its own set of physical properties including composition, appearance, temperature, revolution, size, atmospheres, and moons. They will also learn about Earth's orbit and the intensity of sunlight and how these factors affect the seasons.

Planetary Limerick

A limerick is a simple verse of five lines and is often humorous. When writing a limerick, the last word of lines one, two, and five must rhyme with each other. The final word of the third and fourth lines must rhyme with one another. Using the Think-Pair-Share strategy, have students write a limerick about one planet of their choice. An example, using Mars, is provided below.

There once was a planet named Mars.

Where the water would freeze in jars.

If you tried to live there

With no water and thin air

I am sure you wouldn't get far!

Students complete Investigation 15B before reading section 15.2. In this investigation compare the sizes of other planets to Earth. Students continue to apply their knowledge of scale models from investigation 15A.

Investigation 15B: The Sizes of the Planets

Setup	Materials	Investigation Vocabulary
1. Allow one class period to complete the investigation. 2. Students work in pairs or groups of three.	• metric tape measure (sewing tape) • lacrosse ball • balls of various sizes • simple calculator	diameter

Teaching Tip . . .NASA Educational Resource Centers in California

The Educational Resource Centers (ERC) provide educators access to a wide variety of NASA resources. The ERC assists educators in gathering ideas, doing research, duplicating selections of audiovisual and printed materials. Educators are encouraged to visit the centers to learn how they can incorporate materials produced by NASA into science instruction.

NASA Ames Research Center
Moffett Field, CA
http://amesnews.arc.nasa.gov/erc/erchome.html

NASA Dryden Flight Research Center
Palmdale, CA
http://www.dfrc.nasa.gov/Education/ERC/index.html

NASA Jet Propulsion Laboratory
Pomona, CA
http://learn.jpl.nasa.gov/resources/index.html

Multicultural Experience: Women in Space

Drs. Mae Jemison and Ellen Ochoa were the first African-American and Hispanic women in space, respectively. Born in Decatur, Alabama in 1956, Dr. Jemison served as a science mission specialist aboard the Shuttle *Endeavour*. She logged over 190 hours in space and was co-investigator on the bone cell research experiments during that mission.

Dr. Ochoa was born in Los Angeles, California in 1958. She is a veteran of four space flights and has logged over 978 hours in space as a mission specialist, payload commander, and flight engineer.

Have your students research to find out about Drs. Jemison and Ochoa's current interests and projects.

Addressing Misconceptions: Reflected Sunlight

Students are often unaware that the Moon does not produce its own light. The moonlight humans see is actually light from the sun being reflected on the Moon. The Moon rotates on its axis as it orbits Earth. The same features always face the side of Earth. Since the moon rotates, the sun must rise and set on all sides of the Moon. Therefore, the near side of the moon faces Earth. At times it is illuminated by the sun and at other times it is dark.

Taking a look at the other side - the far side of the Moon faces Earth. In some instances it is illuminated by the Sun and at other times it appears to be dark. It is a misconception to use the term dark side as a meaning for the far side of the Moon because the far side is not always dark. How much of the Moon humans actually see depends on the moon phase. The different shapes of the moon as it orbits Earth are known as the phases of the Moon. There are eight major moon phases which are the result of the changing position of the Moon and Earth and relative to the Sun (Figure 15.1).

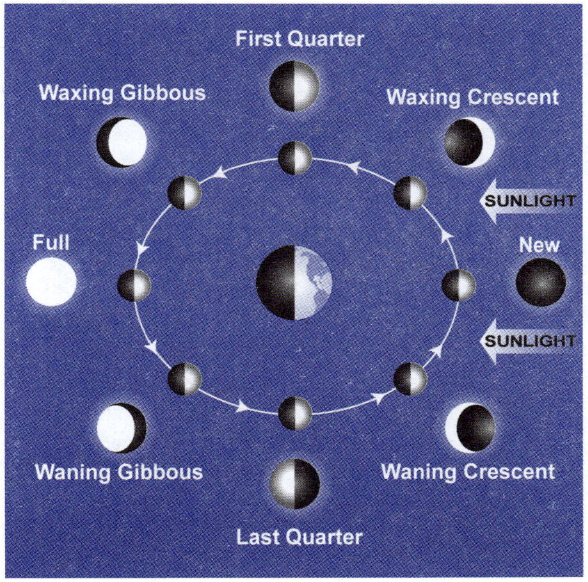

Figure 15.1: Sun and Earth positions relative to the moon's orbit.

Is There a 10th Planet?

An interesting discussion among scientists centers around the possible discovery of a 10th planet in the Kuiper Belt, which is located beyond Neptune. This area is sometimes referred to as the outer solar system. The so-called planet was discovered by scientists Mike Brown, Chad Trujillo, and David Rabinowitz on October 31, 2003. The planet's temporary name is 2003 UB313. Have students do online research to answer the following questions:

1. What is the status of 2003 UB313?
2. What are the physical characteristics of 2003 UB313?
3. Compare and contrast 2003 UB313 to Neptune and Pluto.

Students answer section 15.2 review questions.

Teaching Tip . . .Addressing Misconceptions about Earth's seasons

Most students understand the temperature changes associated with the seasons but they not know why seasons change. Some believe the sole reason is due to Earth's distance from the sun. Earth is closer to the sun in December than it is in June. The seasons are caused by the 23° tilt of Earth's axis relative to its orbit. Earth's northern hemisphere is tilted toward the Sun in June and experiences summer, when the sunlight is more direct and the days are longer. The southern hemisphere is tilted away from the Sun in June and experiences winter. Why? The sunlight is less direct and the days are shorter. Six months later, the northern hemisphere has winter while the southern hemisphere has summer.

Investigation 15B: The Sizes of the Planets

Who remembers what we did in our last investigation?

We found the scale distances of the planets from the sun. We then went outside and built a scale model of our solar system.

You are correct! You built a scale model of our solar system. What do you remember about the distances between the planets and the sun?

They are very large.

This is true. Today, we are going to investigate the sizes of some other very large items in our solar system. We are going to build a model of each of planet and make some comparisons about its size relative to the size of the solar system.

1 Determining scale sizes of the planets compared to the solar system

In the last investigation you used proportions. Just like last time, we are going to use the proportion *100 m = 5,900,000,000 km*. We know diameters of the planets and we know that we are looking for a scale diameter of the planets. Does anyone have a suggestion as to how we can set up our proportion?

Student answers will vary and may be incorrect. Write the equality statement on the board as you describe it.

Let's look at what we know. We know that for every 100 meters, five billion nine hundred million kilometers are represented. We also know that we want to find the number of meters it will take to represent the actual diameter of a planet. We can set this information up as a fraction. If we use both fractions together, we can write an equation that will help us solve for the scale diameter of the planets. Our equation will look like this.

$$\frac{\text{scale diameter (m)}}{\text{diameter of planet (km)}} = \frac{100\text{ m}}{5{,}900{,}000{,}000\text{ km}}$$

Now, let's rearrange the equation so it will be easier to use to solve for the scale diameter of the planet. Does anyone have any ideas or suggestions?

Multiply both sides of the equation by the diameter of the planet in order to isolate scale diameter to one side of the equation by itself.

Very good. Everyone do this.

Write the equation on the board.

Now that we know the equation, how can we find the scale diameter of Mercury?

Use the chart in Part B and find that Mercury's diameter is 4,880 km. Substitute this number into the equation for diameter of planet (km).

Very good. Now that you have the proper equation use your calculators and solve for the scale diameter.

Students solve for the answer which is equal to 0.000083 meters.

15B The Sizes of the Planets

How big are the planets relative to Earth?

Compared to the distances between them, the planets are relatively tiny. Although Earth seems large while standing on it, you could fit almost 12,000 Earths in the space between Earth and the sun! Earth is small compared to some of the other planets, such as Jupiter or Saturn. But other planets are smaller than Earth's moon! This investigation will look at the relative sizes of the planets.

Materials
- Metric tape measure (sewing tape)
- Lacrosse ball
- Other sizes of balls

1 Determining scale sizes of the planets compared to the solar system

What would the planets look like in a 100 meter scale model of the solar system? For example, Mercury has a diameter of 4,880 kilometers. How big would Mercury be in a 100-meter scale model? You can use the same method to determine the scale diameter of Mercury that you used in the last investigation:

$$\frac{x}{4{,}880\text{ km}} = \frac{100\text{ m}}{5{,}900{,}000{,}000\text{ km}}$$

Cross-multiply and rearrange the variables to solve for x:

$$x = \frac{100\text{ m}}{5{,}900{,}000{,}000\text{ km}} \times 4{,}880\text{ km} = 0.000083\text{ m}$$

Based on the example above, the diameter of Mercury in a 100-meter scale solar system would be 0.000083 meters or 0.083 millimeters. This is about the thickness of a human hair! If the distance from the sun to Pluto were 100 meters, Mercury would be smaller than the period at the end of this sentence. The space between the planets is enormous compared to the size of the planets themselves.

2 Calculating the scaled sizes of the planets

Calculate the scaled diameters of the other planets as well as the sun and Earth's moon. Write these values in units of meters in the third column of Table 1.

Table 1: Diameters of the planets, Earth's moon, and sun

Planet	Actual diameter (km)	Scale diameter (m)	Scale diameter (mm)
Sun	1,391,980		
Mercury	4,880	**0.000083**	
Venus	12,100		
Earth	12,800		
Moon	3,475		
Mars	6,800		
Jupiter	142,000		
Saturn	120,000		
Uranus	51,800		
Neptune	49,500		
Pluto	2,300		

70

2 Calculating the scales sizes of the planets

Now that you know how to find the scale diameter of the planets, let's finish off the column of scale diameters in mm on Table 1 in Part 2 of our investigation.

Students complete the data table. A quick review of metric conversions may be necessary as students convert from scale diameter in meters to millimeters.

3 Stop and think

Students complete answers to questions in part three.

4 Modeling the sizes of the planets

Using the information we have calculated in Table 1, do you think that we could build a model of the solar system?

No, our scale is too small.

Our scale is too small to work with, so we will need to increase it to a size that is easier to understand. Suppose we decide that we want a lacrosse ball to represent earth. The diameter of the lacrosse ball is 6.4 cm. How many times does 6.4 go into 12800?

The answer is 2000 times.

Divide each diameter in Table 2 by 2000 in order to determine the scale diameter in cm.

5 Stop and think

Students complete questions. Discuss answers before proceeding with the investigation.

6 Extension: Making a larger scale model of the solar system

You have compared the planets to balls and have made your own construction paper models. Let's use some everyday objects to represent the planets. First, we will need to make a new scale. This time instead of using a lacrosse ball as Earth, we are going to use a globe. Now a globe has a diameter of 30 cm. How many times will 30 go into 12800?

The answer is 426.67 times.

Divide all of the actual diameters in Table 3 by 426.67 to find our new scale diameter.

Students complete calculations.

Now that we have the new scale diameters for the planets, sun and moon, can you think of any object that we could use to act as our model?

Answers will vary.

How many times bigger is 30 cm compared to 0.2 mm?

It is 60 times larger.

Knowing that we are using a ratio of 30 cm to represent 12,800 km on this scale, we figure out what would be the distance between the sun and Pluto. Set the ratio equal to the scale distance between Pluto and the sun over the distance between Pluto and the sun. Cross multiply and solve for the scale distance.

Students complete the calculations and answer questions a through c.

Teaching Tip . . .Stop and think (part five)

If a metric tape measure is not available, have the students measure the circumference with a piece of string. They can then measure the length of the string to find the circumference. Write the formula for calculating circumference ($C= \pi D$) on the board.

Give the student some balls and a tape measure. Allow them time to explore. Whether or not they find balls that can represent the planets will depend on the types of balls you provide.

Pass out the art materials needed to make paper models of the planets. It may be a good idea to put the students in groups of 3 or 4 and let them work together to make the planets. You could assign each group two or three planets to say time and preserve supplies.

Investigation 15B: Data and Answers

1 Determining scale sizes of the planets compared to the solar system

2 Calculating the scales sizes of the planets

Planet	Actual diameter (km)	Scale diameter (m)	Scale diameter (mm)
Sun	1,391,980	0.0236	23.6
Mercury	4,880	0.000083	0.083
Venus	12,100	0.000205	0.205
Earth	12,800	0.002017	0.217
Moon	3,475	0.000059	0.059
Mars	6,800	0.000153	0.153
Jupiter	142,000	0.00241	2.41
Saturn	120,000	0.00203	2.03
Uranus	51,800	0.000878	0.878
Neptune	49,500	0.000839	0.839
Pluto	2,300	0.000039	0.039

3 Stop and think

a. 2.36 cm

b. The Sun's diameter is about 109 times bigger than Earth's. Earth's diameter is about 3.7 times larger than the moon's.

c. All but Mercury and Pluto. Earth may be seen but the moon can not be seen.

d. The size of the planets is amazingly small compared to the size of the solar system.

4 Modeling the sizes of the planets

Planet	Actual diameter (km)	Scale diameter (cm)
Sun	1,391,980	696
Mercury	4,880	2.4
Venus	12,100	6.05
Earth	12,800	6.4
Moon	3,475	1.74
Mars	6,800	3.40
Jupiter	142,000	71.0
Saturn	120,000	60.0
Uranus	51,800	25.9
Neptune	49,500	24.8
Pluto	2,300	1.15

5 Stop and think

a.

Planet	Common spheres
Sun	hot air balloon
Mercury	large glass marble
Venus	tennis ball
Earth	lacrosse ball
Moon	medium glass marble
Mars	ping pong ball
Jupiter	x-large beach ball
Saturn	large beach ball
Uranus	basketball
Neptune	volleyball
Pluto	small glass marble

b. No answer needed.

6 Extension: Making a larger scale model of the solar system

Planet	Actual diameter (km)	Scale diameter of sun or planet (cm)	Representative object
Sun	1,391,980	3262	small mountain
Mercury	4,880	11.4	softball
Venus	12,100	28.4	basketball
Earth	12,800	30	Earth globe
Moon	3,475	8.1	baseball
Mars	6,800	15.9	cantelope
Jupiter	142,000	333	large van
Saturn	120,000	281	van
Uranus	51,800	121	large truck tire
Neptune	49,500	116	large truck tire
Pluto	2,300	5.39	lime

a. 0.20 mm = .020 cm; 30 cm = x(.020 cm); x = 1500, so 30 cm is 1500 times larger than 0.20 mm.

b. This larger scale multiplies everything 1500 times. The proportional distance from the sun to Pluto in investigation 15A was 100 m. With the larger scale, the distance would be 150,000 m, which equals 150 km!

c. It is challenging because the distances from the sun to each planet are so great compared to the size of the planets. The people of Aroostook County in Maine created a model of the solar system along highway route 1 between Presque Isle and Houlton. The planet sizes and distances between are on the same scale. Pluto is 40 miles from the Sun in this model!

Notes:

Chapter 15 Answers

15.1 Section Review

1. No, planets shine by reflected sunlight.
2. Mercury, Venus, Earth, Mars, Jupiter, Saturn, Uranus, Neptune, Pluto
3. An astonomical unit (AU) is the distance from Earth to the Sun (150 million km).
4. distance; mass
5. In order to feel the force of gravity between you and all of the objects around you, it takes an extra-large mass to notice.
6. No, a satellite orbiting Earth is affected by Earth's gravity and inertia.
7. Pluto
8. Answers are:
    a. Largest: JupiterSmallest: Pluto
    b. Strongest: JuupiterWeakest: Pluto
    c. Longest: VenusShortest: Jupiter
    d. Longest: PlutoShortest: Mercury
    e. Most: EarthLeast: Saturn
    f. Saturn
9. A comet revolves around the Sun in highly elliptical orbits. Earth does not always pass by the comet at the same time each year.
10. A meteor is a chunk of burning rock broken off from a comet or asteroid traveling through Earth's atmosphere. A meteroite is a meteor that passes through Earth's atmosphere and strikes the ground.
11. The asteroid belt contains thousands of small, rocky bodies that orbit the Sun. The asteroid belt is located between Mars and Jupiter.
12. The orbits of the planets are not perfect circles because there are slight variations in velocity vectors created during the formation of the solar system.
13. c

15.2 Section Review

1. Mercury
2. Venus
3. a
4. Venus
5. The corrosive atomosphere destroyed it after 23 minutes.
6. Earth's 23 degree axial tilt relative to its rotation.
7. Aerial photos show erosion and patterns of riverbeds similar to those formed by flowing water on Earth.
8. Both have liquid water.
9. Saturn's rings are made up of billions of particles of rock and ice.
10. No, Neptune and Jupiter and Uranus also have rings.
11. Scientists believe all nine planets condensed out of the same cloud of interstellar material as the Sun. The smaller inner planets could not hold onto their ligher gases (hydrogen and helium) and their exposed cores became rocky planets.
12. Pluto has a strongly elliptical orbit so sometimes it passes inside the orbit of Neptune.
13. Pluto, Uranus, and Venus rotate backward.

Connection

1. They measure ground movement, emissions of gases and changes in temperature.
2. Collect pictures, data and other scientific information.
3. Continuous volcanic eruptions release plumes of smoke and deposits of sulfur and sulfur dioxide. The color of the sulfur changes as the temperature goes down which changes the appearance of the surface.
4. Jupiter's magnetic field and intense radiation could destroy the spacecraft.

Activity

a. I am going to design an alien that can live and survive on Jupiter.

b. In order for an organism to live on Jupiter, it must be able to withstand the tremendous gravitational forces on the planet. Therefore, its bone structure must be very strong. My creature will be made out of titanium, one of the strongest and lightest metals. The organism will also be very tiny, probably about the same size as a mosquito on Earth. The reason it has to be so small is that it must be able to navigate through the gaseous surface of Jupiter, and if it were big it would just fall right through and eventually get crushed. Although my organism will be perfectly happy in the frigid temperatures, it will be able to generate heat by taking elements from the atmosphere and either burning them or combining them chemically inside its body (much like a firefly). This light will also serve as its means for illuminating its surroundings, as Jupiter is covered in clouds and is not very light on the surface.

c. My organism will have the ability to use the liquid hydrogen on Jupiter as an energy source. Because Jupiter has such high pressure, my organism would live near the hydrogen lakes and use suction pumps that are built into its body to suck in the liquid hydrogen as its fuel.

In order to move, the organism will have a system of jets that work by expelling the elements it has collected at high speeds. This will propel the creature in the direction it wishes to travel, and it will be able to stay near the surface where it will live. Its jets will be continuously in action. It will be in continuous motion, taking in liquids, burning them, and expelling the gases behind it to move.

Chapter 15 Assessment

Vocabulary

Section 15.1

1. Newton's law of universal gravitation
2. asteroid
3. astonomical unit
4. gas planets
5. meteor
6. solar system
7. comet
8. terrestrial planets
9. meteorite

Concepts

Section 15.1

1. Galileo, using a telescope he built himself, argued that the phases of Venus could not be explained if Earth were at the center of the planets. Secondly, he saw four moons orbiting Jupiter. This meant that not everything in the sky orbited Earth.
2. The Sun is at the center of the solar system because its gravitational force holds the planets in orbit around it. All the planets, moons, asteroids, comets, and meteors are gravitationally bound to the Sun.
3. a
4. The size of the masses; an increase in mass increases the force.
   The distance between objects; an increase in distance decreases the force.
5. Answers are:
   a. Earth reflects
   b. Mars reflects
   c. The Sun emits
   d. The moon reflects
   e. Stars emit
6. The moon shines by reflecting sunlight.
7. We see planets because they reflect light from their stars; the light from their stars is much brighter than the reflected light from the planets which makes them difficult to see.
8. A meteor is a chunk of burning rock broken off from a comet or asteroid traveling through Earth's atmosphere. A meteroite is a meteor that passes through Earth's atmosphere and strikes the ground.
9.

Object	Size	Material	Orbit Shape	Location
Asteroid	most < 1 km	rocky	elliptical	between Mars and Jupiter
Comet	size of Earth mountain	ice and dust	elliptical	orbit around the Sun

10. Some of its ice turns into gas and dust and forms an outer layer called a coma. As the comet approaches the Sun, a tail is formed which faces away from the Sun.
11. no

Section 15.2

12. Jupiter
13. Mars
14. Venus has an atmosphere made of 96% CO_2, which traps heat.
15. b
16. Venus
17. Some of the smaller moons act as "shepherds", keeping the particles confined due to gravity.
18. Venus

19. Venus, Mars, Jupiter, Saturn, Uranus, Neptune
20. a large rotating storm
21. Io is covered with smoking sulfur volcanoes.
22. Sun, Venus, Mars
23. Mars; may be explored by humans

Problems

Section 15.1

1. 8.5 lbs/x = 1.6 N/9.8 N
   weight of rock = 51 lbs
2. Answers are:
   a. 132 lbs
   b. 22 lbs
3. 0.003 AU
4. Answers are:
   a. it would double
   b. it would increase by 4 ×
   c. there would be 1/4 the gravitational attraction
5. Yes
6. The closer to the Sun, the greater the orbital speed
7. b

Section 15.2

8. The Sun's direct rays are shining on the northern hemisphere in summer, but in winter they are shining on the southern hemisphere, therefore causing a lower intensity in the northern hemisphere.

Notes:

Chapter 16: The Sun and Stars

Learning Goals	California Content Standards	Activities and Resources	Assessment
16.1: The Sun and the Stars The student will: • Explore the anatomy of the Sun. • Explain why stars shine. • Identify the different types of stars. • Define light year. • Discuss factors which affect the brightness of stars.	PS 4b PS 4c PS 4d PS 4e	**Laboratory Investigation 16A:** Distance and Brightness **Teaching Illustrations:** Sun Anatomy, Star Color and Temperature, H-R Diagram **Graphic Organizer:** Anatomy of the Sun **Math Skill Sheets:** Scientific Notation **Biographies:** Arthur Walker **Skill and Practice Sheets:** The Sun: A Cross-section (diagram), Understanding Light-years , Calculating Luminosity	**Exam View Test Bank** Chapter Pre-Test **Section 16.1 Review**
16.2: The Life Cycles of Stars The student will: • Explain how stars are born. • Describe the effects of a supernova. • Identify the stages in the life cycle of a star.	PS 4b PS 4d	**Laboratory Investigation 16B:** Spectroscope **Teaching Illustrations:** Forming a Planetary System, Evolution of Sun-Like Star, Spectroscope **Graphic Organizer:** Stars **Connection:** Big Bear Solar Observatory **Chapter Activity:** Sunspots	**Section 16.2 Review** **Chapter Assessment** **Exam View Test Bank**

Estimated Instruction Time	Language Tools	Differentiated Strategies	Materials
SIx 45-minute class periods	**Literary Picks** **Vocabulary:** star, nuclear fusion, light year, brightness, luminosity, main sequence star, white dwarf, red giant, supergiant **Word Origins:** luminosity	**ELL Strategies:** Listed in ELL Ancillary **Learning Strategies:** • Cooperative Learning - Investigation 16A • Discussion - Motivate • Active Research - Explain: What is a Solar Eclipse **Multicultural Experience:** Sun Mythology **Teaching Tip:** Can the Sun Tell Time?	**Investigation 16A:** incandescent lamp, base and 100-watt bulb, digital multimeter, solar cell, meter stick, small cardboard box
Six 45-minute class periods	**Vocabulary:** nebula, protostar, planetary system, binary star, planet nebula, supernova, spectroscope **RAFT Writing:** Extend - Star RAFT **Word Origins:** nebula **Closure Cards**	**Learning Strategies:** • Cooperative Learning - Investigation 16B • Discussion - Motivate: Nebula News **Teaching Tip:** Famous Astronomers, Stargazing	**Investigation 16B:** Project STAR spectrometer, colored pencils, access to light from incandescent and fluorescent light fixtures **Chapter Activity:** cardboard, 3-cm x 3-cm piece of aluminum foil, tape, pin or sharp pointed object, ruler, white paper, solar telescope

16.1 The Sun and the Stars

About the Lesson

In this lesson students learn about many features of the Sun including that it is one of many stars in the Milky Way galaxy. They also learn that different types of stars exist in various sizes, color, brightness, and temperature.

Write the following facts and question on the board. Ask students to think for a moment about each statement or question.

1. Light from the Sun takes about 8.5 minutes to reach Earth.
2. Light travels through space at about 186,000 miles per second.
3. The Sun is 93 million miles from Earth.
4. How long does it take for energy from the Sun's core to reach the photosphere, the Sun's visible surface?

The answer is 150,000 years. Ask the students what this answer really means. Students may be surprised to know the light humans see today was essentially produced in the Sun's core back in the stone age, about 150,000 years ago.

Students complete Investigation 16A before reading section 16.1 of the student text.The investigation answers questions about how factors that affect the brightness of a star. Students also learn how astronomers use light to measure the distances to faraway stars and galaxies.

Investigation 16A: Distance and Brightness

Setup	Materials	Investigation Vocabulary
1. Allow one class period to complete the investigation. 2. Students work in groups three to five.	• incandescent lamp base and 100 watt bulb • digital multimeter • solar cell • meter stick • small cardboard box	brightness luminosity inverse relationship

Literary Picks...

Reading Level	Title
3-5	***Sun, Moon and Stars*** **by Mary Hoffman** Students will truly enjoy this collection of myths and legends dealing with the sun, moon and the stars. Filled with cultural costumes, design, art and some factual information interwoven between the lore.
6-8	***Jumping Off the Planet*** **by David Gerrold** The story of a dysfunctional family with three brothers told from the viewpoint of the 13-year-old middle brother named Charles. The father kidnaps the boys by offering them a trip to the moon on a giant space elevator. Top notch science fiction by an award-winning author.
9-12	***Rendezvous with Rama*** **by Arthur Clarke** This book is a creative piece of science fiction classic by an award-winning author. Details the story of a human encounter with an alien spaceship named Rama. Awarded both the Hugo and Nebula Awards for best novel.

What is a solar eclipse?

A solar eclipse occurs on Earth when the Moon is positioned between the Sun and Earth. The Moon blocks the sunlight and casts a shadow on Earth. A total solar eclipse occurs when a new moon passes directly between the Sun and Earth. The Sun's corona is visible as it (the sun) is completely covered by the Moon.

One can view a total eclipse if he is in its path at the right time. The path may be as wide as 200 miles and cover about one-half of one percent of Earth's surface. Sky watchers who have witnessed an eclipse often describe the experience as awesome and amazing. Have your students view NASA's listing of solar eclipses through 2010, which includes information such as the type of eclipse and regional visibility. Find out more by visiting http://sunearth.gsfc.nasa.gov/eclipse/solar.html.

Multicultural Experience: Sun Mythology

Early civilizations often worshipped the Sun and created interesting cultural stories such as myths, folklore, tales, and legends. In fact, many attempted to explain observations about the Sun, such as why it rises and sets.

Have students research a cultural story, legend, myth or tale about the Sun and create a poster depicting the story. Then have students present their findings to the class. Some examples may include mythology from the Chinese, Hawaiians, and Africans as well as the Aborigines, Hinduism Egyptians, Ancient Greeks, and Romans. You may extend this activity even further by having your students compose their own creative story about the Sun.

Students complete Section 16.1 review questions.

Word Origins...luminosity (from Latin *luminosus* meaning "light")

Many students may watch crime shows on television. A common occurrence is for a forensic investigator to use *luminol* to detect the presence of blood. In other words, luminol brings "to light" the blood that would otherwise be impossible to see.

Teaching Tip . . .Can the Sun Tell Time?

Ask students, "Can you imagine a time when there were no clocks? What if someone said I'll meet you at high noon?" Due to Earth's rotation, the Sun moves across the sky, rising in the east and setting in the west. A sundial may be used to measure the Sun's position in the sky, relative to the position of an observer. A sundial measures apparent solar time from the observer's location.

Your students can construct a simple sundial that allows them to tell time using the Sun based on your zip code and position (latitude). Go to http://www.nasa.gov/audience/forkids/activities/A_Make_a_Sundial.html to make your sundial.

Ancient cultures used a variety of methods for telling time with the Sun. Over 6000 years ago in Egypt, sundials were made out of a stick called a *gnomon* and a face-like dial plate which divided the day into twelve equal hours. Have your students research some of the other ancient methods for using the Sun to tell time.

Investigation 16A: Distance and Brightness

Have you ever been in a parking lot at night?

Most students should answer yes.

What happens to the light around a light pole as it spreads out into the lot?

The farther away from the pole you get, the less light there appears to be.

Good observation. Did you know that you can calculate how far away you are from a light source? All you need is to know the brightness and the luminosity of the object. What do you think is meant by brightness?

Brightness is how much light is given off.

That is a good answer, but not quite correct. *Brightness* is more about how much light is available. That is why when you walk away from the light pole, the light becomes less and less. It is a relationship between the light and the observer. *Luminosity* is the total amount of light a source gives off in every direction. Though the brightness of light may decrease as you move further away from the source, luminosity remains constant. Can you think of a type of scientist who might need to know about brightness and luminosity?

Astronomers.

Good answer. Astronomers may want to calculate the distances to stars and planets. But there are also other components to the universe besides planets and stars. Can anyone name some of those things?

Answers may vary. Sample responses may include galaxies, clusters, black holes, and asteroids.

In today's investigation you will learn how astronomers measure distances to faraway stars and galaxies.

1 Setting up

Turn on your light source.

When all light sources are turned on, turn off the overhead lights in the class room. If there are any windows, make sure they are covered so that the light coming through does not interfere with the investigation

Using your meter stick, place your solar cell 10 cm from the light bulb. Make sure your solar cell is directly facing your light bulb. Check to make sure your electric meter is set to measure milli-amps. Now you are ready to take some brightness and distance measurements.

Walk around and verify that students have the correct settings.

Use the solar cell to measure the light's brightness at the 10 cm. Record your answer in Table 1.

Students continue to take measurements until they have measured the brightness of the bulb 100 cm away. Be sure to demonstrate the proper method for taking measurements.

Investigation 16A *Distance and Brightness*

16A Distance and Brightness

How does light from a star spread out on its way to us?

The brightness of a star depends both the star's distance and on how much light the star actually produces (its luminosity). When astronomers know a star's brightness and luminosity they can calculate the distance to the star. In this Investigation, you will discover the mathematical relationship between how bright a light source appears and it's distance. You will then understand how astronomers are able to use light to measure the distances to faraway stars and galaxies.

Materials

- Incandescent lamp base & 100w bulb
- Digital multimeter
- Solar cell
- Meter stick
- Small cardboard box

1 Setting up

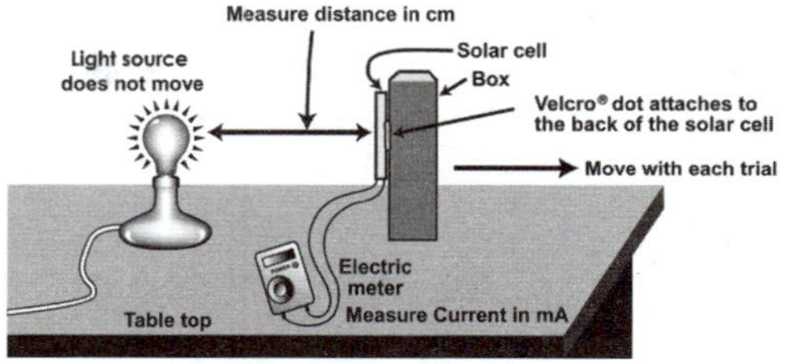

1. Turn on your light source and turn off the overhead lights in the classroom.
2. Place the solar cell exactly 10 centimeters from the light bulb. Make sure the solar cell is directly facing the light bulb and is not at an angle.
3. Make sure the electric meter is set to measure current in milliamps (mA).
4. Measure the light's brightness in mA and record your results in Table 1.
5. Measure the brightness at 10-centimeter intervals and record your data in Table 1.

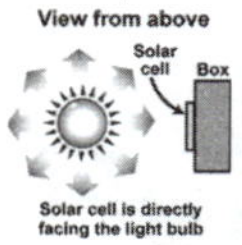

Table 1: Brightness and distance data

Distance (cm)	10	20	30	40	50	60	70	80	90	100
Brightness (mA)										

73

2 Stop and think

Let's think about what you have learned. Does this experiment measure the brightness or luminosity of the light bulb?

It measures the brightness.

How do you know?

We know because as we moved the solar cell farther and farther away, our measurements decreased. Luminosity is constant.

We used a 10 watt light bulb. Do you think that the power rating of 10 watts is a measure of the light's brightness or luminosity? Hoe do you know this?

It is a measure of the light's luminosity. A sample response may be, "I know because the 100 watts would be constant no matter how close or far I am away from the bulb.

Who can describe what happened as you moved the solar cell farther and farther away from the bulb?

The measurements got smaller.

3 Analyzing your data

Students make the graph as described in part three. Discuss the graph trends as students answer questions a-c.

4 Inverse relationships

Have a student volunteer read aloud he definition of an inverse relationship. Use the questions in part four to lead a guided discussion.

5 Identifying the correct inverse relationship

To figure out which equation is correct, you will need to look at the data from part two more closely. Follow each step in part four and enter the data in table 2.

Students may need guidance with the calculations. Set up 3 graphs with the students with distance on the *x*-axis and brightness on the *y*-axis. Be prepared to model how to make a smooth curve graph again.

6 Reaching a conclusion

Students complete questions a and b. Discuss the answers to ensure that students understand the relationship between distance and brightness.

7 Testing your model

Choose a distance you did not actually measure, like 45 cm. Calculate the brightness in mA. Use this as the predicted brightness. Now, move your solar cell to the distance you choose and measure its brightness. This is the actual brightness. Let's see how well the equation worked in predicting the brightness.

Students apply the percent error calculations.

> **Teaching Tip . . .Using the calculator**
>
> Students may need a quick tutorial for using the squared and cubed functions on their calculator as they identify the correct inverse relationship (part five).

Investigation 16A: Data and Answers

1 Setting up

Distance	10	20	30	40	50	60	70	80	90
Brightness (mA)	92.5	28.5	15.2	9.2	6.6	4.8	3.8	3.0	2.0

2 Stop and think

a. This experiment measures the apparent brightness of the light source because it is a measurement of the brightness of the light source at various distances.

b. The power rating of the light bulb is closest to a measurement of the absolute brightness of the light source. This value does not change with distance because it is how much light the bulb is actually emitting.

c. We think that increasing the distance will decrease the brightness of the light source.

3 Analyzing your data

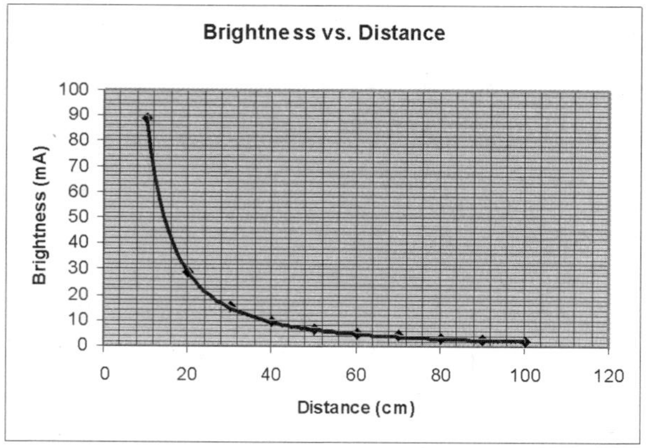

a. The graph is decreasing from left to right.

b. The curve shows that brightness decreases as distance increases by a certain factor. It looks like an exponential relationship.

c. No, I cannot determine the mathematical relationship because the line is a curve. If it were a straight line, I could find the slope of the line and determine the mathematical relationship.

4 Inverse relationships

a. The third equation states that brightness varies with the inverse of the distance cubed.

b. I would need to plug the data into each equation to see which one works for each data point on the graph.

c. I think that $B = 1/D^2$ is correct because I have seen this type of relationship graphed before and the curve looked similar.

5 Identifying the correct inverse relationship

Three graphs:

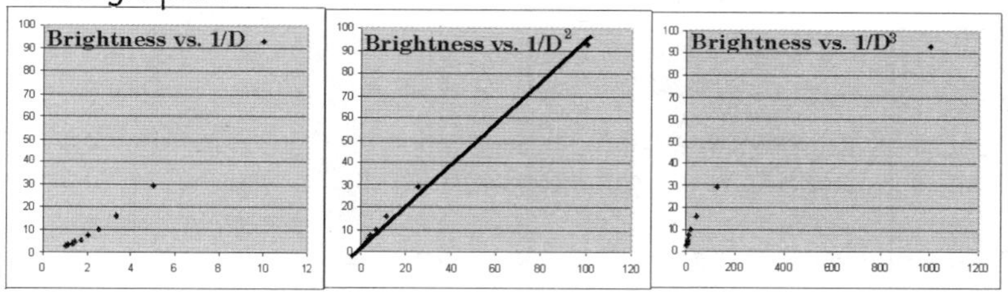

Distance	10	20	30	40	50	60	70	80	90
Distance (cm)	0.10	0.20	0.30	0.40	0.50	0.60	0.70	0.80	0.90
1/D	10	5.0	3.3	2.5	2	1.7	1.4	1.25	1.1
$1/D^2$	100	25.0	11.1	6.25	4.00	2.78	2.04	1.56	1.23
$1/D^3$	1000	125	37.5	15.6	8.00	4.63	2.92	1.95	1.37
Brightness (mA)	92.5	28.5	15.2	9.2	6.6	4.8	3.8	3.0	2.0

6 Reaching a conclusion

a. The graph of brightness vs. $1/D^2$ is the correct relationship because it is the closest one to a straight line.

b. $B = 1/D^2$

7 Testing your model

a. No answer required

b. Prediction for 0.45 meters = 4.9.

c. Actual = 5.5.

d. % error = (4.9 - 5.5)/4.9 = 12.2%.

Notes:

16.2 The Life Cycles of Stars

About the Lesson

In this lesson students learn about the life cycles of stars including how they are born, main sequencing stage, planetary systems, and binary stars. Students are also introduced to other important topics including nuclear fusion, supernovas, and neutron stars.

Nebula News

Ask students, "What do an eagle and a tarantula have in common?" The answer is that both are nebulae. At one time, astronomers referred to any astronomical objects, particularly galaxies beyond the Milky Way, as nebula. In fact, the Andromeda Galaxy is still referred to as the Andromeda Nebula. Today, nebula simply refers to an interstellar cloud of gas and dust such as the Eagle Nebula and Tarantula Nebula. The Eagle Nebula is about 7000 light-years away. However, the Tarantula Nebula is located in the next galaxy over called the Large Magellanic Cloud and is some 160,000 light-years away.

Students complete Investigation 16B before reading section 16.2. In this investigation students use a spectrometer to analyze the light emitted from stars. By doing so, students learn about the chemical composition, speed, temperature, and other properties of stars. Remind students to avoid looking directly into any light source, especially the sun.

Investigation 16B: Spectroscopy

Setup	Materials	Investigation Vocabulary
1. Allow one class period to complete the investigation. 2. Students work in pairs. 3. Examine the spectrometers before class. Calibrate each one ahead of time so s	• Project STAR spectrometer • colored pencils • access to light from incandescent and fluorescent light fixtures	spectroscopy spectrometer spectral lines fluorescent wavelength incandescent

Teaching Tip . . .Famous Astronomers

Have your students learn more about famous astronomers. Two famous astronomers are featured below.

- **Edwin Hubble** (1889-1953)

Hubble is considered one of America's leading modern day astronomers and is renowned for his discoveries of galaxies beyond the Milky Way. Hubble, an intellectual genius and superior athlete was born in Missouri, but raised in Illinois. He studied law as a Rhodes Scholar at Oxford and received his Ph.D at the Yerkes Observatory at the University of Chicago. He moved to California to begin his work at Mt. Wilson Observatory, and stayed there until his death in 1953. He is credited with formulating the empirical Redshift Distance Law of Galaxies, also known as Hubble's Law.

- **Jan Hendrik Oort** (1900-1192)

His ground breaking discoveries ranged from comets to the solar system. By far, no other astronomer has made as many contributions to so many different fields in astronomy. He is recognized as one of the greatest 20th century astronomers with contributions to discoveries of stars, galaxies, clusters, comets, and the Oort Cloud. Oort was one of the few who had the opportunity to observe Halley's Comet on two separate occasions. His first viewing was at age 10 in 1910 and again at age 86 in 1986.

Constellations

Constellations are named groups of stars which form a pattern. This idea of naming star patterns comes from the ancient Greeks. Over time, humans have learned more about the stars in the night sky as well as the patterns they form. Many cultures developed stories and legends based on the patterns they observed.

The constellations we see today are the same ones that were seen by others thousands of years ago. Astronomers have recognized and given names to 88 different constellations to date. Many have been named to honor kings, mythical gods, events, and objects. Today, astronomers use constellations to identify specific regions in the sky.

Word Origins...nebula (from Latin *nebula* meaning "mist")

Survey your students to determine if any of them has asthma. If so, ask a student volunteer to talk about how a nebulizer functions. Then have students look up the meaning of the words nebula and nebulizer and use each in a sentence.

Star RAFT

RAFT writing gives students an opportunity to think critically and creatively about the content of stars. Students are also able to make meaningful connections to prior knowledge as they synthesize new learning into a imaginative piece of writing.

Role: Journalist of the Mars Star Gazette

Audience: People on Mars

Format: Interview

Topic: Select one of the brightest stars as seen from Earth. Examples include Sirius, Canopus, Rigil Kentaurus, Arcturus, Vega, Capella, Rigel, Procyon, Achernar, Betelgeuse, Hadar, Acrux, Altair, Aldebaran, Spica, Pollux, Formalhaut, Becrux, Deneb, Regulus, Adhara, Castor, Gacrux, and Shaula.

Students answer section 16.2 review questions.

Teaching Tip . . .Stargazing

Encourage your students to become observers of the night sky. A stargazer may view about 3,000 stars under optimal conditions with the unaided eye. The best location to choose is away from city lights and smog, which hinder the view. Binoculars and telescopes increase the amount of light your eye can see. It is important to know that the night sky is constantly changing and what you see depends on your location, the time of day, and time of year. Stargazing tips and additional information are available and updated daily on the following astronomy websites sponsored by NASA's Jet Propulsion Lab in California and McDonald Observatory at the University of Texas.

- JPL's Amateur Astronomy Page: NASA Jet Propulsion Lab at http://education.jpl.nasa.gov/erc.html
- StarDate Online - Daily viewing guide for a week with tips, sky almanac, and constellation guide at http://stardate.org/nightsky/weekly.php

Investigation 16B: Spectroscopy

What do you see when you look at the night sky?

Students share observations. Most students will respond by saying the moon, stars, and other stellar objects.

What do the stars look like?

The stars look like tiny specks of light.

Yes, they do. But what appears to be simple and tiny specks of light hold a wealth of information. For instance, the light from a star tells a whole lot about its character. By studying spectra, astronomers are able to learn about specific properties of stars.

1 The spectrometer

A *spectrometer* is used to study spectra, or the particles emitted by light. A spectrometer splits light into a spectrum of colors and displays the different colors of light along a scale. The lower scale measures the wavelength of light in nanometers.

Tell students that the prefix *nano-* means one billionth.

A nanometer is one billionth of a meter. Therefore individual light waves are extremely small. In this investigation you will use a spectrometer compare the spectra of light sources. Hold the spectrometer so the printed side faces up.

Demonstrate the proper technique for holding the spectrometer.

Look in the short end with one eye while you point the slit at a bright light. Direct your eye to the left to see the colors. What do you see?

Students should observe a band of colors in between two rows of numbers. Refer students to the diagram in part one and ask them if this looks like what they see. Students may need to adjust the disk in order to line up the colored band.

The first type of light you will look at is an incandescent light bulb. These bulbs are very bright and should look familiar to you. Point the slit of the spectrometer directly at the bulb as you look through the eyepiece. Where do the lights appear on the scale?

Students use colored pencils to record where the lights appear on the scale.

What does the spectrum look like?

Students should observe that the colors resemble a rainbow.

That is a good observation. Which colors appear on each end of the spectrum?

Students observe blue at one end and red at the other.

The blue light has the highest energy, while red is the lowest. What does the position of each color of light along the spectrum tell you about the relationship between wavelength and energy?

Red light has the longest wavelength and lower energy; and blue light has shorter wavelength and higher energy. The shorter the wavelength, the higher the energy and vice versa.

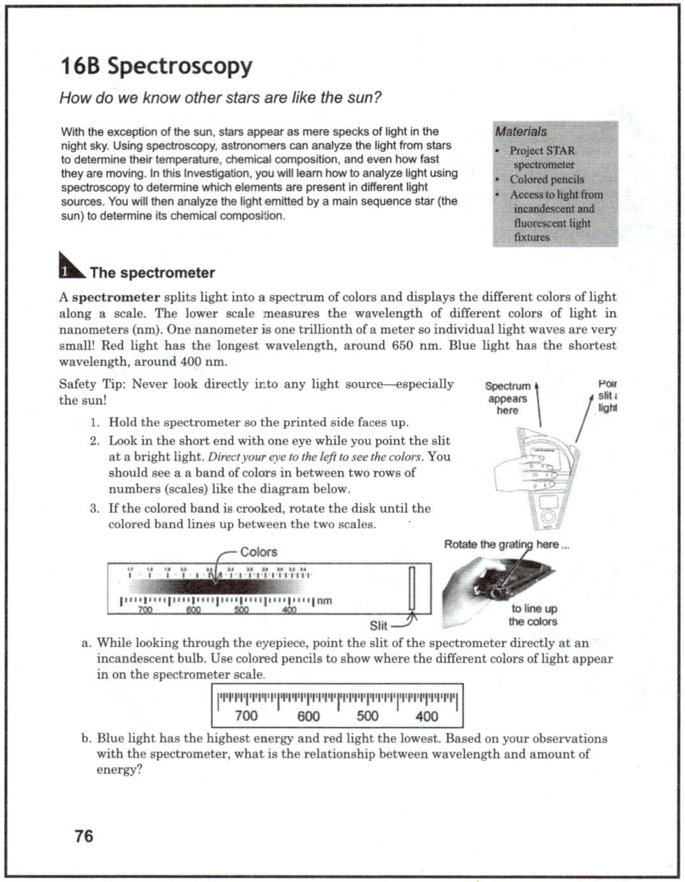

16B Spectroscopy

How do we know other stars are like the sun?

With the exception of the sun, stars appear as mere specks of light in the night sky. Using spectroscopy, astronomers can analyze the light from stars to determine their temperature, chemical composition, and even how fast they are moving. In this Investigation, you will learn how to analyze light using spectroscopy to determine which elements are present in different light sources. You will then analyze the light emitted by a main sequence star (the sun) to determine its chemical composition.

Materials

- Project STAR spectrometer
- Colored pencils
- Access to light from incandescent and fluorescent light fixtures

1 The spectrometer

A **spectrometer** splits light into a spectrum of colors and displays the different colors of light along a scale. The lower scale measures the wavelength of different colors of light in nanometers (nm). One nanometer is one trillionth of a meter so individual light waves are very small! Red light has the longest wavelength, around 650 nm. Blue light has the shortest wavelength, around 400 nm.

Safety Tip: Never look directly into any light source—especially the sun!

1. Hold the spectrometer so the printed side faces up.
2. Look in the short end with one eye while you point the slit at a bright light. *Direct your eye to the left to see the colors.* You should see a a band of colors in between two rows of numbers (scales) like the diagram below.
3. If the colored band is crooked, rotate the disk until the colored band lines up between the two scales.

a. While looking through the eyepiece, point the slit of the spectrometer directly at an incandescent bulb. Use colored pencils to show where the different colors of light appear in on the spectrometer scale.

b. Blue light has the highest energy and red light the lowest. Based on your observations with the spectrometer, what is the relationship between wavelength and amount of energy?

76

2 Using a spectrometer to identify elements in a fluorescent light

Now you will use your spectrometer to look at a different light source, a fluorescent bulb. Fluorescent bulbs are commonly used in homes, businesses, and schools because they are far more efficient than incandescent bulbs.

Point out the source of fluorescent light in your classroom.

Look through the spectrometer at light from a fluorescent bulb. What do you see?

Students should observe colored lines. Allow for students to explore and make notes about what they see.

Do the lines you see look the same as what you observed as you looked at the incandescent bulb?

The lines are different.

The colored lines you see are called *spectral lines*. Examine the line at 546 nanometers on the scale. What color is this line?

The line is green. Some students may not see the green line at the wavelength. If this is so, calibrate students' spectrometers.

Use colored pencils to sketch all the lines you see. Place the lines in the exact position on the scale in part two as you see them in the spectrometer. Then write the color and wavelength of each line in Table 1.

Students sketch spectral lines and record data.

3 What do the lines mean?

Light rays bend as they pass through a prism. How much the rays bend depends on the wavelength and frequency of light. Light rays that have the longest wavelength bend the least; while those with shortest wavelength bend the most. At one end of the visible spectrum is dark red light. It has the longest wavelength. At the opposite end of the spectrum is violet light, with its shorter wavelength.

Draw a line on the board and label one end *red* and the other *violet*. Then label the wavelength at each end of the spectrum as *longest* and *shortest*.

In between the red and violet light are other colors. Which colors belong in between these two colors?

Students may already be familiar with "ROY G BIV" as a way of describing the colors along the spectrum. The colors are red, orange, yellow, green, blue, indigo, and violet.

When elements are heated until they are hot enough to emit light, they produce a pattern of spectral lines. The pattern produced by each element is unique, like a fingerprint.

Point out the spectral lines of the elements shown in part three.

The light produced by a fluorescent source is created when electric current is passed through a gas inside of the tube. This gas, which is made of only one element, absorbs energy, and emits light. Which of the spectral lines shown above looks like what you observed as you looked at the fluorescent light?

Students should identify the spectral line pattern of mercury. Talk about the potential environmental dangers associated with mercury from fluorescent tubes. Encourage students to investigate proper techniques for disposing of fluorescent tubes.

Investigation 16B: Data and Answers

1 The spectrometer

a. Color sketch of spectrometer view.

b. Longer wavelengths have lower energy and shorter wavelengths have higher energy.

2 Using a spectrometer to identify elements in a fluorescent light

a. Color sketch of spectrometer view.

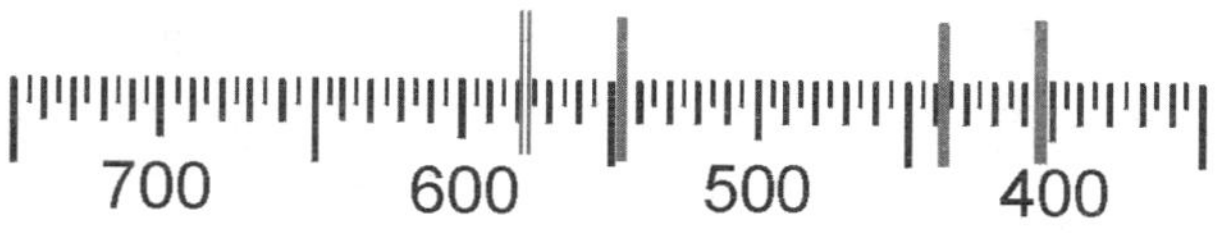

b.

Line number	Spectral line color	Spectral line wavelength (nm)
1	violet	405
2	blue	436
3	green	546
4	yellow (2 close together)	577,579

3 What do the lines mean?

a. The fluorescent light contains mercury.

b. Mercury is a toxic substance that is dangerous for organisms in the environment.

c. They would need to know the exact values of the spectral lines and which elements display each characteristic spectral line.

Notes:

Chapter 16 Answers

16.1 Section Review

1. nuclear fusion
2. photosphere: apparent surface of the sun
   chromosphere: hot layer of plasma above the photosphere
   corona: outermost layer of sun's atmosphere
3. Sunspots are areas of gas that are cooler than the gases around them. Sunspots look like small dark areas.
4. color; temperature; brightness; size
5. Our sun is medium sized.
6. The distance light travels in one year equal to 9.46×10^{12} km
7. The current year - 77
8. You can tell the amount of energy and temperature of the star: red is cooler with lower energy while blue is hotter with higher energy.
9. Brightness decreases by the inverse square of distance away from the source.
10. Luminosity is the total amount of light given off by a star. Brightness is a measure of the amount of light reaching Earth.
11. The star three time farther away will have 1/9 the brightness.
12. (1) Main sequence: stable part of a star's life cycle; on the H-R diagram is found diagonally across from cool/dim to hot/bright.
    (2) White dwarf: cool and dim; found in lower left corner of H-R diagram
    (3) Red giant: cool and bright; found in upper left corner of H-R diagram
    (4) Supergiant: bright red and blue stars; found in very upper portion of H-R diagram
13. False. a white dwarf is much smaller than our Sun.
14. d

16.2 Section Review

1. about 5 billion years
2. nebula
3. gravitational force
4. pressure and temperature both rise
5. about 10 billion years
6. High mass stars burn brighter and hotter and use up their hydrogen fuel faster than low mass stars.
7. Gravity causes the core to contract, raising the temperature. Other nuclear reactions occur and the outer layers of the star get pushed out. The star expands into a red giant.
8. Carbon and oxygen were created by nuclear fusion inside of stars.
9. The massive explosion of a very large star.
10. No, our Sun does not have enough mass to become a supernova.
11. Through spectroscopy, scientists can analyze which elements are present in a star. A star's speed, temperature, rotation rate, and magnetic field can also be determined.
12. Each element has its own unique pattern of spectral lines, just as every person has his or her own unique pattern of fingerprint lines.
13. a
14. iron

Connection

1. The Sun is important to Earth because it is the source of energy, heat, and light. The Sun supports life on Earth and without it, life would not be able to exist. No other star is as close to Earth as the Sun. By studying the Sun, scientists learn a great deal about stars in general.

   Understanding the Sun is also important to people on Earth. Space weather, created by the Sun, can impact Earth. The sun causes radiation storms, magnetic field changes, and movement of energetic particles. Space weather storms can damage satellites, electric power grids, cell phone systems, and Global Positioning Systems. NASA is also interested in knowing about the Sun and events occurring on its surface because space weather can affect astronaut safety.
2. Big Bear Solar Observatory observes the Sun and all types of solar phenomena occurring on the surface of the Sun. Scientists study the solar atmosphere, solar flares, sunspots, and prominences. Powerful telescopes are used to observe the Sun. These telescopes have special cameras and filters designed for safe sun viewing. Specific telescopes observe particular solar events. Two telescopes observe sunspots, flares, and prominences. A smaller telescope watches the Sun from sunrise to sunset, taking images every 30 seconds. A fourth telescope measures Earthshine, which provides information about Earth's temperature, atmosphere, and global warming.

3. Big Bear Solar Observatory is located on Big Bear Lake in Southern California. Big Bear Lake has over 300 sunny days per year. The weather, high altitude, and seasonal changes provide excellent sun viewing opportunities year round. The observatory is also built on a lake, which minimizes convection to produce clear images of the sun. Light winds blowing over the lake also improve image clarity.

Activity

Applying your knowledge: sample answers

Distance from Earth to the Sun: 149,600,000 km

Measured diameter of the Sun: 0.6 cm

Measured distance from viewer to paper: 55 cm

$$\text{Diameter of Sun} = \left(\frac{\text{Diameter of image of Sun}}{\text{Distance from viewer to paper}}\right) \times \text{Distance from Earth to Sun}$$

Diameter of Sun = 1.6 million km

a. 34,000 km
   Diameter of sun from sunspotter: 7.62 cm
   Diameter of one sunspot: .16 cm
   (1,600,000 km/.0000762 km) = (x/.0000016 km)
   x = 34,000 km (sunspots range generally from 100 to 80,000 km)
b. Peaks and valleys are periodic (about every 11 years)
c. 16%

Chapter 16 Assessment

Vocabulary

Section 16.1

1. sunspots
2. nuclear fusion
3. stars
4. light year
5. supergiant
6. main sequence
7. red giant

Section 16.2

8. binary star
9. nebula
10. spectroscopy
11. white dwarf
12. protostar
13. supernova

Concepts

Section 16.1

1. b
2. b; d; a; c
3. hydrogen atoms are fused together to form helium; mass is lost and converted to energy
4. extremely high temperature; extremely high density; extremely high mass
5. As a material is heated, it begins to glow first red, then yellow, then orange, and then white. When it reaches its highest temperature, blue light is dominant visible color.
6. c
7. Hydrogen and helium are the two most common elements in the universe. The two most common elements in Earth's crust are oxygen and silicon. Helium is rare on Earth.
8. Compared to other stars in the galaxy, the sun is average in all categories.
9. An H-R diagram is useful to astronomers because it helps them to categorize stars into groups.

10. That mass is roughly the minimum mass needed to create the gravitational pressure required to ignite fusion reactions in the core.
11. Even though its mass is about the mass of the Sun, the density of the materials in the star is much lower, and the red giant has a much larger volume.

Section 16.2

12. Answers are:
    a. A spectrometer is a tool used by astronomers to analyze electromagnetic waves emitted from glowing objects. A spectrometer splits visible light, for example, into spectral lines of bright colors and displays the bright lines.
    b. Each element has its own unique pattern of lines, much like a human fingerprint. This enables astronomers to determine the identity of the materials in stars and other distant glowing bodies.
13. The lifespan of a star mostly depends upon its mass.
14. nebula, protostar, sun-like star, red giant, white dwarf, (planetary nebula forms during white dwarf stage) black dwarf .
15. It will move into the red giant phase, then move into the white dwarf phase, then possibly form a planetary nebula.
16. No, stars eventually the core of a star runs out of hydrogen for fuel.
17. As the sun formed, it was surrounded by a spinning cloud of gas and dust. Gravity caused the cloud/dust shape to flatten into a disk shape along the axis of its rotation.

Problems

Section 16.1

1. 1 light year = 9.46×10^{12} km
   2.3 million light years = (2.3 million ly × 9.46×10^{12} km/ly) = 2.18×10^{19} km
2. The Sun, because the larger the mass of a star the shorter its lifespan.
3. The star that is 200 ly away will look 1/4 as bright as the star that is 100 ly away. The star that is 300 ly away will look 1/9 as bright as the star that is 100 ly away.
4. 169,000 years before it was seen.
5. Answers are:
    a. A white dwarf is dimmer and hotter than a supergiant.
    b. main sequence
    c. red giant

Section 16.2

6. Answers are:
    a. nuclear fusion in a star
    b. supernova
    c. nuclear fusion in a star
    d. early universe
    e. nuclear fusion in a star

Chapter 17: Galaxies and the Universe

Learning Goals	California Content Standards	Activities and Resources	Assessment
17.1: Galaxies The student will: • Define galaxy. • Identify different types and shapes of galaxies. • Discuss the black hole theory. • Describe the evidence which supports the Big Bang theory.	PS 2g PS 4a PS 4b PS 4c	**Laboratory Investigation 17A:** The Milky Way Galaxy **Teaching Illustrations:** Types of Galaxies, Big Bang Theory **Graphic Organizer:** Galaxies **Connection:** A Spectacular Show of Lights **Chapter Activity:** Galaxies and the Universe	**Exam View Test Bank** Chapter Pre-Test **Section 17.1 Review** **Chapter Assessment** **Exam View Test Bank**

Estimated Instruction Time	Language Tools	Differentiated Strategies	Materials
Eight 45-minute class periods	**Literary Picks** **Vocabulary:** galaxy, Milky Way galaxy, Big Bang Theory	**ELL Strategies:** Listed in ELL Ancillary **Learning Strategies:** • Cooperative Learning - Investigation 17B • Active Research - Motivate: The Milky Way Galaxy • Active Learning - Explain: Hubble Space Telescope • Think-Pair-Share - Extend: Reflective Closure **Teaching Tip:** Galaxy Transparencies	**Investigation 17B:** **Chapter Activity:** 100-watt light bulb, solar cell, electric meter, box, Velcro dots

17.1 Galaxies

About the Lesson

In this lesson, students learn about the characteristics of galaxies and explore the postulates of the Big Bang theory.

Investigation 17A: The Milky Way Galaxy

(see answers on page 282)

This investigation provides students an opportunity to find out what scientists have learned about the Milky Way galaxy. Meet with your school's media specialist ahead of time to reserve a day for your students to visit the library. Inform the specialist of the topic in advance. In this way, he or she can prepare appropriate books and magazines or reserve computers for your students. The research and discussion questions are given below.

- Are there other stars with planets in our galaxy?
- Suppose only one in every one million stars has a solar system with a planet like Earth. How many earth like planets would that make in the Milky Way galaxy?
- Something unusual is happening at the very core of the galaxy. What do astronomers think is hiding there?
- How long does it take the Milky Way galaxy to make one complete rotation?
- Roughly how old is the Milky Way galaxy?
- The fastest spacecraft built by humans goes 30 km/sec. How long would this spacecraft take to cross the galaxy?

Students complete Investigation 17B after reading section 17.1 of the student text.The investigation answers questions about the amount of time needed for starlight to reach Earth. Students will need to use the library for additional research in part three of the investigation.

Investigation 17B: The Time Machine

Setup	Materials	Investigation Vocabulary
1. Allow one class period to complete the investigation. 2. Students work in groups three to five.	• astronomy reference materials • simple calculator	speed of light light-year

Literary Picks...

Reading Level	Title
3-5	***The Universe*** **by Seymour Simon** A spectacular book of photographs that includes brief explanations of galaxies, nebulas, black holes, and stars. The book is easy to understand and covers a wide range of topics such as star formation, Big Bang theory, and quasars.
6-8	***Maria Mitchell: The Soul of an Astronomer*** **by Beatrice Gormley** Selected as a NSTA Outstanding Science Trade Books for Students, 2004. This is truly an excellent biography depicting America's world-renowned female astronomer. The story highlights to the contributions she made to advancing women in science.
9-12	***Turn Right at Orion*** **by Mitchell Begelman** A narrative scientific fiction story, Astrophysicist Begelman takes us on a journey to the center of the Milky Way some sixty million years in the future. Considered a textbook in disguise, this book is really a great introduction to basic space science.

Hubble Space Telescope

The Hubble Space Telescope is a Low Earth Orbiting (LEO) satellite which orbits around 375 miles above Earth's surface. Hubble completes an orbit around Earth approximately every 95 minutes, traveling at about 5 miles per second. This is an amazing 18,000 miles per hour! At that speed it would only take around 10 minutes to travel from New York City to Los Angeles, California. Launched in 1990, Hubble's maintains four major scientific objectives.

1. To determine the physical characteristics and dynamics of celestial bodies.
2. To determine history and evolution of the universe.
3. To see if the laws of nature are universal.
4. To find out if the nature of processes which occur in extreme conditions exist in and between astronomical objects.

To date, Hubble has provided information from as far away as the Ultra Deep Field, which is 12 to 13 billion light-years away. Scientists have had a glimpse of over 10,000 galaxies at different evolutionary stages. Hubble's cutting edge mission has taken us into deep space, above and beyond our wildest dreams.

Students can track the Hubble Space Telescope at NASA's Hubble Operations site. Visit *http://hubble.nasa.gov/operations/tracking.php* to learn more.

Teaching Tip . . .Galaxy Transparencies

Imagine The Universe is a service of the High Energy Astrophysics Science Archive Research Center at NASA's Goddard Space Flight Center

Follow the "Teachers's Corner" link at *http://imagine.gsfc.nasa.gov/docs/teachers/galaxies/transparencies/* to download files of printable transparencies. Some examples include *Types of Galaxies*, *Galaxy Classification Images*, *Deep Survey Images*, and *Hubble Fork Diagrams*. These are excellent images of the various types of galaxies including: spiral, elliptical, barred spiral, peculiar, and irregular.

Reflective Closure

Using Think-Pair-Share have students answer the following closure questions based on all 17 chapters of the book:

1. Looking back at all that you have learned in this book, describe the two topics you like best.
2. Which additional topics would you like to study?

Students complete Section 17.1 review questions.

Investigation 17B: The Time Machine

Make a Power Point or photo presentation of some past events. Pictures from a variety of decades would be best. Show the class the photos and explain each photo. It would be helpful to point out the different styles of fashion, hair cuts, eyeglasses and so forth.

Today I want to share some photographs with you. When we look at photos, what are we really looking at?

Answers will vary. Steer students toward responses like the past, or history.

Is there anything else we could watch that would show us the past?

Answers will vary. Responses may include old movies and old TV programs.

Those are very good answers. What would you think if I told you that every time you look up into the sky and see a star you are looking into the past? Just think for a moment. We know that there are very great distances between the stars and earth. It is going to take a long time for the light from a distant star to reach Earth. So, what does it mean to us timewise when we look up at a star?

It means that when we see the light from a star, we are really looking at light that has left the star a long time ago.

Exactly. And if the light left the star a long time ago, then we are really seeing the star as it used to be. The light that the star is giving off at this moment won't reach Earth for quite some time. We may see the star as it is this moment in a few years from now.

1 How fast is fast?

Light travels really fast. Does anyone know how fast light travels?

It travels at 300,000 km/s.

You are correct. But how fast is that really? In order to find out, let's compare some objects and their speeds to the speed of light. We can do this by finding out how fast the object moves in km/sec and then finding its percent speed of light. The first thing we need to do is to figure out how we can convert speed in km/hr to km/sec. Does anyone have any ideas?

Yes, first divide the speed by 60 to convert to minutes. Then divide again by 60 to convert to seconds.

Let's take a look at Table 1. It says that a diving falcon has a speed of 259 km/hr. Take the 250 and divide it by 60, and then divide it by 60 again. What does this equal?

The answer is 0.069 km/sec.

Very good. Go ahead and complete the rest of the conversions from km/hr to km/sec.

Students complete the conversions.

Now let's find out how long it will take these objects to fly to the moon. Take the moon's distance away, the 384,400 km and divide it by the speed in km/hr. That will give us how many hours it will take. If we divide the hours by 24 it will give us how many days it will take. Let's work one together.

17B The Time Machine

How long does it take for starlight to reach Earth?

Light travels very fast at 300,000 kilometers per second. However, the universe is very large. Light that reaches your eyes from distant stars or galaxies has been traveling for a long time. Suppose you look through a telescope and see a galaxy that is one million light-years away. Because it takes one million years for the light to reach you, you are actually seeing the galaxy as it was one million years ago! Looking out into the universe is like looking back in time. The farther away you look in space, the farther back you see in time.

Materials

- Astronomy reference materials

1 How fast is fast?

When you travel in a car or a jet, your speed is usually measured in miles per hour or kilometers per hour. It takes about six hours to fly across the country on a modern airliner, which travels about 1000 km/hr. This is about as fast as a person can travel today, but it is nothing compared to how fast light travels.

Complete Table 1 below and compare the speeds of these fast moving objects to the speed of light (150,000,000 km/sec). To calculate the percent of speed of light;

$$\frac{\text{Speed of object}}{\text{Speed of light}} \times 100 = \text{Percent of speed of light}$$

Table 1: Comparing speeds

Object	Speed in km/hr	Speed in km/sec	Time to reach the moon 384,400 km away (hours and days)	% of speed of light
Diving falcon	250			
Mag-Lev Train	430			
Top fuel dragster	535			
Commercial jet	1000			
Space shuttle	28,000			
Comet	200,000			

a. Earth is about 150,000,000 km from the sun. How long does it take the light from the sun to reach Earth?

b. Jupiter is 778,000,000 km from the sun. How long does it take the light from the sun to reach Jupiter?

c. Pluto is 5,900,000,000 km away from the sun, the farthest planet away. How long does it take light from the sun to reach Pluto?

Take the distance to the moon, 384,400 km and divide it by the 250 km/hr the falcon dives. Then divide it by 24 hours. The answer is 64.07 days. Students complete the remaining conversions.

Now we can figure out the percentage of the speed of light that specific objects travel. We will use the equation speed of the object divided the speed of light times 100 to solve for the percentage of the speed of light. Let's figure out what percent of the speed of light the diving falcon is traveling. We will use the speed of the falcon in km/sec. So take the 900,000km/sand divide it by the speed of light which is 300,000 km/sec. Then multiply it by 100. What is the answer?

The answer is.000023%. Students complete the remaining calculations.

Now that we are finished, would anyone like to make an observation about the speed of light?

Light is FAST!

Light is fast. From our previous investigations, we have learned that Earth is 150,000,000 km from the sun. Using what we just learned in calculating the time it took for our objects to reach the moon, do you think there is a way to calculate how long it would take light from the sun to reach Earth?

Yes. We can use the same procedure as before. Take the distance from the sun to Earth and divide it by the speed of light. Then divide it by 24 to find out how many days the light travels. Students complete the remaining calculations for questions a through c.

2 How far is far?

Because the distances are so large and it takes so much time for the light to travel from a star to a planet, scientists like to use the measurement *light-year*. By using the measurement, light-year, scientists have a number that is easier to work with, especially since we saw that the distance light can travel adds up very quickly. We know how far light travels in one second. Let's find out how far light travels in other measurements of time. Let's all take a look at Table 2 on page 84. The first distance is done as an example for us.

Model the initial calculation for students. Then students complete the data table in part two.

3 Looking back in time

Wow! Light can travel 66 quadrillion 225 billion 600 millions kilometers in one year. Talk about amazing. How would you like to work with numbers that large?

This number is very large. Most students would think it is too confusing to work with.

Exactly. Numbers that large can be confusing and hard to work with. That is why scientists like to describe the large distances in space as light years. One light year is 66,225,600,000,000 km. Which number would you rather use in a calculation?

The one light year.

Now, we are going to the media center and research how far sky objects are from earth. Then I want you to tell me how far back in time you are looking when you see these objects.

Reserve a day for your students to continue their research in the library or your school's media center.

Investigation 17A: Data and Answers

1 The Milky Way galaxy

a. Yes, there have been several discovered in recent years.

b. Around 200,000.

c. A supermassive black hole surrounded by an unusually compact and powerful radio signal producing accretion disk.

d. Over 200,000,000 years.

e. Around 13.6 billion years, give or take 800 million years.

f. About 10,000,000,000 years (ten billion).

Notes:

Investigation 17B: Data and Answers

1 How fast is fast?

Object	Speed in km/hr	Speed in km/sec	Time to reach the moon 384,400 km away (hrs and days)	% speed of light
Diving falcon	250	0.07	1,537.6 hours or 64 days	0.00002
Mag-Lev Train	430	0.12	894 hours or 37 days	0.00004
Top fuel dragster	535	0.15	718.5 hours or 30 days	0.00005
Commercial jet	1000	0.28	384 hours or 16 days	0.00009
Space Shuttle in orbit	28,000	8.0	13.7 hours	0.00266
Comet	200,000	55.5	1 hour 55 min	0.0185

a. About 8 and a half minutes.

b. About 43 hours.

c. About 5 and a half hours.

2 How far is far?

Unit	Speed of light (in km/sec)		Time in seconds	Distance (in km)
1 light second	300,000	x	1	300,000
1 light minute	300,000	x	60	18,000,000
1 light hour	300,000	x	3,600	1,080,000,000
1 light day	300,000	x	86400	2.59×10^{10}
1 light week	300,000	x	604,800	1.8144×10^{11}
1 light year	300,000	x	31,536,000	9.4608×10^{12}

3 Looking back in time

Object	Distance from Earth (light years)	Years looking back in time
Sirius	8.5	8.5
Vega	23.5	23.5
Arcturus	37	37
Canopus	316	316
Polaris	430	430
Betelgeuse	430	430
Rigel	775	775
Andromeda Galaxy (M31)	2.5 million	2.5 million

Chapter 17 Answers

17.1 Section Review

1. galaxy
2. the Milky Way
3. d
4. (1) He found galaxies outside our own, including Andromeda. (2) He discovered that the further away a galaxy is, the faster it was moving away from Earth.
5. Any four of the following: spiral; barred spira; elliptical; lenticular; irregular
6. c
7. One way is to find a star whose luminosity is known and apply the inverse square law to find the distance. Another way is to compare the size and type of galaxy with closer galaxies to estimate the luminosity of the farther ones.
8. black hole
9. light cannot escape it
10. Spectral lines were shifted toward the red end of the spectrum
11. That the further away a galaxy was, the faster it was moving away from Earth.
12. a single point, smaller than an atom
13. 1 billion years
14. cosmic background radiation left from Big Bang

Connection

1. An aurora is caused by charged particles and energy that penetrate Earth's upper atmosphere. Charged electrons from the sun travel down Earth's magnetic field lines that make up the magnetosphere. When the electrons collide with atmospheric gases, light particles are released creating aurora lights.
2. The Space Environment Center is the national and international warning center that monitors the space environment. It is located in Boulder, Colorado and provides real-time information about solar events and space weather. The POES travels over the poles gathering data and images. The role of the SEC is to provide timely data to people and organizations that need accurate space weather information. These interested parties may include the U.S power grid, NASA, airline industry, cell phone companies, satellite missions, and users of GPS.
3. The best conditions for observing an aurora are:
   a. Increased geomagnetic activity
   b. A higher latitude location
   c. Sky must be clear and dark
   d. No cloud cover
   e. No moonlight, sunlight, or city lights

Chapter Activity

Sample Data Table

Distance (cm)	Brighness (mA)
10	92.5
20	28.5
30	15/2
40	9.2
50	6.6
60	4.8
70	3.8
80	3.0
90	2.5
100	2.0

(graph is on next page)

Applying your knowledge: sample answers

a. We think increasing the distance will decrease the brightness of the light.
b. The graph is decreasing from left to right.
c. No, I cannot determine the mathematical relationship because the line is a curve. If it were a straight line, I could find the slope of the line and determine the mathematical relationship.

Brightness vs. Distance graph of sample data:

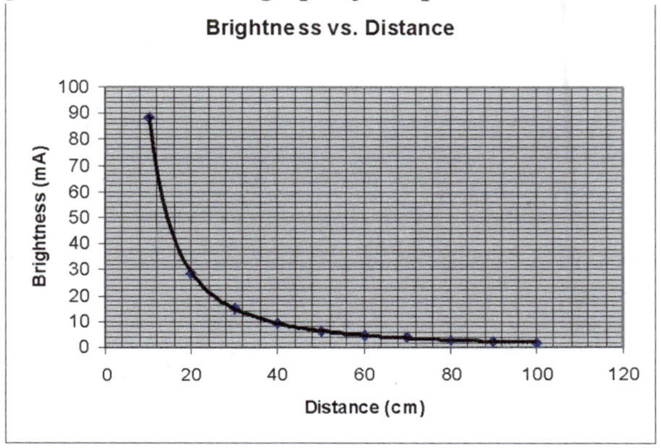

Chapter 17 Assessment

Vocabulary

Section 17.1

1. galaxy
2. Big Bang theory

Concepts

Section 17.1

1. Spiral: central, dense area surrounded by spiraling arms
   Barred spiral: bar-shaped structure in the center
   Elliptical: central portion of aspiral galaxy without the arms
   Lenticular: lens-shaped with a smooth, even distribution of stars; no central, denser area
   Irregular: peculiar shapes
2. Milky Way; elliptical
3. Answers are:
   a. 26,000 light years from the center
   b. 100,000 light years across and 3,000 light years thick
   c. black hole
4. The idea that the universe exploded outward from a sngle point smaller than an atom into the vast expanse of galaxies and space we see today.
5. (1) galaxies expanding away from each other, (2) microwave background radiation, (3) ratios of the elements
6. 10-20 billion years old
7. The star is moving away from Earth
8. c
9. The universe is expanding.
10. a
11. Helium was formed in the Sun; the Sun was formed about 4 billion years ago and helium was formed about 4 minutes after the Big Bang.
12. At the end of star life cycles
13. c
14. Its gravity is so strong that the escape velocity equals or exceeds the speed of light.

Problems

Section 17.1

1. Earth is the smallest, then our Sun, the black hole, the Miky Way galaxy, the universe
2. Our moon is closest, then the Sun, Sirius, Andromeda galaxy

Notes:

Reference Materials

Meeting the Needs of all Learners

The middle school classroom is a diverse learning community of students with various abilities, backgrounds, and learning styles. An effective teacher modifies his or her teaching style to meet the needs of all learners. In doing so, the teacher creates an engaging learning environment in which all students achieve.

Learning Modalities

All students can learn, but each student has a unique learning style. Many factors contribute to how students learn including personality, strengths, weaknesses, cultural experiences, and learning modalities. *Learning modality* refers to how students process information. Students tend to prefer their strongest modality; but they often apply a combination of modalities as they learn. The three major types of learning modalities are visual, auditory, and kinesthetic or tactile. Common characteristics of students who tend to embrace a specific modality are described below.

Visual Learners

- prefer to acquire knowledge by observing the surroundings
- enjoy reading illustrated books, doodling, interactive (online) lessons, active research, and teacher demonstrations
- memorize new material when accompanied by a diagram
- pay attention to details
- benefit from graphic organizers and vocabulary cards

Auditory Learners

- enjoy group discussions and activities such as *Think-Pair-Share*
- have excellent memory of facts, figures, and faces and often apply mnemonic strategies
- are successful when teachers employ guided practice
- enjoy being read to or listening to music

Kinesthetic/Tactile Learners

- thrive when completing "hands-on" activities like investigations and experiments
- excel at problem solving or may have profound mechanical skills
- enjoy working cooperatively or at stations and centers
- volunteer to participate in active learning and research
- are outgoing

Effective teachers use a variety of modality based learning strategies to meet the needs of all learners. The more students are exposed to various strategies, the more they tend to modify their own learning style. The balance between a student's natural strengths and the development of new and alternative modalities is enhanced when teachers vary their approach and incorporate a blend of learning strategies.

Learning Strategies

Research suggests that a blended approach to employing learning strategies improves overall student performance in all classroom settings. The CPO Science teaching model highlights many of these strategies, which are explained in the next few pages.

Active Learning

Active learning directly engages students in the learning process through activities such as problem-solving exercises, games, simulations, role-playing, presentations, and debates. Active learning is student-centered and teacher facilitated. This approach directly involves students in meaningful content-based learning. Active learning translates into increased mastery and retention of science content because students enjoy the learning process more. Although active learning may be time consuming it provides helpful feedback to gauge student comprehension and is effective for multiple modalities. Teacher planning and monitoring are essential for successful active learning experiences.

Active Research

Active research is student-centered. It includes learning experiences such as conducting interviews, Internet and library research, writing reports, developing presentations, creating posters, and building models. Active research may be done independently, in pairs, or as group work. The teacher must clearly state expectations or provide grading rubrics at the onset of the activity. In this manner the desired outcomes are clear.

Closure Cards

Closure provides necessary feedback to help teachers determine if students have mastered major content themes. A significant component of closure involves students summarizing and internalizing new knowledge. Students may complete closure cards at the end of a lesson, chapter, or unit. Use a 3x5 index card have the students answer the following questions:

1. Write down two new facts you have learned.
2. What else would you like to learn?

Cooperative Learning

Cooperative learning is a small-group strategy which involves students of varying abilities working together to achieve a common measurable goal. Cooperative learning fosters academic achievement. An effective cooperative learning activity assigns a specific role and responsibility to each group member while reinforcing the importance of teamwork.

Demonstration

Teacher demonstrations are used in many ways to enhance student learning and to link observation with comprehension. Demonstrations may be used to introduce a new concept or to supplement instruction of a difficult topic. They are also applicable when materials or time are limited, or when an activity involves the use of dangerous materials. Teachers should always practice and perfect a demonstration before presenting it to students.

Discussion

Whole group and small group discussions invoke meaningful learning through creative inquiry and active student participation. Whole class discussions may be used to introduce new topics and concepts or as refreshers for past learning experiences. Small group discussions are often used to build upon and extend a new concept and to assess student progress. Discussions promote active listening skills and an array of supported opinions.

Graphic Organizers

Graphic organizers are visual tools that stimulate, guide, and enhance student learning. They promote learning by association, patterns, and connecting key concepts, ideas, and terms. Three basic types of graphic organizers (including one example of each) are discussed below.

Task specific organizers explore meaningful content learning and basic skills. Examples include Venn diagrams, story maps, sorting trees, and concept circles. The Venn diagram shown below compares and contrasts the qualities of bacteria and viruses.

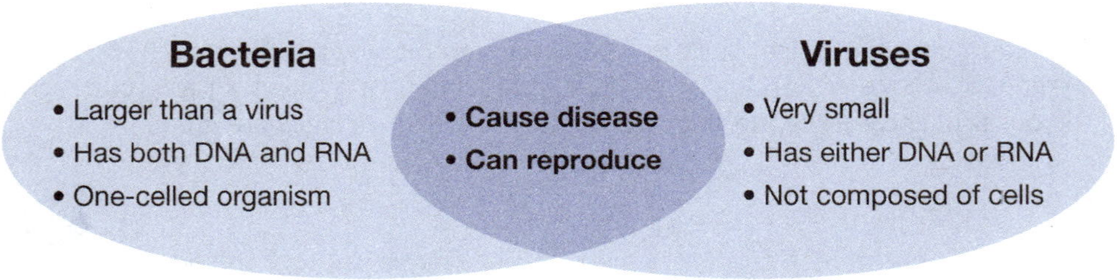

Thinking process maps and charts enhance cognitive development and critical thinking skills and include concept maps, K-W-L charts, and thinking maps such as compare and contrast charts. The K-W-L chart shown below may be used to assess prior knowledge and prediction skills, address any misconceptions, and to determine what students have learned. The acronym (K-W-L) stands for "What I **K**now," "What I **W**ant to Know," and "What I **L**earned."

KNOW	WANT	LEARNED

Instructions for creating a K-W-L chart

1. Create 3 columns with the following headings: **K**now, **W**ant, and **L**earned.
2. Engage students in a discussion to determine what they know about the topic. Record student responses in the *Know* column.
3. Then have the students predict what they will learn by asking, "What do you want to know about this topic?" Record responses in the *Want* column. This step helps to set the purpose for learning and encourages students to take an active role in the process.
4. After the concept has been taught then ask, "What did you learn?" Record the responses in the *Learned* column. This step clarifies any misconceptions, assesses the overall knowledge gained, and brings closure to the lesson.

Brainstorming Webs use brainstorming to encourage individual and group creativity. Some examples are clustering, mind mapping, brainstorming wheels and webs.

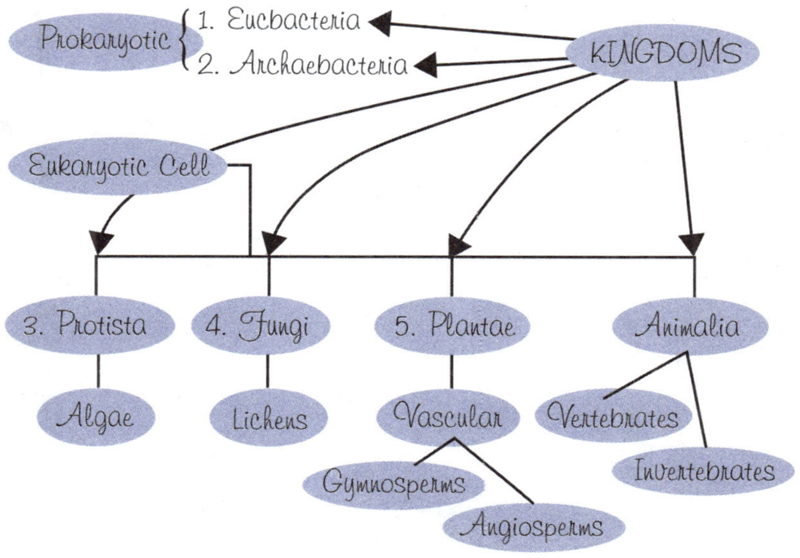

Relating Bloom's Taxonomy to Types of Graphic Organizers

Level I Knowledge	Level II Comprehension	Level III Application	Level IV Analysis	Level V Synthesis	Level VI Evaluation
• K-W-L Chart • Spider Map	• Main Idea • Action Maps • Story Board • Sorting Tree	• Flow Charts • Concepts Maps • Herringbone • Mind Mapping • Cause and Effect	• Compare/Contrast Thinking Maps • Classification Tree • Network Tree	• Idea Maps • Problem/Solution Charts	• Venn Diagrams • Comparison Matrix

Guided Practice

Guided practice is useful when students are introduced to a new task, topic, or concept but are unable to work independently because they have limited prior knowledge. The teacher facilitates student inquiry after introducing the topic. The purpose of guided practice is to guide student learning, correct misconceptions, model and reteach tasks, and provide feedback.

Mnemonics

Mnemonics enhance student memory and recall. They are powerful cognitive tools for visual and auditory learners which help to connect existing and newly acquired knowledge. Three types of mnemonic strategies are discussed below.

Mnemonic Strategies

	Letter	Keyword	Pegword
DESCRIPTION	Creating a word or phrase out of the first letters of a list.	Identify a term, provide the definition of the term, create a keyword that goes along with the term, and link to an imagine.	Creating rhyming words that represent numbers or a specific order.
EXAMPLE	"My Very Educated Mother Just Served Us Nine Pizzas" *This mnemonic represents the planets in order of increasing distance from the Sun.*	**Term:** VEIN **Keyword:** IN *Visualize a car driving into a parking lot with a sign that reads "IN" written inside of a heart.* **Definition:** The blood flows into the heart through the vein.	**One = Bun** **Two = Shoe** **Three = Tree** **Four = Door** **Five = Tribe** **Six = Sticks** **Seven = Heaven** *The atomic number for Aluminium is "**a Bun in a Tree**" or 13.*

Paired Learning

Middle schoolers learn effectively while interacting with other students. Paired learning activities encourage students to exchange information, develop ideas, solve problems. These exercises develop social and team building skills while reinforcing instruction. Effective paired learning activities have clearly stated objectives and are teacher-monitored.

RAFT

RAFTs are a creative writing strategy used to demonstrate content mastery. The format is flexible and focuses on individual student views and selected roles. The chart below explain the specific elements of the RAFT. Examples of each component are shown in blue.

Role	What role will the student take? (writer, reporter, character, scientist, doctor, judge, eyewitness, inventor, juror, observer)
Audience	Who will read what you write? (teacher, peers, parents, fictional character, politician)
Format	What is the best way to present the writing? (letter, newspaper, article, report, journal, interview, poem, song, biography)
Topic	Who and/or what is the subject of your writing? (personal interest, inventions, aliens, food, animals, books, vacation, sports)

Stations/Centers

Stations and centers may be used to address many different learning objectives at once. It is a student-centered strategy which promotes hands-on learning. Effective stations and centers will do the following:

- Promote independence and responsibility
- Allow learning through self-discovery
- Manage limited resources
- Allow for one-on-one time with students
- Promote cooperative and hands-on learning
- Increase student decision-making skills
- Allow teachers to observe, assess, and correct misconceptions

Think-Pair-Share

The "Think-Pair-Share" strategy is designed to stimulate thinking, discussion, and information processing. It is a four-step process that starts with a thought-provoking question. Think-Pair-Share may be adapted to any content area and helps to promote sharing of ideas in a non-threatening context.

Step 1	The teacher poses a topic question to the students.
Step 2 THINK	Students are given a limited amount of time to ponder the question and come up with a response.
Step 3 PAIR	Students are paired together to discuss and analyze their individual response, and then formulate a single or combined response.
Step 4 SHARE	In the final step the paired students share their response with the teacher and the entire class. Visual presentations may be included to emphasize the response.

Vocabulary Cards

Traditional vocabulary cards are useful for all students and are a structured method for learning new words. There are many types of vocabulary cards, however the most common is a 3x5 file card with the vocabulary word written on the front side. The back side of the card may contain any of the following: definition, picture, the word used in a proper sentence, pronunciation, parts of speech, synonyms, or antonyms. The cards may be hole punched and held together with a shower hook or placed in a binder.

BLOOM'S TAXONOMY

Bloom's Taxonomy is the most widely used and well known system for classifying educational objectives in the cognitive domain. Bloom categorized cognitive learning outcomes into a hierarchal system from the most basic level of learning (knowledge) to the highest level of learning (evaluation).

	Level I: Knowledge	Level II: Comprehension	Level III: Application	Level IV: Analysis	Level V: Synthesis	Level VI: Evaluation
COGNITIVE DOMAIN	*The remembering of previously learned information and material*	*Understanding the meaning of material*	*Using of learned material in new and concrete circumstances*	*Breaking down material into smaller parts and understanding organized structure*	*Linking and combining informational parts together to form a whole*	*Judging the value of material for a given purpose or standard*
PROCESS VERBS	collect, define, describe, examine, identify, label, list, match, name, outline, quote, read, recall, record, repeat, reproduce, retell, select, state, tabulate	associate, classify, cite, convert, describe, defend, distinguish, discuss, estimate, explain, express, extend, generalize, give, group example, infer, order, predict, restate, rewrite, summarize, trace, understand	administer, arrange, articulate, assess, apply, calculate, chart, classify, collect, compute, construct, demonstrate, develop, determine, discover, establish, experiment, manipulate, modify, predict, prepare, produce, report, relate, show, solve, teach, transfer, use	break down, connect, correlate, diagram, differentiate, discriminate, divide, focus, identify, illustrate, infer, order, outlines, prioritize, recognize, select, separate, subdivide	adapt, anticipate, categorize, combine, compile, compose, create, devise, design, explain, facilitate, generate, modify, negotiate, organize, plan, rearrange, reconstruct, reinforce, relate, reorganize, revise, rewrite, structure, summarize, tell, validate, write	appraise, compare, contrast, conclude, confront, convince, criticize, critique, decide, defend, grade, interpret, judge, justify, measure, persuade, rank, support, test, translate
INSTRUCTIONAL OBJECTIVES	*The student knows*: • common terms • specific facts • methods and procedures • basic concepts • principles	*The student will*: • interpret charts and graphs • interpret verbal material • translate from one medium to another • justify methods and procedures • estimate future consequences implied in data • understand facts and principle	*The student will*: • apply concepts and principles to new circumstances • apply laws and theories • solve mathematical problems • construct charts and graphs • demonstrate proper procedures and methods	*The student will*: • distinguish between facts and inferences • recognize unstated assumptions • recognize logical errors in reasoning • evaluate the pertinency of data • analyze the organizational structure of work	*The student will*: • design an experiment • Link learning from different areas into a plan for problem solving • devise a new plan for classifying • create and write an organized theme • write a creative short story • give an organized speech	*The student judges*: • logical consistency of written material • adequacy with which conclusions are supported by data • value of a work by use of internal criteria • value of a work by use of external standards of excellence
ACTION	memorizing recalling identification recognizing descriptive	describing interpreting translating organizing	problem solving applying implementing	separating dissecting identifying subdividing	combining creating integrating formulating	decision making resolving assessing concluding

Pre-Lab Support for English Language Learners (ELLs)*

Connect to Prior Knowledge and Experience Determine what students already know about the concepts being investigated. Connect the current inquiry to the educational and cultural background of ELLs. Tap into their prior knowledge and experience through open-ended questions and guided discussion. Students use graphic organizers to record and analyze their thoughts in this process.

Monitor for Misconceptions of Science Discover and correct any misconceptions students have about the scientific information being investigated in the lab. First help students to organize and link the information they already know, then clarify and elaborate the scientific concepts. Encourage students to become problem solvers, fostering a scientific view of the world.

Post steps to The Scientific Method Create and post a visual aid to remind ELLs of the steps to the scientific method. Students can create posters to be used for this purpose at the beginning of the year when the scientific method is first taught.

Identify Materials For each investigation, take time to display the lab materials and teach the names for each item. Students can record this information in a "Science Tools" picture glossary so that they can refer to it at any time throughout the school year. Encourage students to use the correct scientific vocabulary for all equipment used in class.

Teach & Post Relevant Vocabulary Carefully choose and teach the vocabulary that ELLs will need to be successful in reaching the learning objectives for each investigation. Prior to the investigation, employ learning activities that allow students to listen to, pronounce, write and understand the words. Provide translations of the terms or allow students to use bilingual dictionaries. Post the words clearly in the classroom, with illustrations whenever possible. Give students many opportunities before, during and after the lab to use these words in authentic communication, particularly in discussion with peers in small groups or pairs.

Use Skill Sheets CPO Science Learning Systems offers a number of skill sheets on the Teacher Resource CD-ROM. ELLs use these sheets for additional practice of certain skills related to the content of CPO texts and investigations. The sheets can be assigned as group work or homework.

Speak Clearly Speak clearly while giving students directions for the investigation. Slow your rate of speech and enunciate clearly. Use props, actions, visual aids, gestures and facial expression to help convey meaning. Model the correct usage of new vocabulary and grammar. Utilize cognates whenever possible. Avoid using idioms.

Check for Comprehension Stop frequently to check for student comprehension. Use hand signals, questions, surveys or other methods to allow students to express their level of understanding. Repeat, clarify, or paraphrase information that is not clear. Create an environment where students feel comfortable to ask for clarification or help whenever they need it.

***See ELL ancillary for additional strategies**

Assisting ELLs during the Investigation*

Model Procedures Model proper safety precautions, lab procedures and clean up. Model how to fill in charts, graphs and other lab charts. Demonstrate how to complete a lab report. Post clear examples of all types of written reports in the classroom.

Provide Supportive Materials ELLs need supportive materials that allow them to organize and analyze new information. Use visual materials such as graphic organizers, charts and graphs. Use scaffolding to move students from dependence to independence in all investigative activities. When introducing a new procedure, provide graphic organizers with more support, such as a graph with the titles for the x and y axes already filled in. As student competence increases, gradually remove the support.

Develop Academic Language Well-supported investigations give students hands-on experience with both science and language. Allow students to talk throughout the investigations. As you circulate around the room listening to students, interact with them and help them to adjust their everyday English to academic language. Whenever needed, give students the vocabulary and language structures they will need to speak or write about their observations. For example, demonstrate the use of the past tense in sentences in a lab report. "We observed that…."

Support for ELLs after the Investigation*

Provide and Post Lab Report Format Model a clear example of a correct lab report, as you expect it to be submitted. Take students through each step and provide language support whenever needed. Post a copy of the example in the classroom so that students can continually refer to it. The report can be scaffolded, providing a lot of support when students need it and gradually removing it as students become more capable of completing it on their own.

Provide Opportunities to Evaluate Learning Give students opportunities to reflect on their own learning process. Use discussion, journals or any variety of graphic organizers to provoke students to think not only about what they have learned but how. Students can become problem solvers in their own education by learning more about their own strengths, weaknesses and use of strategies.

Provide Opportunities to Expand Learning Give students opportunities to extend their learning by giving them further experiences with the science they encounter in the investigations. If possible, find ways to help students to apply what they have learned in the investigation to new situations. Encourage students to make the connection between science and their everyday world.

*See ELL ancillary for additional strategies

Family Involvement in Science Education

Family involvement in science education is a critical component for ensuring student success. Many research studies have shown that a student's performance in science class is enhanced when families are involved. By encouraging their student to apply skills and concepts acquired in class, families promote a genuine interest in learning more about science. Below are suggestions to increase family participation in science education.

Inform families about the program

Send a letter home to inform the family about the science program their student is using. A sample letter is given on the next page.

Be proactive

Open the lines of communication before a problem arises. Encourage parents or guardians to write a note or to call the teacher when they have information to share.

Reinforce science skills at home

Send home a list of science skills that parents can help develop and reinforce through activities at home. Observing, asking questions, predicting, measuring, comparing, exploring, testing ideas, and trying new experiences are among skills families can emphasize in daily activities.

Start a parent newsletter

Get students, faculty, and administrators involved in creating the newsletter. Designate a portion of the newsletter for parents or guardians to express their concerns.

Employ community resources

Create a list of community resources which support science learning such as recycling centers, museums, parks, gardens and nurseries. Include some ideas of what to look for and connections to the classroom for parents. Distribute them during open house or conferences for potential weekend family outings.

Develop a Science Backpack

Create a backpack for each major unit in your course. Fill it with instructions and materials for students to do science at home with their families. Choose activities that make use of common household materials. Develop a system for each student to "check-out" the backpack.

Discuss current events in science

Use science in the news to foster discussion. Ask students to discuss or critique an article about a science event with their family.

Have a "Family Science Night"

Plan a family science night to bring families together to enjoy and participate in science education. Consider the following tips to ensure your family night is a success.

- Have snacks, a raffle, and door prizes to entice people to come. Ask local businesses or the PTO for donations, write a small grant, or ask student council to help subsidize the night.
- Get students volunteers to help put up posters, set up tables, and work the snack table. Have students create activities for their families.
- Provide a baby-sitting service somewhere in the building for parents to drop off little ones while they work with older children.
- Advertise the *Science Night* through the school newsletter, community newspaper, posters in the building, and phone calls to parents. Let people know something exciting and free is going to take place.
- Have enough stations and activities to keep people engaged. Choose activities that are easy to set up and tear down, are not overly messy, and engage family members. Have some students on duty for maintenance of the station throughout the night.
- Make it a theme night. Pick one concept and develop difference stations around that concept.
- Have an integrated approach where you incorporate stations around Earth, life, and physical science.
- Send thank-you notes to everyone who attended and assisted.

Dear Parent or Guardian:

This year your student will study science using the CPO FOCUS ON PHYSICAL SCIENCE program published by CPO Science. This new program, recently adopted by the State Board of Education, teaches all of the California Science Standards using a variety of methods. The general approach is guided-inquiry where students do two major investigations in each chapter. This will account for about percent of the instructional time in science. The CPO program is supported by a package of high quality equipment, technology components, a special web site (www.cpo.com) and a variety of teacher's materials all designed to ensure success.

Students work in small groups to complete investigations. They will experiment, discuss their observations, and share their findings through written reports and presentations. This cooperative approach to learning encourages team-building skills and prepares students for life beyond high school. Business leaders have stated that learning to work as part of a team is critical to success. Students will be highly motivated because they enjoy actively doing science, not just reading about it. Special materials and teaching components ensure that all students have an equal opportunity to achieve.

We believe that your student will enjoy science and will master the objectives stated in the California Science Standards which are provided in the student text, on the CPO Science web site, and on the California Department of Education web site at www.cde.ca.gov. The focus in grade eight is on physical science. However, the emphasis is on life science and earth science in grades seven and six respectively. This three-year middle school series will properly prepare your student for any state testing and high school science classes.

We look forward to a very enjoyable and productive year in science. If you have any questions concerning our science class and how you might get involved please let us know. The best way to contact me is ______________________________.

Respectfully,

Building Science Vocabulary

Prefixes, roots, and suffixes are the building blocks of language.

- A **prefix** is a word part added to the beginning of a word or root.
- A **root** is a word base, the part that gives a word its core meaning. A root can appear at the beginning, middle, or end of a word.
- A **suffix** is a word part added to the end of a word or root.

The most common sources of English word parts are Latin and Greek words. Sometimes both ancient languages contribute roots with the same meaning. For example, the word *solar* is based on the Latin word for "sun," *sol*; the word *helium* is based on the Greek word for "sun," *helios*. Knowing word parts and how they are combined can help your students figure out the meanings of new science vocabulary words.

Prefixes

Prefix	Meaning	Example	Add Your Example...
a-	without	abiotic	
ab-	away	absent	
ad-	toward	adhere	
alt-	high	altitude	
bi-	two	binoculars	
circum-	around	circumference	
co-, com-	with, together	compare	
de-	down, from, off	descend	
dia-	through	diameter	
dis-	not, apart	disappear	
endo-	within	endoskeleton	
equ-	equal	equinox	
ex-, exo-	off, outside	exhale	
hemi-	half	hemisphere	
hetero-	different	heterogeneous	
homo-	same	homogeneous	
hyper-	over	hyperventilate	
hypo-	under	hypodermic	
im-, in-	not	impossible	
inter-	between	interrupt	
iso-	equal	isobar	

Prefix	Meaning	Example	Add Your Example...
mega-	large	megaton	
meta-	change	metamorphosis	
micro-	small	microorganism	
multi-	many	multiply	
non-	not	nonliving	
omni-	all	omnivorous	
paleo-	ancient	paleontology	
para-	beside	parasite	
per-	through	permeable	
poly-	many	polygon	
post-	after	postpone	
pre-	before	prehistoric	
prim-	first	primary	
pro-	forward	produce	
re-	again, back	reflect	
semi-	half	semiconductor	
strat-	to spread or stretch	stratus	
sub-	under	submarine	
super-	above	superior	
tele-	distant, far	telescope	
trans-	across	transfer	
un-	not	unlike	
uni-	one	universe	

Roots

Root	Meaning	Example	Add Your Example...
anthrop	human	anthropology	
aqua	water	aquarium	
aster, astro	star	astronomy	
bio	life	biology	
celer	fast	accelerate	
cept	to take or hold	accept	
chron	time	chronology	
clin	to lean or bend	inclined	
cosm	world, universe	cosmic	
cycl	circle	cycle	
dict	to say or speak	predict	
dorm	to sleep	dormant	
duc, duct	to lead or draw	conduct	
fac	to make	artifact	
flect, flex	to bend	flexible	
flu	to flow	fluid	
form	to shape	formation	
fract	to break	fracture	
geo	earth, land	geology	
graph	to write	biography	
grav	heavy	gravity	
hab, habit	to live	habitat	
helio	sun	helium	
hydro	water	hydrometer	
ject	to throw	reject	
litho	rock	lithosphere	
log	word, speech	dialog	
luna	moon	lunar	

Root	Meaning	Example	Add Your Example...
mar	sea	marine	
miss, mit	to send	transmit	
morph	form	metamorphic	
noct, noc	night	nocturnal	
nucl	nut, kernel	nucleus	
pend	to hang	suspend	
phon	sound	telephone	
photo	light	photosphere	
port	to carry	portable	
rad	ray	radiant	
sci	to know	science	
scope	to see	microscope	
scrib, script	to write	manuscript	
sed	to settle	sediment	
sens, sent	to feel	sense	
sol	sun	solar	
spec	to look	specimen	
sphere	ball, globe	atmosphere	
spir	breath	respiration	
stella	star	stellar	
struct	to build	construct	
tend	to stretch	extend	
terra	earth, land	terrestrial	
therm	heat	thermal	
tract	to pull	attract	
trib	to give	tributary	
vert	to turn	invert	
vid, vis	to sewe	visible	
volv	to roll	revolve	
vor	to eat	carnivore	

Suffixes

Suffix	Meaning	Example	Add Your Example...
able-, -ible	capable of	movable	
-aceous	of or pertaining to	herbaceous	
-al	of, relating to	global	
-ance, -ence	action, state, or condition	dependence	
-ary, -arium	a place for something	aquarium	
-ase	forms names of enzymes	amylase	
-ate	to act on or change	saturate	
-cule	small	molecule	
-cyte	cell, hollow container	lymphocyte	
-d, -ed	past tense	formed	
-elle	small	organelle	
-emia	blood	anemia	
-en	made of	wooden	
-er, -or	person who is or does	teacher	
-escent	becoming	effervescent	
-ful	full of, having	careful	
-gram	a written record	seismogram	
-graph	something written or drawn	autograph	
-ic	related to	seismic	
-ify, -fy	to make or cause to become	solidify	
-ism	the act, state, or theory of	magnetism	
-ist	person who is or does	scientist	
-ity, -ty	degree of	acidity	
-ive	forms adjectives from verbs	active	
-ize	forms verbs from nouns	hypothesize	
-less	without	harmless	
-logy	study of	meteorology	
-ly	in the manner of	slowly	

Suffix	Meaning	Example	Add Your Example...
-ment	act or result of	measurement	
-mer	part	polymer	
-meter	to measure	thermometer	
-oid	like, resembling	asteroid	
-oma	mass, tumor	melanoma	
-ose	forms names of sugars	glucose	
-phyll	leaf	chlorophyll	
-s, -es	plural, more than one	earthquakes	
-sion, -tion	state of	erosion	
-zoa	animal	protozoa	

Additional Materials

The following pages list materials used in the Investigations in addition to the kits provided. For ease of ordering and buying, the list is divided into two categories: Non-Consumables and Consumables.

Non-Consumables

Item	Amount
Metric and English ruler	1
Meter stick	1
Digital or triple-beam balance	1
Scissors	1
250 mL beaker	1
100 mL graduated cylinder	1
Tire pressure gague	1
steel hex nuts 3/8" or ½" size	6
Bicycle pump	1
Spot plate	1
Pipette	10
Washers	30
Key ring	1
Loops of string	3
Track and field tape measure (25 m or more)	1
Lacrosse ball	1
Assorted balls (basketball, volleyball, baseball, etc.)	8
Earth globe (30 cm)	1

Item	Amount
Incandescent lamp base	1
100 watt bulb	1
Alcohol thermometer	1
Digital multimeter	1

Additional Physical Science Materials

Consumables

Item	Amount
Small cardboard box which may be sealed	2
A selection of small objects to fit into the box	1-5
16 oz. foam cups	10
Uncooked rice	¼ cup
Cellophane tape	1 roll
Plastic cups	6
Glycerin	30 mL
Corn syrup	30 mL
Molasses	30 mL
Vegetable oil	30 mL
Food coloring	4 colors
Modeling clay	2.5 sticks
Granulated sugar	40 mL
Sugar cube	2
Red litmus paper	22 pieces
Blue litmus paper	22 pieces
Red cabbage juice	50 mL
Lemon juice	10 mL
Vinegar	20 mL
Seltzer	10 mL

Item	Amount
Baking soda solution	10 mL
Bar soap solution	10 mL
Ammonia	20 mL
Green tea	10 mL
Antibacterial cleaner	10 mL
Apple juice	10 mL
Hydrogen peroxide	10 mL
Epsom salts	5 grams
Raw potato	1 slice
Baking soda	10 grams
Phenol red	20 mL
Calcium chloride	10 grams
Glow stick	1 (any variety)
Heat pack	1
Effervescent tablet	1
Small paper cup	1
Reference book of organic compounds	1
Licorice sticks	2
Round toothpicks	8
Colored gum drops	20

CPO Focus on Physical Science Skill Sheets

The Investigation Skill sheets are found in the front of the *CPO Focus of Physical Science Investigation Manual*.

Math, Reading, Biographies, Skill and Practice sheets are found on the Teacher Resource CD in a folder labeled Skill and Practice.

Chapter # and Title	Investigation Skills	Math Skill Sheets	Reading Skill Sheets	Biographies	Skill and Practice
Chapter 1: What Physics and Chemistry are About					
• Section 1.1	• Lab Safety • Lab Report Format		• Using Your Textbook		
• Section 1.2	• Recording Observations	• Stopwatch Math • SI Units • SI-English conversions • Dimensional Analysis	• SQ3R- Reading and Study Method		
Section 1.3			• Study Notes	• Joule	
Chapter 2: Experiments and Variables					
Section 2.1	• Measuring Mass • Measuring Length • Measuring Temperature	• Creating line graphs • Measuring angles • Solving equations with 1 variable • What's the scale?	• Internet Research		
Section 2.2		• Averaging	• Bibliographies		• Scientific processes • Percent Error • Significant Differences in Measurement
Chapter 3: Key Concepts in Physical Science					
Section 3.1	• Measuring Temperature		• Science Vocabulary		
Section 3.2					• Temperature Scales • Specific Heat

Chapter # and Title	Investigation Skills	Math Skill Sheets	Reading Skill Sheets	Biographies	Skill and Practice
Chapter 4: Density and Buoyancy					
Section 4.1	• Calculating Volume • Measuring Volume				• Calculating Slope From a Graph • Density
Section 4.2				• Archimedes	• Archimedes Principle • Buoyancy
Chapter 5:The Structure of Matter					
Section 5.1					
Section 5.2					
Chapter 6: The Atom					
Section 6.1				• Ernest Rutherford • Niels Bohr • Marie and Pierre Curie • Rosalyn Yalow	• Atoms, Isotopes, and Ions
Section 6.2				• Albert Einstein	• Structure of the Atom
Chapter 7: The Periodic Table					
Section 7.1					
Section 7.2					
Chapter 8: Molecules and Compounds					
Section 8.1					• Sketching Molecules and Compounds
Section 8.2					• Dot Diagrams • Chemical Formulas
Chapter 9: Water and Solutions					
Section 9.1					• Calculating Concentration
Section 9.2					

Chapter # and Title	Investigation Skills	Math Skill Sheets	Reading Skill Sheets	Biographies	Skill and Practice
Chapter 10: Chemical Reactions					
Section 10.1					• Chemical Equations
Section 10.2				• Lise Meitner	• Predicting Chemical Equations
Chapter 11: The Chemistry of Living Organisms					
Section 11.1					• Classifying Reactions
Section 11.2					
Chapter 12: Distance, Time, and Speed					
Section 12.1					• Position on the Coordinate Plane
Section 12.2					• Velocity
Section 12.3					• Analyzing Graphs of Motion with Numbers • Analyzing Graphs of Motion with and without Numbers • Calculating Slope From a Graph
Chapter 13: Forces					
Section 13.1				• Chien-Shiung Wu	• Mass vs Weight
Section 13.2					• Equilibrium
Chapter 14: Newton's laws of motion					
Section 14.1				• Isaac Newton	
Section 14.2					• Applying Newton's Laws • Acceleration • Newton's Second Law

Chapter # and Title	Investigation Skills	Math Skill Sheets	Reading Skill Sheets	Biographies	Skill and Practice
Chapter 15: The Planets					
Section 15.1		• Ratios/Proportions		• Copernicus • Galileo Galilei • Kepler • Benjamin Banneker	• Touring the Solar System • Gravity Problems
Section 15.2					
Chapter 16: The Sun, Stars, and Galaxies					
Section 16.1		• Scientific Notation		• Arthur Walker	• The Sun: A Cross-section (diagram) • Understanding Light-years • Calculating Luminosity
Section 16.2					
Chapter 17: Galaxies and the Universe				• Hubble	
Miscellaneous					
				•	• Blank x-y coordinates graph

Notes:

Teaching Illustrations

Chapter	Section	Title
1	1.2	Metric Prefixes
		Converting Between Units of Length
	1.3	Steps in Learning Through Inquiry
2	2.1	Relationships Between Variables
		Direct and Inverse Relationships
		Dependent and Independent Variables
		Basic Math Operations Chart
		Formula for Volume of a Rectangular Box
3	3.1	Converting Grams to Kilograms
		Atoms in a Solid, Liquid, and Gas
	3.2	Energy Use in a Typical Car
		Celsius and Fahrenheit Temperature Scales
		Converting Between Fahrenheit and Celsius
		Specific Heat Values of Common Materials
		Converting Energy
4	4.1	Density Range of Common Materials
		Steps to Solving a Problem
		Solving Density Problems
	4.2	Example of Buoyant Force
		Average Density Explanation
5	5.1	Intermolecular Forces
		Temperature and Time Graph
		Composition of Earth's Atmosphere
	5.2	Thermal Conductors and Insulators

Chapter	Section	Title
6	6.1	Units of Elementary Charge
		Size and Structure of an Atom
		Electron, Proton, and Neutron Table
	6.2	Diagram: Spectral Lines and Energy
		Bohr's Model of the Atom
		Orbital
		Energy Levels
7	7.1	Periodic Table of Elements
		Reading the Periodic Table
		Groups in the Periodic Table
		Energy Levels in the Periodic Table
8	8.1	Chemical Formulas and Structure Diagrams
		Chlorine and Sodium Bonds
	8.2	Periodic Table and Reactivity
		Valence Electrons and Bonds
		Lewis Dot Diagrams
		Example of Oxidation Numbers Adding to Zero
		Periodic Table and Electron Acceptors and Donors
		Predicting a Chemical Formula
9	9.1	Solid Dissolving
		Water: A Polar Molecule
		Supersaturated and Unsaturated Sugar Graph
	9.2	The pH Scale
		pH Ranges in Nature
		Neutralization
		pH Scale with Acids and Bases
10	10.1	The Products of the Reaction
		Chemical Equation
		Balancing Chemical Equations
	10.2	Exothermic Reactions
		Comparing Nuclear and Chemical Reactions

Chapter	Section	Title
11	11.1	Composition of the Human Body Photosynthesis Reaction ADP/ATP
	11.2	Enzymes
12	12.1	X and Y Coordinates
	12.2	Forms of Speed Equation
	12.3	Graphs: Speed vs. Time and Position vs. Time Drawing a Speed vs. Time Graph
13	13.1	Four Elementary Forces Calculating Weight Friction Airplane Forces
	13.2	Free-Body Diagram
14	14.1	Newton's First Law Newton's Third Law
	14.2	Example of a Newton Units of Acceleration Acceleration Formula Newton's Second Law Equation for Newton's Second Law Object in Free Fall

Chapter	Section	Section
15	15.1	The Solar System Planet Size Relative to the Sun Planet Overview
	15.2	Mercury Venus Earth Mars Jupiter Saturn Uranus and Neptune Pluto
16	16.1	Sun Anatomy Star Color and Temperature H-R Diagram
	16.2	Forming a Planetary System Evolution of Sun-Like Stars Spectroscope
17	17.1	Types of Galaxies Big Bang Theory

Notes:

Notes:

Notes:

Notes:

Notes:

Notes:

Notes:

Notes:

Notes:

Notes:

Notes:

Notes:

Notes: